Soft Magnetic Materials and Their Application

Soft Magnetic Materials and Their Application

Editor

Shenglei Che

Basel • Beijing • Wuhan • Barcelona • Belgrade • Novi Sad • Cluj • Manchester

Editor
Shenglei Che
Research Center of Magnetic
and Electronic Materials,
College of Materials Science
and Engineering
Zhejiang University of
Technology
Huzhou
China

Editorial Office
MDPI
St. Alban-Anlage 66
4052 Basel, Switzerland

This is a reprint of articles from the Special Issue published online in the open access journal *Materials* (ISSN 1996-1944) (available at: www.mdpi.com/journal/materials/special_issues/7GBC3BZUN9).

For citation purposes, cite each article independently as indicated on the article page online and as indicated below:

Lastname, A.A.; Lastname, B.B. Article Title. *Journal Name* **Year**, *Volume Number*, Page Range.

ISBN 978-3-7258-0616-4 (Hbk)
ISBN 978-3-7258-0615-7 (PDF)
doi.org/10.3390/books978-3-7258-0615-7

© 2024 by the authors. Articles in this book are Open Access and distributed under the Creative Commons Attribution (CC BY) license. The book as a whole is distributed by MDPI under the terms and conditions of the Creative Commons Attribution-NonCommercial-NoDerivs (CC BY-NC-ND) license.

Contents

About the Editor vii

Preface ix

Shenglei Che
Special Issue: "Soft Magnetic Materials and Their Applications"
Reprinted from: *Materials* **2024**, *17*, 89, doi:10.3390/ma17010089 1

Yao Ying, Linghuo Hu, Zhaocheng Li, Jingwu Zheng, Jing Yu, Wangchang Li, et al.
Preparation of Densified Fine-Grain High-Frequency MnZn Ferrite Using the Cold Sintering Process
Reprinted from: *Materials* **2023**, *16*, 3454, doi:10.3390/ma16093454 5

Yajing Shang, Fan Luo and Zhongxia Duan
Influence of MnZn Ferrite Homogeneous Fibers on the Microstructure, Magnetic, and Mechanical Properties of MnZn Ferrite Materials
Reprinted from: *Materials* **2023**, *16*, 209, doi:10.3390/ma16010209 17

Danyue Ma, Xiujie Fang, Jixi Lu, Kun Wang, Bowen Sun, Yanan Gao, et al.
Analysis and Measurement of Differential-Mode Magnetic Noise in Mn-Zn Soft Ferrite Shield for Ultra-Sensitive Sensors
Reprinted from: *Materials* **2022**, *15*, 8704, doi:10.3390/ma15238704 26

Wenjie Yang, Junjie Liu, Xiangfeng Yu, Gang Wang, Zhigang Zheng, Jianping Guo, et al.
The Preparation of High Saturation Magnetization and Low Coercivity Feco Soft Magnetic Thin Films via Controlling the Thickness and Deposition Temperature
Reprinted from: *Materials* **2022**, *15*, 7191, doi:10.3390/ma15207191 39

Siyuan Wang, Jingwu Zheng, Danni Zheng, Liang Qiao, Yao Ying, Yiping Tang, et al.
Low Core Losses of Fe-Based Soft Magnetic Composites with an Zn-O-Si Insulating Layer Obtained by Coupling Synergistic Photodecomposition
Reprinted from: *Materials* **2022**, *15*, 8660, doi:10.3390/ma15238660 48

Zhengqu Zhu, Jiaqi Liu, Huan Zhao, Jing Pang, Pu Wang and Jiaquan Zhang
Study of the Soft Magnetic Properties of FeSiAl Magnetic Powder Cores by Compounding with Different Content of Epoxy Resin
Reprinted from: *Materials* **2023**, *16*, 1270, doi:10.3390/ma16031270 59

Peng Wu, Shengyu Yang, Yuandong Huang, Guowu Wang, Jinghao Cui, Liang Qiao, et al.
A Study on an Easy-Plane FeSi$_{3.5}$ Composite with High Permeability and Ultra-Low Loss at the MHz Frequency Band
Reprinted from: *Materials* **2023**, *16*, 5133, doi:10.3390/ma16145133 70

Jinchao Du, Pei Gong, Xiao Li, Shaoqi Ning, Wei Song, Yuan Wang, et al.
Effect of Al and La Doping on the Structure and Magnetostrictive Properties of Fe$_{73}$Ga$_{27}$ Alloy
Reprinted from: *Materials* **2023**, *16*, 12, doi:10.3390/ma16010012 81

Jinji Sun, Yan Lu, Lu Zhang, Yun Le and Xiuqi Zhao
A Method to Measure Permeability of Permalloy in Extremely Weak Magnetic Field Based on Rayleigh Model
Reprinted from: *Materials* **2022**, *15*, 7353, doi:10.3390/ma15207353 91

Yuan Cheng, Yaozhi Luo, Ruihong Shen, Deyu Kong and Weiyong Zhou
Testing and Analysis Method of Low Remanence Materials for Magnetic Shielding Device
Reprinted from: *Materials* **2023**, *16*, 681, doi:10.3390/ma16020681 . . . . **101**

Anqi Wang, Baozhi Tian, Yulin Li, Shibo Xu, Lubin Zeng and Ruilin Pei
Transient Magnetic Properties of Non-Grain Oriented Silicon Steel under Multi-Physics Field
Reprinted from: *Materials* **2022**, *15*, 8305, doi:10.3390/ma15238305 . . . . **110**

Xu Zhang, Yu Sun, Bin Yan and Xin Zhuang
Correlation of Magnetomechanical Coupling and Damping in $Fe_{80}Si_9B_{11}$ Metallic Glass Ribbons
Reprinted from: *Materials* **2023**, *16*, 4990, doi:10.3390/ma16144990 . . . . **122**

Zhiye Li, Yuechao Ma, Anrui Hu, Lubin Zeng, Shibo Xu and Ruilin Pei
Investigation and Application of Magnetic Properties of Ultra-Thin Grain-Oriented Silicon Steel Sheets under Multi-Physical Field Coupling
Reprinted from: *Materials* **2022**, *15*, 8522, doi:10.3390/ma15238522 . . . . **134**

Deji Ma, Baozhi Tian, Xuejie Zheng, Yulin Li, Shibo Xu and Ruilin Pei
Study of High-Silicon Steel as Interior Rotor for High-Speed Motor Considering the Influence of Multi-Physical Field Coupling and Slotting Process
Reprinted from: *Materials* **2022**, *15*, 8502, doi:10.3390/ma15238502 . . . . **151**

Xiujie Fang, Danyue Ma, Bowen Sun, Xueping Xu, Wei Quan, Zhisong Xiao, et al.
A High-Performance Magnetic Shield with MnZn Ferrite and Mu-Metal Film Combination for Atomic Sensors
Reprinted from: *Materials* **2022**, *15*, 6680, doi:10.3390/ma15196680 . . . . **167**

Haoying Pang, Feng Liu, Wengfeng Fan, Jiaqi Wu, Qi Yuan, Zhihong Wu, et al.
Analysis and Suppression of Thermal Magnetic Noise of Ferrite in the SERF Co-Magnetometer
Reprinted from: *Materials* **2022**, *15*, 6971, doi:10.3390/ma15196971 . . . . **178**

Ye Liu, Hang Gao, Longyan Ma, Jiale Quan, Wenfeng Fan, Xueping Xu, et al.
Study on the Magnetic Noise Characteristics of Amorphous and Nanocrystalline Inner Magnetic Shield Layers of SERF Co-Magnetometer
Reprinted from: *Materials* **2022**, *15*, 8267, doi:0.3390/ma15228267 . . . . **188**

About the Editor

Shenglei Che

Prof. Dr. Shenglei Che currently works at the College of Material Science and Engineering, Zhejiang University of Technology, and serves as the director for Research Center of Magnetic and Electronic Materials. Prof. Che's current research interest focuses on magnetic composites, high-frequency soft magnetic materials, bio-applications of nano-magnetic powders, and low-resource-consumption rare earth permanent magnets. He has published over 170 papers and over 200 invention patents in the field of functional materials. Prof. Dr. Shenglei Che obtained his PhD at the Tokyo Institute of Technology, Japan, in 1997 and then worked at Sumitomo Metal Mining Co. Ltd. and TDK Corporation as a material scientist before he moved to Zhejiang University of Technology in 2011.

Preface

Soft magnetic materials normally show no magnetic properties outside of a magnetic field but can be easily magnetized and demagnetized within magnetic fields. Most soft magnetic materials have small coercive field strengths, in the same order as the Earth's magnetic field (about 40 A/m) or lower.

Soft magnetic materials are widely used in electronics, energy conversion, information processing, and many other application scenarios. Indeed, it is safe to assume that whenever you use electricity, soft magnetic materials are working in the background.

Recently, technological progress in soft magnetic materials has focused on the processing of rapidly quenched amorphous and nanocrystalline materials (either in the form of ribbons or powders), as well as improvements in the magnetic properties of soft magnetic composites (SMCs) and soft ferrites to fit the requirements of high-frequency devices driven by Wide-Bandgap Semiconductors (SiC/GaN). Moreover, new soft magnetic devices have also been successively designed for the rapid development of fabrication methods and new applications.

The aim of this Special Issue was to present the latest achievements in the synthesis, fundamentals, characterization, and applications of soft magnetic materials, including soft ferrites, alloys, and composites, together with their processing technology, characterization, and applications in electronics, motors, wave absorbing, sensors, communications, etc.

Seventeen papers are published in this Special Issue. For convenience of reading, these papers can be divided into four categories according to the contents of their studies: I. Soft Ferrite Materials; II. Soft Magnetic Alloys and Composites; III. Characterization of Soft Magnetic Materials; and IV. Applications of Soft Magnetic Materials.

We congratulate all the authors published in this collection for their remarkable results presented in these papers. We truly believe that these works will help the research community to enhance their understanding of the present status and trends in soft magnetic materials and their applications.

Shenglei Che
Editor

Editorial

Special Issue: "Soft Magnetic Materials and Their Applications"

Shenglei Che [1,2]

[1] College of Materials Science and Engineering, Zhejiang University of Technology, Hangzhou 310014, China; cheshenglei@zjut.edu.cn
[2] Research Center of Magnetic and Electronic Materials, Zhejiang University of Technology, Hangzhou 310014, China

1. Introduction

Soft magnetic materials normally show no magnetic properties outside of a magnetic field but can be easily magnetized and demagnetized within magnetic fields. Most soft magnetic materials have small coercive field strengths, in the same order as the Earth's magnetic field (about 40 A/m) or lower.

Soft magnetic materials are widely used in electronics, energy conversion, information processing, and many other application scenarios. Indeed, it is safe to assume that whenever you use electricity, there are soft magnetic materials working for you.

Recently, technological progress in soft magnetic materials has been focused on the processing of rapidly quenched amorphous and nanocrystalline materials (either in the form of ribbons or powders), as well as the improvement in magnetic properties of soft magnetic composites (SMCs) and soft ferrites to fit the requirements of high-frequency devices driven by Wide Bandgap Semiconductors (SiC/GaN). Moreover, new soft magnetic devices have also been successively designed for the rapid development of fabrication methods and new applications.

The aim of this Special Issue is to present the latest achievements in the synthesis, fundamentals, characterization, and applications of soft magnetic materials, including soft ferrites, alloys, and composites, together with their processing technology, characterization, and applications in electronics, motors, wave absorbing, sensors, communications, etc.

Seventeen papers are published in this Special Issue. For the convenience of introduction, these papers can be divided into four categories according to the contents of their studies.

2. Soft Ferrite Materials

Soft ferrites mainly include MnZn ferrites and NiZn ferrites. As a kind of ceramics, ferrites are normally prepared via a sintering process at 900–1450 °C. To prepare soft ferrites with small grain sizes and better high-frequency properties, Ying et al. applied cold sintering technique to sinter MnZn ferrites at around 300 °C and obtained a densified, fine-grained, high-frequency MnZn ferrite bulk [1]. The relative density could be increased to 97.2% by annealing the samples at 950 °C for 6 h. Annealing also reduced the power loss at high frequencies.

Shang et al. synthesized homogeneous MnZn ferrite fibers using the solvothermal method and used them as a reinforcing phase to prepare homogeneous-fiber-reinforced MnZn ferrite materials [2]. The best magnetic and mechanical properties were obtained at a fiber content of about 2 wt%.

Ma et al. measured the differential-mode magnetic noise in MnZn soft ferrites for magnetic shielding and analyzed and optimized the structural parameters of the shield on the differential-mode magnetic noise; some useful results for suppressing magnetic noise and breaking through the sensitivity of the magnetometer were obtained [3].

Citation: Che, S. Special Issue: "Soft Magnetic Materials and Their Applications". *Materials* 2024, *17*, 89. https://doi.org/10.3390/ma17010089

Received: 13 December 2023
Accepted: 21 December 2023
Published: 23 December 2023

Copyright: © 2023 by the author. Licensee MDPI, Basel, Switzerland. This article is an open access article distributed under the terms and conditions of the Creative Commons Attribution (CC BY) license (https://creativecommons.org/licenses/by/4.0/).

3. Soft Magnetic Alloys and Composites

High saturation magnetization and low coercivity FeCo soft magnetic thin films with different thicknesses were prepared via magnetron sputtering by controlling the thickness and deposition temperature in a paper by Yang et al. [4]. This FeCo thin film with high saturation magnetization and low coercivity could be an ideal candidate for high-frequency electronic devices.

Wang et al. reported a novel photodecomposition method to create a ZnO insulating layer on FeSiAl powders [5]. Combined with conventional coupling treatment processes, a thin and dense Zn-O-Si insulating layer was coated on the surface of iron powder in situ. Treating the iron powder before coating by photodecomposition led to a synergistic effect, significantly reduced the core loss, while the effective permeability only decreased slightly.

Zhu et al. studied the effect of epoxy resin (EP) on the insulated coating and pressing effect of FeSiAl magnetic powders by changing the content of EP and characterized the soft magnetic properties of the powders and SMPCs and revealed that best overall performance can be obtained when the EP content was 1 wt.% [6].

Wu et al. reported an FeSi3.5 easy-plane composite with high permeability and ultra-low loss at the MHz frequency band [7]. Through loss measurement and separation, they found that the real reason why magnetic materials do not work properly at MHz due to overheat is dramatical increase of the excess loss and the easy-plane composite can greatly reduce the excess loss. This makes the easy-plane FeSi3.5 composite become an excellent soft magnetic composite and it is possible for magnetic devices to operate properly at higher frequencies, especially at the MHz band and above.

Du et al. studied the changes of microstructure, magnetostriction properties and hardness of the Fe73Ga27−xAlx alloy and (Fe73Ga27−xAlx)99.9La0.1 alloy (x = 0, 0.5, 1.5, 2.5, 3.5, 4.5) by doping Al into the Fe73Ga27 and (Fe73Ga27)99.9La0.1 alloy, respectively [8]. The results indicated that trace La doping can improve the magnetostriction properties and deformation resistance of Fe-Ga alloy, which provides a new design idea for the Fe-Ga alloy, broadening their application in the field of practical production.

4. Characterization of Soft Magnetic Materials

In order to solve the problem that the relative permeability of the permalloy is missing and difficult to measure accurately in an extremely weak magnetic field (EWMF, <1 nT), Sun et al. proposed a method to measure the permeability in EWMF based on the Rayleigh model [9]. This method can obtain the relative permeability in any EWMF and avoid test errors caused by extremely weak magnetization signals.

Cheng et al. proposed a novel method of measuring the remanence of materials in a magnetic shielding cylinder, which prevents interference from the Earth's magnetic field and reduces the measurement error [10]. This method is used to test concrete components, composite materials, and metal materials commonly applied in magnetic shielding devices and determine the materials that can be used for magnetic shielding devices with 1 nT, 10 nT, and 100 nT as residual magnetic field targets.

Wang et al. built a test system that combined temperature, stress, and electromagnetic fields along with other fields at the same time [11]. It can accurately simulate the actual complex working conditions of a motor and explore the dynamic characteristics of non-grain oriented (NGO) silicon steel. The rationality of this method was verified by checking the test results of the prototype, and the calculation accuracy of the motor model was found to be improved.

Zhang et al. investigated the magnetomechanical coupling factors (k) and damping factors (Q−1) of FeSiB amorphous ribbons annealed in air at different temperatures [12]. The k and Q−1 of FeSiB-based epoxied laminates with different stacking numbers show that a −3 dB bandwidth and Young's modulus are expressed in terms of the magnetomechanical power efficiency for high lamination stacking.

5. Applications of Soft Magnetic Materials

As the core part of an electrical driving system, the electrical machine faces the extreme challenge of maintaining a high power density and high efficiency output under complex working conditions. The research and development of new soft magnetic materials has an important impact on solving the current bottlenecks of electrical machines. Li et al. studied the variation trends of magnetic properties of ultra-thin grain-oriented electrical silicon steel (GOES) under thermal–mechanical–electric–magnetic fields and explored its possible application in motors [13]. They verified that grain-oriented silicon steel has great application prospects in the drive motors (IPMs) of electric vehicles, and it is an effective means to break the bottleneck of current motor design.

High silicon steel has low loss and high mechanical strength, making it extremely suitable as a high-speed motor rotor core material. Ma et al. investigated the feasibility of using high silicon steel as the material of an interior rotor high-speed motor [14] and verified the results via theoretical analysis and experimental characterization.

Fang et al. proposed a high-performance magnetic shielding structure composed of MnZn ferrite and mu-metal film [15]. The use of the mu-metal film with a high magnetic permeability restrains the decrease in the magnetic shielding coefficient caused by the magnetic leakage between the gap of magnetic annuli. This proposed combined magnetic shielding is of great significance to further promoting the performance of atomic sensors sensitive to magnetic fields.

Ferrite magnetic shields are widely used in ultra-high-sensitivity atomic sensors because of their low noise characteristics. However, their noise level varies with temperature and affects the performance of the spin-exchange relaxation-free (SERF) co-magnetometer. Pang et al. established the thermal magnetic noise model of a ferrite magnetic shield and calculated the thermal magnetic noise of ferrite more accurately by testing the low-frequency complex permeability at different temperatures [16]. The experimental results demonstrated the effectiveness of the proposed method.

Magnetic shields are an important part of electronic equipment, ultra-sensitive atomic sensors, and in basic physics experiments. Particularly in spin-exchange relaxation-free (SERF) co-magnetometers, the magnetic shield is an important component for maintaining the SERF state. Liu et al. applied different amorphous and nanocrystalline materials as the innermost magnetic shielding layers of an SERF co-magnetometer and analyzed their magnetic noise characteristics [17]. The experimental results show that compared with an amorphous material, using a nanocrystalline material as the inner magnetic shield layer can effectively reduce the magnetic noise and improve the sensitivity and precision of the rotation measurement.

The Guest Editors would like to congratulate all the authors published in this Special Issue for the remarkable results presented in these papers. We truly believe that these works will help the research community to enhance their understanding of the present status and trends in soft magnetic materials and their applications.

Funding: This work was supported by the National Key Research and Development Program (Grant no. 2022YFE0109800).

Acknowledgments: The Guest Editors would like to thank all the authors and reviewers for their work for this special issue.

Conflicts of Interest: The author declares no conflict of interest.

References

1. Ying, Y.; Hu, L.; Li, Z.; Zheng, J.; Yu, J.; Li, W.; Qiao, L.; Cai, W.; Li, J.; Bao, D.; et al. Preparation of Densified Fine-Grain High-Frequency MnZn Ferrite Using the Cold Sintering Process. *Materials* **2023**, *16*, 3454. [CrossRef] [PubMed]
2. Shang, Y.; Luo, F.; Duan, Z. Influence of MnZn Ferrite Homogeneous Fibers on the Microstructure, Magnetic, and Mechanical Properties of MnZn Ferrite Materials. *Materials* **2023**, *16*, 209. [CrossRef] [PubMed]
3. Ma, D.; Fang, X.; Lu, J.; Wang, K.; Sun, B.; Gao, Y.; Xu, X.; Han, B. Analysis and Measurement of Differential-Mode Magnetic Noise in Mn-Zn Soft Ferrite Shield for Ultra-Sensitive Sensors. *Materials* **2022**, *15*, 8704. [CrossRef] [PubMed]

4. Yang, W.; Liu, J.; Yu, X.; Wang, G.; Zheng, Z.; Guo, J.; Chen, D.; Qiu, Z.; Zeng, D. The Preparation of High Saturation Magnetization and Low Coercivity FeCo Soft Magnetic Thin Films via Controlling the Thickness and Deposition Temperature. *Materials* **2022**, *15*, 7191. [CrossRef] [PubMed]
5. Wang, S.; Zheng, J.; Zheng, D.; Qiao, L.; Ying, Y.; Tang, Y.; Cai, W.; Li, W.; Yu, J.; Li, J.; et al. Low Core Losses of Fe-Based Soft Magnetic Composites with an Zn-O-Si Insulating Layer Obtained by Coupling Synergistic Photodecomposition. *Materials* **2022**, *15*, 8660. [CrossRef]
6. Zhu, Z.; Liu, J.; Zhao, H.; Pang, J.; Wang, P.; Zhang, J. Study of the Soft Magnetic Properties of FeSiAl Magnetic Powder Cores by Compounding with Different Content of Epoxy Resin. *Materials* **2023**, *16*, 1270. [CrossRef]
7. Wu, P.; Yang, S.; Huang, Y.; Wang, G.; Cui, J.; Qiao, L.; Wang, T.; Li, F. A Study on an Easy-Plane FeSi3.5 Composite with High Permeability and Ultra-Low Loss at the MHz Frequency Band. *Materials* **2023**, *16*, 5133. [CrossRef] [PubMed]
8. Du, J.; Gong, P.; Li, X.; Ning, S.; Song, W.; Wang, Y.; Hao, H. Effect of Al and La Doping on the Structure and Magnetostrictive Properties of Fe73Ga27 Alloy. *Materials* **2023**, *16*, 12. [CrossRef]
9. Sun, J.; Lu, Y.; Zhang, L.; Le, Y.; Zhao, X. A Method to Measure Permeability of Permalloy in Extremely Weak Magnetic Field Based on Rayleigh Model. *Materials* **2022**, *15*, 7353. [CrossRef] [PubMed]
10. Cheng, Y.; Luo, Y.; Shen, R.; Kong, D.; Zhou, W. Testing and Analysis Method of Low Remanence Materials for Magnetic Shielding Device. *Materials* **2023**, *16*, 681. [CrossRef] [PubMed]
11. Wang, A.; Tian, B.; Li, Y.; Xu, S.; Zeng, L.; Pei, R. Transient Magnetic Properties of Non-Grain Oriented Silicon Steel under Multi-Physics Field. *Materials* **2022**, *15*, 8305. [CrossRef] [PubMed]
12. Zhang, X.; Sun, Y.; Yan, B.; Zhuang, X. Correlation of Magnetomechanical Coupling and Damping in Fe80Si9B11 Metallic Glass Ribbons. *Materials* **2023**, *16*, 4990. [CrossRef] [PubMed]
13. Li, Z.; Ma, Y.; Hu, A.; Zeng, L.; Xu, S.; Pei, R. Investigation and Application of Magnetic Properties of Ultra-Thin Grain-Oriented Silicon Steel Sheets under Multi-Physical Field Coupling. *Materials* **2022**, *15*, 8522. [CrossRef] [PubMed]
14. Ma, D.; Tian, B.; Zheng, X.; Li, Y.; Xu, S.; Pei, R. Study of High-Silicon Steel as Interior Rotor for High-Speed Motor Considering the Influence of Multi-Physical Field Coupling and Slotting Process. *Materials* **2022**, *15*, 8502. [CrossRef] [PubMed]
15. Fang, X.; Ma, D.; Sun, B.; Xu, X.; Quan, W.; Xiao, Z.; Zhai, Y. A High-Performance Magnetic Shield with MnZn Ferrite and Mu-Metal Film Combination for Atomic Sensors. *Materials* **2022**, *15*, 6680. [CrossRef] [PubMed]
16. Pang, H.; Liu, F.; Fan, W.; Wu, J.; Yuan, Q.; Wu, Z.; Quan, W. Analysis and Suppression of Thermal Magnetic Noise of Ferrite in the SERF Co-Magnetometer. *Materials* **2022**, *15*, 6971. [CrossRef] [PubMed]
17. Liu, Y.; Gao, H.; Ma, L.; Quan, J.; Fan, W.; Xu, X.; Fu, Y.; Duan, L.; Quan, W. Study on the Magnetic Noise Characteristics of Amorphous and Nanocrystalline Inner Magnetic Shield Layers of SERF Co-Magnetometer. *Materials* **2022**, *15*, 8267. [CrossRef] [PubMed]

Disclaimer/Publisher's Note: The statements, opinions and data contained in all publications are solely those of the individual author(s) and contributor(s) and not of MDPI and/or the editor(s). MDPI and/or the editor(s) disclaim responsibility for any injury to people or property resulting from any ideas, methods, instructions or products referred to in the content.

Article

Preparation of Densified Fine-Grain High-Frequency MnZn Ferrite Using the Cold Sintering Process

Yao Ying [1,2,*], Linghuo Hu [1,2], Zhaocheng Li [1,2], Jingwu Zheng [1,2], Jing Yu [1,2], Wangchang Li [1,2], Liang Qiao [1,2], Wei Cai [1,2], Juan Li [1,2], Daxin Bao [3] and Shenglei Che [1,2,*]

[1] College of Materials Science and Engineering, Zhejiang University of Technology, Hangzhou 310014, China
[2] Research Center of Magnetic and Electronic Materials, Zhejiang University of Technology, Hangzhou 310014, China
[3] Hengdian Group DMEGC Magnetics Co., Ltd., Dongyang 322118, China
* Correspondence: yying@zjut.edu.cn (Y.Y.); cheshenglei@zjut.edu.cn (S.C.)

Abstract: The densified MnZn ferrite ceramics were prepared using the cold sintering process under pressure, with an acetate ethanol solution used as the transient solvent. The effects of the transient solvent, the pressure and annealing temperature on the density, and the micromorphology and magnetic properties of the sintered MnZn ferrites were studied. The densified MnZn ferrite was obtained using the cold sintering process and its relative density reached up to 85.4%. The transient solvent and high pressure are essential to the cold sintering process for MnZn ferrite. The annealing treatment is indispensable in obtaining the sample with the higher density. The relative density was further increased to 97.2% for the sample annealed at 950 °C for 6 h. The increase in the annealing temperature reduces the power loss at high frequencies.

Keywords: MnZn ferrite; cold sintering process; power loss; high frequency

Citation: Ying, Y.; Hu, L.; Li, Z.; Zheng, J.; Yu, J.; Li, W.; Qiao, L.; Cai, W.; Li, J.; Bao, D.; et al. Preparation of Densified Fine-Grain High-Frequency MnZn Ferrite Using the Cold Sintering Process. *Materials* 2023, 16, 3454. https://doi.org/10.3390/ma16093454

Academic Editor: Antoni Planes

Received: 13 March 2023
Revised: 11 April 2023
Accepted: 19 April 2023
Published: 28 April 2023

Copyright: © 2023 by the authors. Licensee MDPI, Basel, Switzerland. This article is an open access article distributed under the terms and conditions of the Creative Commons Attribution (CC BY) license (https://creativecommons.org/licenses/by/4.0/).

1. Introduction

MnZn ferrites are the core materials in power transformers, sensors and other electronic devices due to their excellent soft magnetic performance, such as high initial permeability (μ_i), high flux density (B_s), low power loss (P_{cv}) and relatively high Curie temperature (T_C) [1–6]. With the continuous development of 5G communication technology and third generation wide band gap (WBG) semiconductors, electronic devices are developing toward miniaturization and high efficiency [7–9]. When the magnetic core in the electronic device is subjected to an AC magnetic field, some of the power is lost in the core and dissipated as heat and sometimes noise. The lost power is termed as power loss. Low power loss is the crucial factor for high efficiency in electronic devices. Increasing the operating frequency is the essential way to achieve miniaturization. However, P_{cv} increases dramatically when the frequency increases to above 500 kHz, due to the dramatic increase in eddy current loss (P_{ev}) and residual loss (P_{rv}). In order to reduce the P_{ev} and P_{rv}, it is necessary to refine the grain of the MnZn ferrite to nearly a single domain, namely smaller than 3 μm.

To achieve the purpose of grain refinement, an effective method is to reduce the sintering temperature [10,11]. In recent years, the cold sintering process (CSP), as an energy-efficient sintering technology, has emerged. The CSP is a technique according to which the densified ceramics or metals are sintered at low temperatures, typically below 300 °C and high pressure, with the aid of a transient solvent [12–15]. In the typical cold sintering process, the introduction of a certain amount of liquid is indispensable. The ceramic powder is uniformly wetted with a small amount of solvent, and the particle surface is decomposed and/or partially dissolved in the solvent and forms a supersaturated solution in the system. The application of high pressure at an appropriate temperature triggers the dissolved solute to precipitate in the interstices between the particles. This process is usually referred to as "dissolution precipitation" [12,16–18]. To date, some ceramics and metals with high density, ranging from mono-, di-, ternary, quaternary and pentameric compounds, have been successfully prepared using CSP [19,20]. Zinc ferrite was successfully prepared using CSP and the sample density reached 93% of the theoretical density after annealing treatment [21]. Mingming Si [22] prepared zinc oxide/polyetheretherketone (PEEK) composites with excellent performance using the cold sintering method.

In order to refine the grain of MnZn ferrite to nearly a single domain with a grain size of less than 3 μm, the cold sintering process was employed in this work. The cold sintering process can inhibit the growth of grains during the ceramic sintering process and obtain a densified ceramic with uniform and fine-grain size. The cold sintering process can effectively reduce the volatilization of the elements, which occurs in the traditional sintering process due to the high sintering temperature. To focus on the effect of the cold sintering process during the preparation of the fine-grain MnZn ferrite, grain boundary additives were not added to this series of samples. The effects of the transient solvent, the pressure and annealing temperature on the density, and the micromorphology and magnetic properties of the sintered MnZn ferrite were studied. The densified MnZn ferrite was successfully prepared. The relative density of the as-CSPed sample reached up to 85.4%. After annealed at 950 °C for 6 h, the relative density of the sample was further increased to 97.2%.

2. Experimental

MnZn ferrite powders were firstly prepared by solid state reaction. The raw materials, namely Fe_2O_3, Mn_3O_4 and ZnO, with a purity of 99% were weighed, according to the stoichiometry of $Mn_{0.888}Zn_{0.049}Fe_{2.063}O_4$. This composition was also used to investigate the effect of low-temperature sintering on MnZn ferrites in our previous work [11]. It was beneficial for us to compare the different properties of the MnZn ferrites sintered by these two processes. The mixtures were calcined at 970 °C in the nitrogen atmosphere. After that, the obtained MnZn ferrite was ball milled into powders. The flow chart of the CSP for MnZn ferrite is shown in Figure 1. Firstly, the calcined MnZn ferrite powder (50 g) and ethanol solution acetic acid (150 g) with a concentration of 4 M were mixed in a beaker and then stirred for 24 h. After pretreatment, the beaker was placed in an oven at 80 °C for 6 h to obtain the dried powder. Then 12 wt% acetate ethanol at a concentration of 2 M was added to the dried powder and they were ground together in a mortar for 30 min. The cold sintering process was carried out in a steel mold with an outer diameter of 16 mm and an inner diameter of 8 mm, which was surrounded by a heating element with a temperature control system. Then, 2 g mixtures were placed in the steel mold and heated to 330 °C at a rate of 4 °C/min. Meanwhile, uniaxial pressure was applied to the mold vertically for 2 h. Finally, the samples obtained by cold sintering were annealed for 6 h at higher temperatures (850 or 950 °C) in an atmosphere of nitrogen.

The crystal structures of the samples were examined from 10 to 80° with a scanning speed of 20°/min using an X-ray diffractometer (XRD) (Ultima IV, Rigaku, Japan) with Cu K_α radiation. The microstructure and cross-sectional morphology of the sintered samples were observed using a scanning electron microscope (SEM) (SU1510, Hitachi, Japan). Prior to the SEM measurements, the cross-sections were sanded using sandpaper of different sizes, from 400 to 7000 mesh, and then etched in 4% HF acid solution for 90 s. The average grain size (D) was estimated. The infrared (IR) spectra were measured using an infrared spectrometer (Nicolet 6700, Thermo, USA). The thermal gravimetric (TG) curves were measured in the temperature range from 25 to 900 °C using a thermal gravimetric analyzer (TG209 F1, Netzsch, Germany). The sample density (ρ) was measured using the Archimedes drainage method. The P_{cv} was measured using an AC BH curve analyzer (SY-8218, IWATSU, Japan).

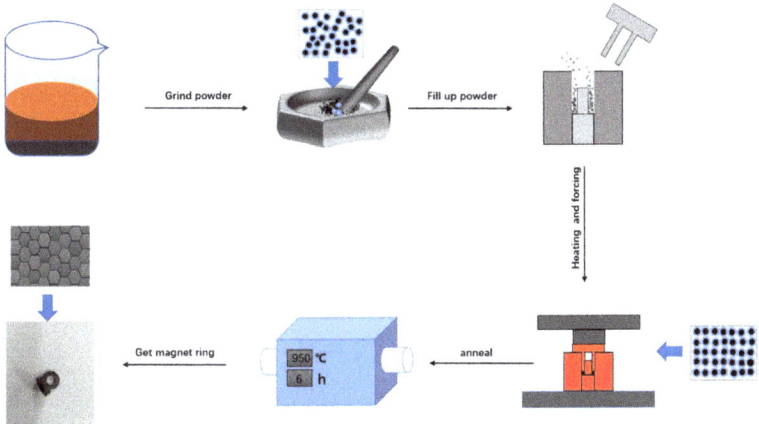

Figure 1. Flow chart of the cold sintering process for MnZn ferrite.

3. Results and Discussion

3.1. Effect of Solvents on the Cold Sintering

Figure 2 shows the XRD patterns of the calcined, as-CSPed and annealed MnZn ferrite samples. The diffraction peaks of the calcined sample are basically consistent with those of the MnZn ferrite, but two impure diffraction peaks of MnO were found at around 41.1° and 60°. The MnO peaks disappeared in the as-CSPed and annealed samples. During the calcination process, the raw materials of Fe_2O_3, Mn_3O_4 and ZnO undergo a solid-state reaction and, consequently, form spinel phase MnZn ferrite with releasing oxygen. The equilibrium oxygen partial pressure of the solid-state reaction is higher at a high temperature. In this work, the calcination was carried out in nitrogen, so the reaction atmosphere was reductive compared to the equilibrium oxygen partial pressure. Therefore, Mn_3O_4 partially reverted to MnO during the calcination. It was also found that the diffraction peaks of the as-CSPed sample were wider than those of the calcined sample. In CSP, the acetic acid treatment leads to the formation of an amorphous layer on the surface of the powder, which weakens the crystallinity of the powder. Therefore, the XRD peaks become wider.

To detect the valency change of the Mn ions, the infrared (IR) spectra were measured for the calcined, as-CSPed and annealed MnZn ferrite samples, as shown in Figure 3. It was found that three absorption peaks appeared at 549, 669 and 875 cm^{-1}. These peaks hardly changed for the calcined, as-CSPed and annealed MnZn ferrite samples. The absorption peak for the Mn^{2+} ion in the manganese ferrite ranged from 545–550 cm^{-1} [23]. Therefore, the 549 cm^{-1} peak in this series of samples was ascribed to Mn^{2+} ion. In Mn_2O_3, the characteristic absorption peaks appeared at 668 and 870 cm^{-1} [24]. Therefore, the 669 and 875 cm^{-1} peaks were ascribed to Mn^{3+} ion. It is obvious that the 669 and 875 cm^{-1} peaks were very weak and the 549 cm^{-1} peak was much stronger than these two peaks in this series of samples. Thus, Mn ions in this series of samples mainly existed in the form of Mn^{2+}, and the number of Mn^{3+} was very small.

Since MnZn ferrite is partially soluble in acetic acid, an ethanol solution of acetate was chosen as the transient solvent for the cold sintering process [21]. During the cold sintering process, Mn, Zn and Fe dissolve in the transient solvent in the ionic state. With an increase in pressure at high temperature, MnZn ferrite grains begin to contact each other and form necks at the contact points, and Mn^{2+}, Zn^{2+} and Fe^{3+} ions also gradually diffuse to the vicinity of the grain neck with the solvent and precipitate. Thus, the diffraction peaks for MnO disappear in the XRD pattern of the as-CSPed sample. The diffraction peaks of the annealed sample are narrower than those of the as-CSPed sample, as shown in Figure 2. It may be due to the fact that the gain size of the annealed samples becomes larger.

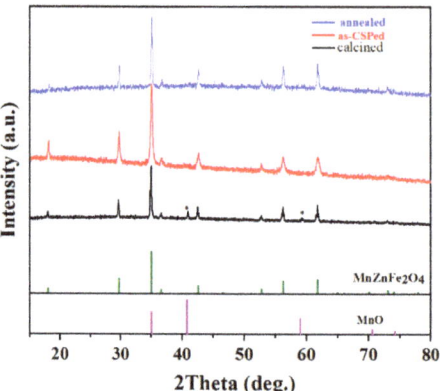

Figure 2. XRD of the calcined, as-CSPed and annealed MnZn ferrite samples.

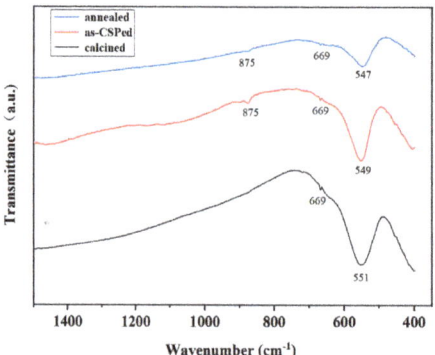

Figure 3. Infrared spectra taken in the frequency range 400–1500 cm^{-1} for the calcined, as-CSPed and annealed MnZn ferrite samples.

Figure 4 shows the SEM fractural morphology of the samples without the ethanol solution of acetate and with the 12 wt% ethanol solution of acetate after the cold sintering process. It can be seen that the grain adhesion without the ethanol solution of acetate is poor and there are many pores. The sintering traces are not obvious. The sample with the ethanol solution of acetate, as a transient solvent, has better particle adhesion and less pores, exhibiting better sintering. By measuring the sample density, the density of the as-CSPed sample without the ethanol solution of acetate was 3.53 g/cm^3, while that with the ethanol solution of acetate increased to 4.08 g/cm^3. Thus, the introduction of the moderate ethanol solution of acetate is beneficial for increasing the densification in the CSP.

The experimental setup can be regarded as a well-closed system. Due to the slow heating rate, the transient solvent can be retained for a long time. Therefore, the system is in a non-equilibrium state before the solvent completely evaporates. The acetic acid solution has a wetting effect on the grains, which reduces the friction between grains. At the same time, the Mn^{2+}, Zn^{2+} and Fe^{2+} ions on the grain surface dissolve, and defects are formed in the MnZn ferrite grain surface. Under the action of high stress, the local pressure between the two grains is very high, which enhances the dissolution of the contact interface between the solid and the liquid. Due to the greater chemical potential in the contact area between the particles, the particles diffuse through the liquid phase and precipitate at locations away from the force contact area. During the holding process, the solid transfer accelerates, the supersaturation increases significantly as the transient solvent evaporates, and small particles begin to grow, which leads to larger grains [25,26].

Figure 4. Fractural SEM images of the sample after cold sintering: (**a**) without ethanol acetate solution and (**b**) with 12 wt% ethanol acetate solution.

3.2. Effect of Pressure on Cold Sintering

Figure 5 shows the fractural morphology of the sample after the cold sintering under different pressures. When the pressure is lower than 1000 MPa, most of the grains exist individually and the pores are obvious. As the pressure exceeds 1000 MPa, some grains connect, and the pores between the grains become smaller. This indicates that high pressure promotes the cold sintering.

In the conventional sintering process, the driving force for powder consolidation is provided by thermal energy at the high sintering temperature. In the cold sintering process, the driving force is mainly provided by the mechanical energy originating from the large external pressure, and the applied pressure increases the solubility of the particles with sharp edges. In this system, a relatively high temperature of 330 °C and a high pressure exceeding 1000 MPa promotes the solubility of the MnZn ferrite and, meanwhile, the evaporation of the liquid phase, which results in the formation of a supersaturated liquid and mass transport via diffusion.

Due to the synergy of the external applied pressure and the capillary stress exerted by the liquid, the atoms diffuse through the liquid from the higher chemical potential areas to the regions of lower chemical potential, where precipitation occurs [18]. The contact areas are found at high chemical potential, and precipitation occurs instead where the chemical potential is lower, away from the contact areas. This is demonstrated by the morphological change of the sample, as shown in Figure 5. As the pressure increases, the original pores between the grains are filled with substances, linking the grains and the grains become tighter.

Figure 6 shows the densities and relative density of the samples after cold sintering at different pressures. The density of sample increases with the increasing pressure and reaches 4.31 g/cm^3. The relative density reaches 85.4%. This also implies that the applied pressure improves the CSP densification.

A sintering body with a relative density of over 85% can be considered to be almost dense enough in some cases of usage, but it is still lower than the normal relative density of ferrite cores, i.e., 90%. Subsequently, an annealing treatment was investigated.

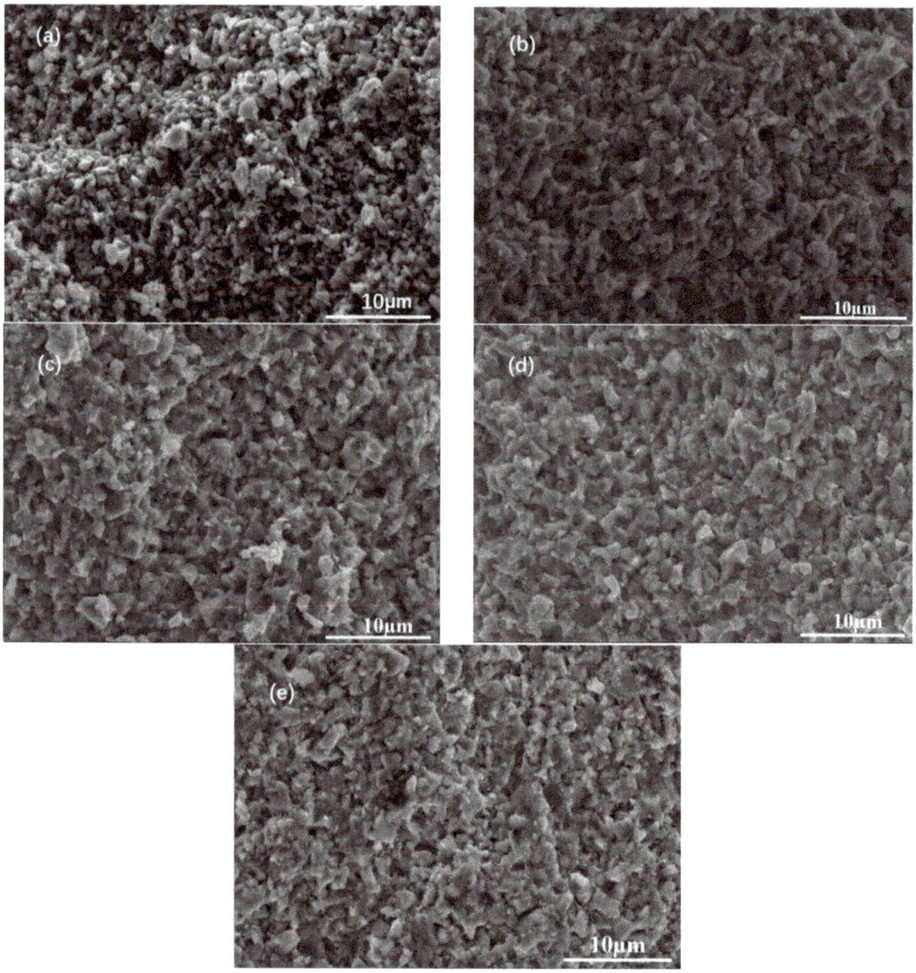

Figure 5. Morphology of the MnZn ferrites after cold sintering under the pressure: (**a**) 600, (**b**) 800, (**c**) 1000, (**d**) 1200 and (**e**) 1400 MPa at 330 °C for 2 h.

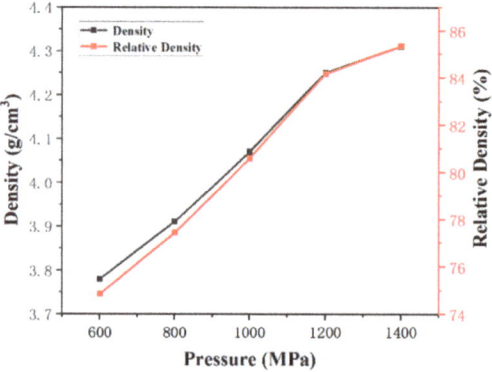

Figure 6. Densities and relative densities of the samples after cold sintering under different pressures.

3.3. Effect of Annealing Treatment

A fully densified structure may not be achieved for the MnZn ferrite using just a one-step CSP, due to the limited dissolution and formation of the grains. In this case, a post-annealing process is necessary to further increase the crystallinity and the density. Figure 7 shows the fractural morphology of the MnZn ferrite samples after annealing at high temperatures of 850 and 950 °C for 6 h. After being annealed at 850 °C, the sample is not dense and many pores exist, which means that 850 °C is not high enough for densification. As the annealing temperature increases to 950 °C, the grains grow, the pores decrease and the sample exhibits good densification.

We further measured the TG curve of the as-CSP sample (Figure 8). When the temperature exceeds 850 °C, the TG curve starts to decrease. This indicates the further solid-phase reaction and the densification process. When the sample is annealed at 850 °C, the solid-phase reaction between the powders is not sufficient, there still exist many pores in the sample and the degree of densification is not enough. When increasing the annealed temperature, the solid-phase reaction is promoted, grains are grown and the pores are removed, so the density increases.

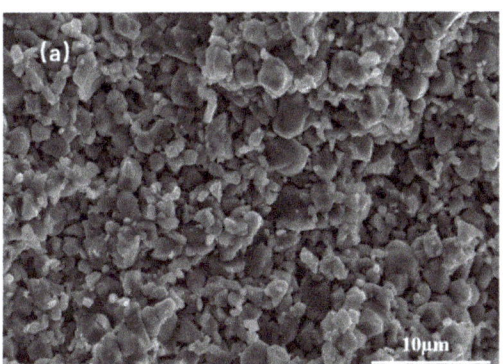 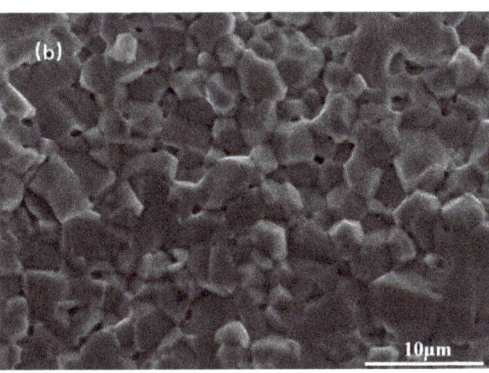

Figure 7. SEM images of the MnZn ferrite samples after annealing at: (**a**) 850 °C and (**b**) 950 °C for 6 h.

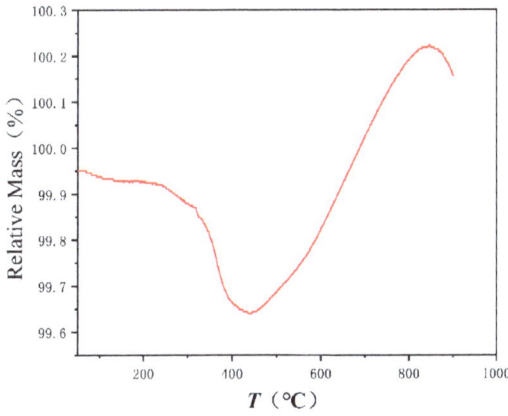

Figure 8. TG curve of the as-CSPed sample.

Figure 9 shows the CSP pressure dependence of the density of the samples before and after annealing at 950 °C for 6 h. The density of the samples with different pressures before annealing was 3.78–4.31 g/cm^3, and the density increased to 4.80–4.91 g/cm^3 after annealing at 950 °C.

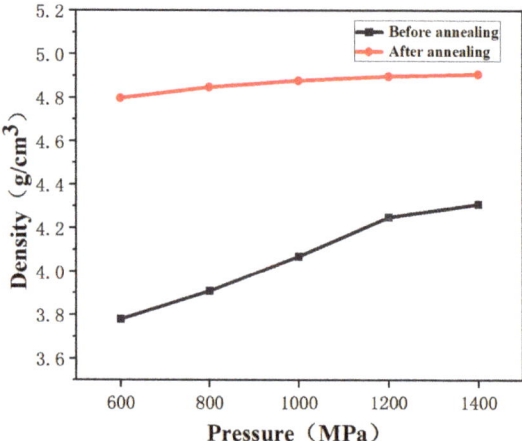

Figure 9. Densities of the samples with different CSP pressures after annealing at 950 °C.

Therefore, the annealing treatment significantly enhanced the density of the sample. Figure 10 shows the relative densities of the samples with different CSP pressures after annealing. The relative density increased remarkably after annealing. The highest relative density increased up to 97.2%, which is compared with the densities of most MnZn ferrites prepared using the conventional high-temperature sintering process. The conventional high-temperature sintering process requires the use of temperatures above 1100 °C to densify the MnZn ferrite.

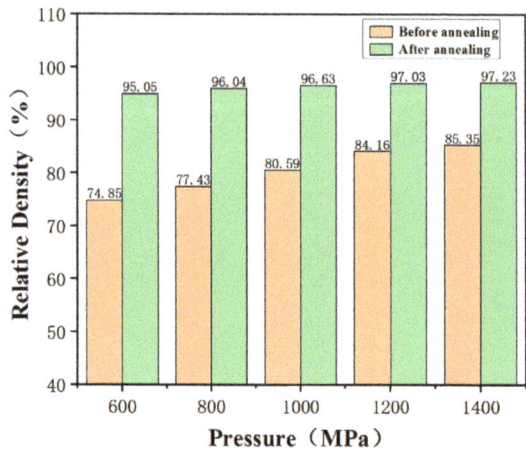

Figure 10. Relative densities of the samples with different CSP pressures after annealing at 950 °C.

After the treatment of grinding and etching, the fractural images of the samples at different pressures after annealing at 950 °C were measured by SEM. Figure 11 shows the corresponding SEM images of the samples. The statistical histogram of the particle size is shown in the corresponding illustration. The grain size was calculated and is shown in Table 1. For the sample with the CSP pressure below 1000 MPa, there are some pores between the grains and the sample is not very compact. When the pressure increases up to above 1200 MPa, the pores decrease and the sample becomes more compact. Therefore, the increase in pressure promotes densification of the sample after annealing. At the same time, the average grain size also increases from 2.67 to 2.90 μm. This is attributed to the fact that the pressure during the cold sintering stage promotes the growth of crystal grains. Due to the no grain boundary phase formed in these samples, there also exist some pores caused by the small grain abscission in the etching process. The average grain size in this series of samples was less than 3 μm, which was also less that the critical size (3.8 ± 0.7 μm) of a single domain in MnZn ferrites. Reducing the grain size below the single-domain size is beneficial to reduce the power loss at high frequencies [27].

Table 1. Average grain size of the MnZn ferrites sintered at different CSP pressures.

Pressure (MPa)	600	800	1000	1200	1400
D_{ave} (μm)	2.67	2.72	2.80	2.87	2.90

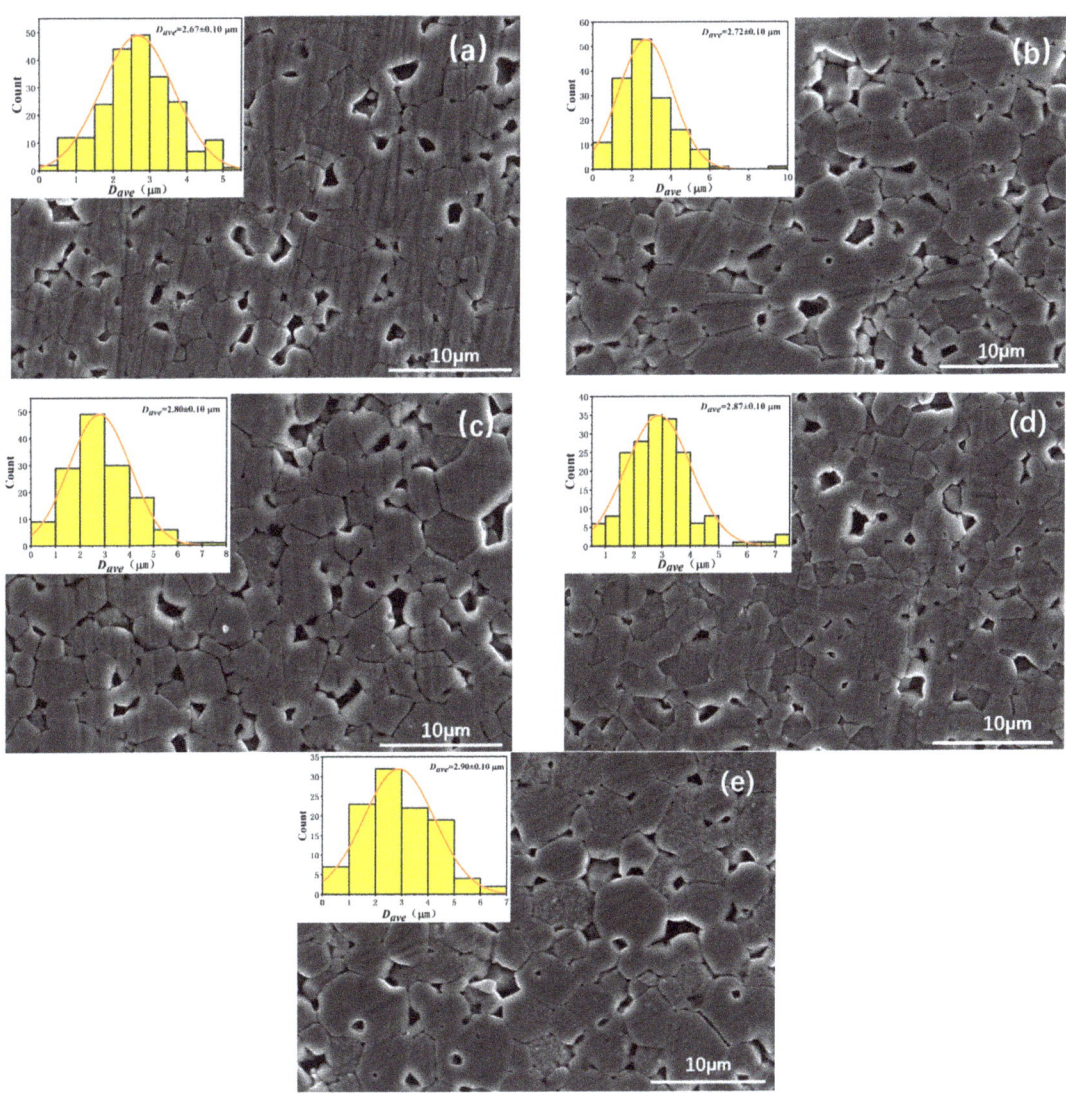

Figure 11. SEM images of samples with different CSP pressure: (**a**) 600, (**b**) 800, (**c**) 1000, (**d**) 1200 and (**e**) 1400 Mpa after annealing at 950 °C.

3.4. Magnetic Properties

Figure 12 shows the temperature dependence of the P_{cv} [$P_{cv}(T)$] for the samples after annealing at 850 and 950 °C for 6 h, under the conditions of 500 kHz/10 mT and 1 MHz/10 mT. The power loss of MnZn ferrite increases with temperature and frequency. The P_{cv} of MnZn ferrite annealed at 850 °C was 342 and 520 kW/cm^3 at 500 kHz/10 mT and 1 MHz/10 mT, 25 °C, respectively. When the annealing temperature increased to 950 °C, the P_{cv} was greatly reduced to 75 and 108 kW/m^3 at 500 kHz/10 mT and 1 MHz/10 mT, respectively. Figure 13 shows the $P_{cv}(T)$ curves for the annealed sample after annealing at 950 °C for 6 h, under the conditions of 500 kHz/30 mT and 1 MHz/30 mT. The $P_{cv}(T)$ curves show similar behavior to those under 10 mT. When the magnetic flux density increased from 10 to 30 mT, the P_{cv} increased.

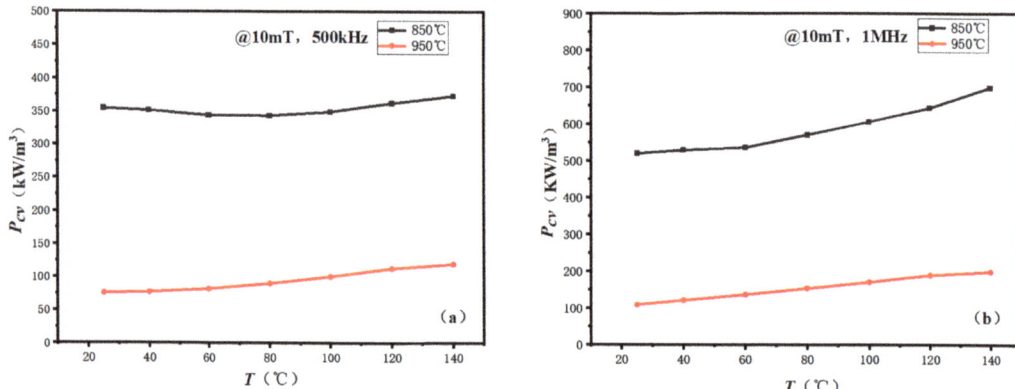

Figure 12. The $P_{cv}(T)$ curves of the CSP samples at different frequencies: (**a**) 500 kHz and (**b**) 1 MHz after annealing for 6 h.

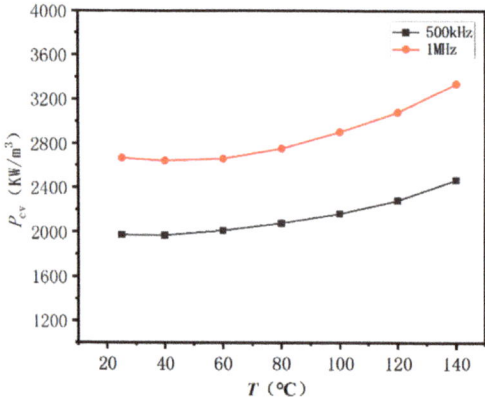

Figure 13. Temperature dependence of P_{cv} for the sample annealed at 950 °C under 500 kHz/30 mT and 1 MHz/30 mT.

As the annealing temperature increases, the density of the sample increases and the porosity decreases. Thus, the power loss decreases. However, compared to the traditional sintering process, the power loss of the CSP samples was higher. The grain boundary of MnZn ferrites can only be formed above 1000 °C [28]. In the conventional sintering process, additives such as CaCO$_3$, SiO$_2$ and V$_2$O$_5$ are generally added to form a high-resistance glass phase on the grain boundary [29–33]. The formation of the high-resistance grain boundary will remarkably reduce the P_{cv}. In this series of samples, the high-resistance grain boundary was not formed. In the next work, the grain boundary additives and sintering aid will be co-added to improve the formation of the high-resistance grain boundary and to reduce the power loss. Through the construction of the high-resistance grain boundary in the fine-grained MnZn ferrite, the power loss at a high frequency may be reduced dramatically and make it a good candidate material with a magnetic core for the high-frequency switching power supply.

4. Conclusions

In this study, MnZn ferrites with fine grains and high density were prepared using the cold sintering process. It is found that the transient solvent and high pressure are the key factors for the CSP process for MnZn ferrite. The as-CSPed MnZn ferrite with a relative density of up to 85% was obtained. After being annealed at 950 °C for 6 h, the relative density was further increased to 97%, while the average grain size was maintained at less than 3.0 μm. The increase in the annealing temperature reduces the P_{cv}s at high frequencies of 500 kHz and 1 MHz.

This work confirms that it is feasible to prepare densified fine-grain MnZn ferrite using CSP, which is a different process from the traditional low-temperature sintering process. The grains of the sample prepared by CSP are more uniform and the average grain size was smaller than the critical size of a single domain. The average grain size in this series of samples was also smaller than that in the sample prepared by the low-temperature sintering process [11]. This work lays a solid foundation for the follow-up development of low-P_{cv} MnZn ferrite by CSP for high frequency applications.

Author Contributions: Conceptualization, Y.Y. and S.C.; Methodology, L.H., Z.L., J.Z., J.Y., L.Q. and J.L.; Software, W.L. and W.C.; Validation, Y.Y.; Formal analysis, Z.L. and J.Y.; Investigation, Y.Y. and L.H.; Resources, D.B., J.Z. and J.L.; Data curation, L.H.; Writing—original draft, Y.Y. and L.H.; Writing—review & editing, Y.Y.; Supervision, Y.Y. and S.C.; Project administration, D.B. and S.C.; Funding acquisition, S.C. All authors have read and agreed to the published version of the manuscript.

Funding: This work was supported by the Intergovernmental International Science, Technology and Innovation Cooperation Key Project of the National Key R&D Programme (No. 2022YFE0109800) and the National Natural Science Foundation of China (NSFC) (No. U1809215).

Institutional Review Board Statement: Not applicable.

Informed Consent Statement: Not applicable.

Data Availability Statement: The data presented in this study are available on request from the corresponding author. The data are not publicly available due to privacy.

Conflicts of Interest: The authors declare no conflict of interest.

References

1. Praveena, K.; Sadhana, K.; Bharadwaj, S.; Murthy, S.R. Development of nanocrystalline Mn–Zn ferrites for high frequency transformer applications. *J. Magn. Magn. Mater.* **2009**, *321*, 2433–2437. [CrossRef]
2. Ott, G.; Wrba, J.; Lucke, R. Recent developments of Mn-Zn ferrites for high permeability applications. *J. Magn. Magn. Mater.* **2003**, *254*, 535–537. [CrossRef]
3. Jiang, K.; Li, K.; Peng, C.; Zhu, Y. Effect of multi-additives on the microstructure and magnetic properties of high permeability Mn–Zn ferrite. *J. Alloy. Compd.* **2012**, *541*, 472–476. [CrossRef]
4. Jeong, W.H.; Song, B.M.; Han, Y.H. Analysis of Power Losses in Mn–Zn Ferrites. *Jpn. J. Appl. Phys.* **2002**, *41*, 2912–2915. [CrossRef]
5. Liu, Y.; He, S. Development of low loss Mn–Zn ferrite working at frequency higher than 3 MHz. *J. Magn. Magn. Mater.* **2008**, *320*, 3318–3322. [CrossRef]
6. Tokatlidis, S.; Kogias, G.; Zaspalis, V.T. Low loss MnZn ferrites for applications in the frequency region of 1–3 MHz. *J. Magn. Magn. Mater.* **2018**, *465*, 727–735. [CrossRef]
7. Sai, R.; Ralandinliu Kahmei, R.D.; Shivashankar, S.A.; Yamaguchi, M. High-Q On-Chip C-Band Inductor With a Nanocrystalline MnZn-Ferrite Film Core. *IEEE Trans. Magn.* **2019**, *55*, 2801304. [CrossRef]
8. Yi, S.; Bai, G.; Wang, X.; Zhang, X.; Hussain, A.; Jin, J.; Yan, M. Development of high-temperature high-permeability MnZn power ferrites for MHz application by Nb_2O_5 and TiO_2 co-doping. *Ceram. Int.* **2020**, *46*, 8935–8941. [CrossRef]

9. Wu, T.; Ju, D.; Wang, C.; Huang, H.; Li, C.; Wu, C.; Wang, C.; Liu, H.; Jiang, X.; Ye, K.; et al. Ferrite materials with high saturation magnetic induction intensity and high permeability for magnetic field energy harvesting: Magnetization mechanism and Brillouin function temperature characteristics. *J. Alloy. Compd.* **2023**, *933*, 167654. [CrossRef]
10. Papazoglou, P.; Eleftheriou, E.; Zaspalis, V.T. Low sintering temperature MnZn-ferrites for power applications in the frequency region of 400 kHz. *J. Magn. Magn. Mater.* **2006**, *296*, 25–31. [CrossRef]
11. Ying, Y.; Xiong, X.B.; Wang, N.C.; Zheng, J.W.; Yu, J.; Li, W.C.; Qiao, L.; Cai, W.; Li, J.; Huang, H.; et al. Low temperature sintered MnZn ferrites for power applications at the frequency of 1 MHz. *J. Eur. Ceram. Soc.* **2021**, *41*, 5924–5930. [CrossRef]
12. Rajan, A.; Solaman, S.K.; Ganesanpotti, S. Cold Sintering: An Energy-Efficient Process for the Development of $SrFe_{12}O_{19}$–Li_2MoO_4 Composite-Based Wide-Bandwidth Ferrite Resonator Antenna for Ku-Band Applications. *ACS Appl. Electron. Mater.* **2021**, *3*, 2297–2308. [CrossRef]
13. Maria, J.P.; Kang, X.Y.; Floyd, R.D.; Dickey, E.C.; Guo, H.Z.; Guo, J.; Baker, A.; Funihashi, S.; Randall, C.A. Cold sintering: Current status and prospects. *J. Mater. Res.* **2017**, *32*, 3205–3218. [CrossRef]
14. Funahashi, S.; Kobayashi, E.; Kimura, M.; Shiratsuyu, K.; Randall, C.A. Chelate complex assisted cold sintering for spinel ceramics. *J. Ceram. Soc. Jpn.* **2019**, *127*, 899–904. [CrossRef]
15. Liu, Y.L.; Sun, Q.; Wang, D.W.; Adair, K.; Liang, J.N.; Sun, X.L. Development of the cold sintering process and its application in solid-state lithium batteries. *J. Power Sources* **2018**, *393*, 193–203. [CrossRef]
16. Galotta, A.; Sglavo, V.M. The cold sintering process: A review on processing features, densification mechanisms and perspectives. *J. Eur. Ceram. Soc.* **2021**, *41*, 1–17. [CrossRef]
17. Yu, T.; Cheng, J.; Li, L.; Sun, B.; Bao, X.; Zhang, H. Current understanding and applications of the cold sintering process. *Front. Chem. Sci. Eng.* **2019**, *13*, 654–664. [CrossRef]
18. Vakifahmetoglu, C.; Karacasulu, L. Cold sintering of ceramics and glasses: A review. *Curr. Opin. Solid State Mater. Sci.* **2020**, *24*, 100807. [CrossRef]
19. Guo, H.; Guo, J.; Baker, A.; Randall, C.A. Hydrothermal-Assisted Cold Sintering Process: A New Guidance for Low-Temperature Ceramic Sintering. *ACS Appl. Mater. Interfaces* **2016**, *8*, 20909–20915. [CrossRef]
20. Guo, J.; Berbano, S.S.; Guo, H.; Baker, A.L.; Lanagan, M.T.; Randall, C.A. Cold Sintering Process of Composites: Bridging the Processing Temperature Gap of Ceramic and Polymer Materials. *Adv. Funct. Mater.* **2016**, *26*, 7115–7121. [CrossRef]
21. Kamani, M.; Yourdkhani, A.; Poursalehi, R.; Sarraf-Mamoory, R. Studying the cold sintering process of zinc ferrite as an incongruent dissolution system. *Int. J. Ceram. Eng. Sci.* **2019**, *1*, 125–135. [CrossRef]
22. Si, M.; Hao, J.; Zhao, E.; Zhao, X.; Guo, J.; Wang, H.; Randall, C.A. Preparation of zinc oxide/poly-ether-ether-ketone (PEEK) composites via the cold sintering process. *Acta Mater.* **2021**, *215*, 117036. [CrossRef]
23. Katsnelson, E.Z.; Karoza, A.G.; Meleshchenko, L.A.; Pankov, V.V.; Khavkin, B.G. Detailed Study of IR Absorption Spectra of Manganese–Zinc Ferrites. *Phys. Status Solidi (b)* **1987**, *141*, 599–609. [CrossRef]
24. Mohamed, I.; Zaki, M.A.H. Lata Pasupulety and Kamlesh Kumari. Bulk and surface characteristics of pure and alkalized TG, Mn_2O_3: IR, XRD, XPS, specific adsorption and redox catalytic studies. *New J. Chem* **1998**, *22*, 875–882.
25. Ndayishimiye, A.; Sengul, M.Y.; Bang, S.H.; Tsuji, K.; Takashima, K.; Hérisson de Beauvoir, T.; Denux, D.; Thibaud, J.-M.; van Duin, A.C.T.; Elissalde, C.; et al. Comparing hydrothermal sintering and cold sintering process: Mechanisms, microstructure, kinetics and chemistry. *J. Eur. Ceram. Soc.* **2020**, *40*, 1312–1324. [CrossRef]
26. Ivakin, Y.D.; Smirnov, A.V.; Kurmysheva, A.Y.; Kharlanov, A.N.; Solis Pinargote, N.W.; Smirnov, A.; Grigoriev, S.N. The Role of the Activator Additives Introduction Method in the Cold Sintering Process of ZnO Ceramics: CSP/SPS Approach. *Materials* **2021**, *14*, 6680. [CrossRef]
27. Van der Zaag, P.J. New views on the dissipation in soft magnetic ferrites. *J. Magn. Magn. Mater.* **1999**, *196*, 315–319. [CrossRef]
28. ShengleiChe, M.W.; Mori, K.; Sato, N.; Aoki, T.; Murase, T. Dissolution-segregation behavior of Ca during the sintering process of MnZn ferrite. *J. Jpn. Soc. Powder Powder Metall.* **2009**, *56*, 111–115.
29. Shokrollahi, H.; Janghorban, K. Influence of additives on the magnetic properties, microstructure and densification of Mn–Zn soft ferrites. *Mater. Sci. Eng. B* **2007**, *141*, 91–107. [CrossRef]
30. Wang, S.-F.; Chiang, Y.-J.; Hsu, Y.-F.; Chen, C.-H. Effects of additives on the loss characteristics of Mn–Zn ferrite. *J. Magn. Magn. Mater.* **2014**, *365*, 119–125. [CrossRef]
31. Zhang, L.; Li, X.L.; Liu, D.; Ying, Y.; Jiang, L.Q.; Che, S.L. Influence of Particle Size and Ca Addition on the Magnetic Performance and Microstructure of MnZn Ferrite. *Mater. Sci. Forum* **2014**, *787*, 362–367. [CrossRef]
32. Xu, J.; Bai, G.; Fan, X.; Zhang, Z.; Liu, X.; Yan, M. Two-step doping of SiO_2 and CaO for high-frequency MnZn power ferrites. *J. Eur. Ceram. Soc.* **2023**, *43*, 2469–2478. [CrossRef]
33. Janghorban, K.; Shokrollahi, H. Influence of V_2O_5 addition on the grain growth and magnetic properties of Mn–Zn high permeability ferrites. *J. Magn. Magn. Mater.* **2007**, *308*, 238–242. [CrossRef]

Disclaimer/Publisher's Note: The statements, opinions and data contained in all publications are solely those of the individual author(s) and contributor(s) and not of MDPI and/or the editor(s). MDPI and/or the editor(s) disclaim responsibility for any injury to people or property resulting from any ideas, methods, instructions or products referred to in the content.

Article

Influence of MnZn Ferrite Homogeneous Fibers on the Microstructure, Magnetic, and Mechanical Properties of MnZn Ferrite Materials

Yajing Shang, Fan Luo and Zhongxia Duan *

Institute of Electrical Engineering, Chinese Academy of Sciences, Beijing 100190, China
* Correspondence: duanzhongxia@mail.iee.ac.cn

Abstract: MnZn ferrite homogeneous fibers were synthesized via a simple solvothermal method and they were used as a reinforcing phase to prepare homogeneous-fiber-reinforced MnZn ferrite materials. The effects of MnZn ferrite homogeneous fibers (0 wt% to 4 wt%) doping on the microstructure, magnetic, and mechanical properties of MnZn ferrite materials were studied systematically. The results showed that MnZn ferrite homogeneous fibers exhibited high purity, good crystallinity, and smooth 1D fibrous structures, which were homogeneous with MnZn ferrite materials. Simultaneously, a certain content of MnZn ferrite homogeneous fibers helped MnZn ferrite materials exhibit more uniform and compact crystal structures, less porosity, and fewer grain boundaries. In addition, the homogeneous-fiber-reinforced MnZn ferrite materials possessed superior magnetic and mechanical properties such as higher effective permeability, lower magnetic loss, and higher Vickers hardness compared to ordinary MnZn ferrite materials. In addition, the magnetic and mechanical properties of homogeneous-fiber-reinforced MnZn ferrite materials first increased and then gradually decreased as the homogeneous fiber content increased from 0 wt% to 4 wt%. The best magnetic and mechanical properties of materials were obtained as the fiber content was about 2 wt%.

Keywords: MnZn ferrite homogeneous fibers; homogeneous-fiber-reinforced MnZn ferrite; microstructure; magnetic properties; mechanical properties

Citation: Shang, Y.; Luo, F.; Duan, Z. Influence of MnZn Ferrite Homogeneous Fibers on the Microstructure, Magnetic, and Mechanical Properties of MnZn Ferrite Materials. *Materials* 2023, 16, 209. https://doi.org/10.3390/ma16010209

Academic Editor: Shenglei Che

Received: 11 November 2022
Revised: 20 December 2022
Accepted: 22 December 2022
Published: 26 December 2022

Copyright: © 2022 by the authors. Licensee MDPI, Basel, Switzerland. This article is an open access article distributed under the terms and conditions of the Creative Commons Attribution (CC BY) license (https://creativecommons.org/licenses/by/4.0/).

1. Introduction

MnZn ferrite is an important member of the soft magnetic materials, which has been widely used in network communication, information storage, automatic control, medical diagnosis, and various apparatuses from household appliances to scientific equipment [1–3]. Moreover, it has superior characteristics such as high magnetic permeability, high electrical resistivity, high stability, large magnetic induction, and low magnetic loss [4–6]. Thus, it is crucial to develop MnZn ferrite exhibiting better performances with the rapid development of modern science and technology.

After decades of studies, it is well known that doping a small amount of oxides is one of the most effective strategies for improving the performances of MnZn ferrite. Common oxides doped in MnZn ferrite have been systematically researched and can be divided into three types: First, dopants such as SiO_2 [7], CaO [8], and HfO_2 [9] tend to change the chemical compositions and further adjust the resistivity of grain boundaries. The first kind of dopants can enhance the resistivity and lower the loss of MnZn ferrite, but the segregation of the above cations at the grain boundaries may bring down the initial permeability of the material due to magnetic dilution. Second, soluble oxides involving TiO_2 [10], SnO_2 [11], NiO [12], and Co_2O_3 [13] are easy to dissolve into the lattices and alter the performance. The second kind of dopants can effectively adjust the intrinsic magnetic properties of MnZn ferrite. However, they may also reduce partial magnetic properties of the material. Third, dopants with low melting points such as V_2O_5 [14], Bi_2O_3 [15], and MoO_3 [16] can modify the microstructure of the MnZn ferrite during the process of sintering. The third kind of dopants is critical for improving the permeability and reducing the loss of MnZn ferrite. Larger amounts of non-magnetic dopants will damage the microstructure and reduce the properties of the material. Accordingly, it can be found that the addition of the above dopants can improve some performances of MnZn ferrite, but often lead to the deterioration of other performances. Therefore, the MnZn ferrite homogeneous-fiber- and particle-co-reinforced structure is proposed to prepare the MnZn ferrite with excellent microstructure, magnetic, and mechanical properties.

In this work, MnZn ferrite homogenous fibers synthesized by the solvothermal method were used as the reinforcing phase to prepare homogeneous-fiber-reinforced MnZn ferrite materials. The microstructure, phase composition, magnetic, and mechanical properties of materials were characterized by scanning electron microscopy (SEM), X-ray diffraction (XRD), an LCR meter, a magnetic automatic test device, and a hardness tester. Furthermore, the effect of homogeneous fiber content (0 wt% to 4 wt%) on the homogeneous-fiber-reinforced MnZn ferrite materials was investigated in detail.

2. Experimental Details

2.1. Synthesis of MnZn Ferrite Homogenous Fibers

$MnSO_4 \cdot 4H_2O$ (≥99%), $ZnSO_4 \cdot 7H_2O$ (≥99%), $FeSO_4 \cdot 7H_2O$ (≥99%), $H_2C_2O_4$ (≥98%), ethylene glycol (EG, ≥99%), and deionized water were used as raw materials. In a typical process, 0.5 mmol $MnSO_4 \cdot 4H_2O$, 0.5 mmol $ZnSO_4 \cdot 7H_2O$, and 2 mmol $FeSO_4 \cdot 7H_2O$ were dissolved in a mixture of EG and deionized water with a volume of 30 mL (EG:water = 3:1). Then, 3 mmol $H_2C_2O_4$ was added into the above mixture under continuous stirring for 60 min. Subsequently, the obtained mixture was transferred into a Para polyphenyl-lined autoclave of 50 mL capacity and heated at 120 °C for 24 h. After that, the products were centrifuged to separate solids, washed with ethanol and deionized water for several times, and dried in an oven at 80 °C for 12 h. Finally, the as-prepared precursor fibers were annealed in air atmosphere at 500 °C for 2 h. The MnZn ferrite homogenous fibers, as a nominal composition of $Mn_{0.5}Zn_{0.5}Fe_2O_4$, were collected by the solvothermal method.

2.2. Synthesis of Homogeneous-Fiber-Reinforced MnZn Ferrites

MnO (≥99%), ZnO (≥99%), Fe_2O_3 (≥99%), and MnZn ferrite homogenous fibers ($Mn_{0.5}Zn_{0.5}Fe_2O_4$, made by our laboratory) were used as raw materials. The preparation process was as follows: First, MnO, ZnO, and Fe_2O_3 were evenly mixed for 120 min at a speed of 300 r/min in the planetary mill. After that, the evenly mixed powders were dried at 90 °C for 10 h in the oven. Subsequently, the dried powders were pre-sintered at 850 °C for 2.5 h in an air atmosphere and cooled to room temperature. Next, the pre-sintered materials were evenly mixed for 120 min at a speed of 300 r/min, and then dried at 90 °C for 10 h. Additionally, the dried materials were mixed with 0.85 wt% polyvinyl alcohol for granulation, followed by 50 mesh screening to obtain the MnZn ferrite powders. Then, the as-prepared MnZn ferrite homogeneous fibers of 1 wt%, 2 wt%, 3 wt%, and 4 wt%, together with MnZn ferrite powders, were evenly mixed and dried to obtain the 1 wt%, 2 wt%, 3 wt%, and 4 wt% homogeneous-fiber-reinforced MnZn ferrite composite powders. After that, the composite powders were pressed into toroidal shapes with the dimensions of OD (outer diameter) 20 mm, ID (inner diameter) 10 mm, and h (height) 8 mm under an external load of 350 MPa for 15 min. Finally, the annular samples were sintered at 1280 °C for 6 h in a Blank equilibrium atmosphere-controlled sintering system to obtain the required 1 wt%, 2 wt%, 3 wt%, and 4 wt% homogeneous-fiber-reinforced MnZn ferrite samples. In addition, the ordinary MnZn ferrite sample was prepared with 0 wt% MnZn ferrite homogeneous fibers by the above preparation process.

2.3. Characterizations

The surface morphology of the samples was characterized by SEM (Supra 35VP, ZEISS, Jena, Germany). The phase composition and crystallographic structure of the samples were analyzed by XRD (D8 Advance, Bruker, Karlsruhe, Germany) over the 2θ range of 10° and 80° with Cu Kα irradiation (λ = 0.15418 nm) at 40 kV and 40 mA. The LCR meter (IM3570A988-06, HIOKI E.E. CORPORATION, Hioki, Japan) was used to measure the effective permeability of the samples at room temperature with a frequency range of 10 kHz to 1000 kHz. The magnetic automatic test device (MATS-3000SA, Hunan Linkjoin Technology Co., Ltd., Loudi, China) was used to measure the magnetic loss of the samples at room temperature in the frequency range from 10 kHz to 1000 kHz. The Vickers hardness of the samples was determined on a hardness tester (HV-1000ZDT, Shanghai Jvjing Precision Instrument Manufacturing Co., Ltd., Shanghai, China). The load was 200 g and the loading time was 15 s.

3. Results and Discussion

3.1. Characterization of the MnZn Ferrite Homogeneous Fibers

The general SEM image of the as-prepared MnZn ferrite homogeneous fibers is shown in Figure 1a. It can be seen that the typical MnZn ferrite homogeneous fibers exhibited a 1D fibrous structure with a large aspect ratio and had a smooth surface area. The XRD pattern of the corresponding MnZn ferrite homogeneous fibers is shown in Figure 1b. It can be observed that the major diffraction peaks appearing at 18.1°, 29.8°, 35.1°, 36.7°, 42.6°, 52.9°, 56.3°, 61.8°, and 73.3° were assigned to the (111), (220), (311), (222), (400), (422), (511), (440), and (622) planes, respectively, and matched well with the standard powder diffraction data of MnZn ferrite, which indicated the high purity and good crystallinity of MnZn ferrite homogeneous fibers. In addition, the MnZn ferrite homogeneous fibers had the same chemical composition as the MnZn ferrites. Therefore, it can be known that the MnZn ferrite homogeneous fibers prepared by the solvothermal method had high purity, good crystallinity, and smooth 1D fibrous structures, which were homogeneous with MnZn ferrites.

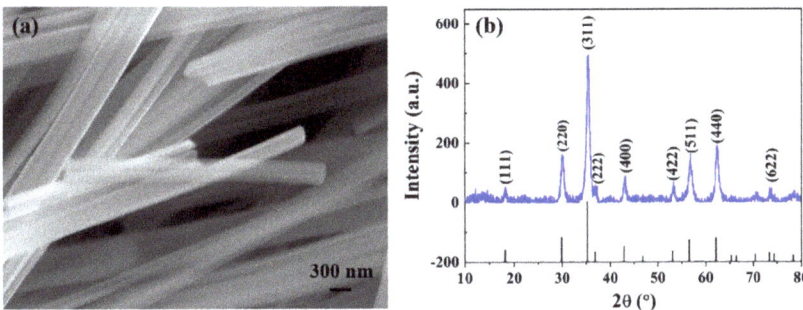

Figure 1. (a,b) SEM image and XRD spectrum of the MnZn ferrite homogeneous fibers.

3.2. Microstructure Characterization Analysis of MnZn Ferrite

The SEM images of typical MnZn ferrite samples are shown in Figure 2. Figure 2a,b display a cross-sectional morphology of ordinary MnZn ferrite with 0 wt% MnZn ferrite homogeneous fibers. Figure 2c,d reveal a representative cross-sectional morphology of homogeneous-fiber-reinforced MnZn ferrite with 2 wt% MnZn ferrite homogeneous fibers. It can be clearly seen from Figure 2a,b that the 0 wt% ordinary MnZn ferrite sample had a relatively good crystal structure, while there were small pores between the grains of the sample. With the introduction of MnZn ferrite homogenous fibers, it can be noticed from Figure 2c,d that the 2 wt% homogeneous-fiber-reinforced MnZn ferrite sample had a more uniform and compact crystal structure. Meanwhile, the sample exhibited less porosity and fewer grain boundaries.

Figure 2. SEM images of typical MnZn ferrite samples. (**a,b**) 0 wt% ordinary MnZn ferrite; (**c,d**) 2 wt% homogeneous-fiber-reinforced MnZn ferrite.

This was because the introduction of MnZn ferrite homogenous fibers took the bridge between the grains, which could induce grain crystallization and promote the homogeneity of the materials. In addition, the nanoscale MnZn ferrite homogeneous fibers with finer grain size than the MnZn ferrite particles could make the composite powders rearrange at the low-temperature sintering stage, and the homogeneous fibers can be filled into the small pores between the grains, gradually eliminating the pores and grain boundaries of materials. Thus, the homogeneous-fiber-reinforced MnZn ferrite exhibited a better microstructure compared with the ordinary MnZn ferrite.

3.3. Magnetic Properties Analysis of the Homogeneous Fiber Reinforced MnZn Ferrite

To understand the function of MnZn ferrite homogeneous fibers on the magnetic properties of the homogeneous-fiber-reinforced MnZn ferrites, the effective permeability and magnetic loss of the homogeneous-fiber-reinforced MnZn ferrite samples prepared with different fiber contents (0 wt% to 4 wt%) were measured by an LCR meter and magnetic automatic test device, respectively. The effective permeability of the homogeneous-fiber-reinforced MnZn ferrite samples with different fiber contents (0 wt% to 4 wt%) versus frequency at room temperature is shown in Figure 3. It can be seen that the effective permeability of all samples first decreased and then gradually increased with the increase in the frequency range from 10 kHz to 1000 kHz. At the same frequency, the effective permeability of the 1 wt%, 2 wt%, 3 wt%, and 4 wt% homogeneous-fiber-reinforced MnZn ferrite samples was higher than that of the ordinary MnZn ferrite sample with 0 wt% homogeneous fibers. In addition, it can be found that the effective permeability of the homogeneous-fiber-reinforced MnZn ferrite samples increased with the increase in fiber content as the content of MnZn ferrite homogeneous fibers was less than 2 wt%, and then gradually decreased with the increase in fiber content as the fiber content was larger than 2 wt%. The effective permeability reached a maximum as the fiber content was 2 wt%.

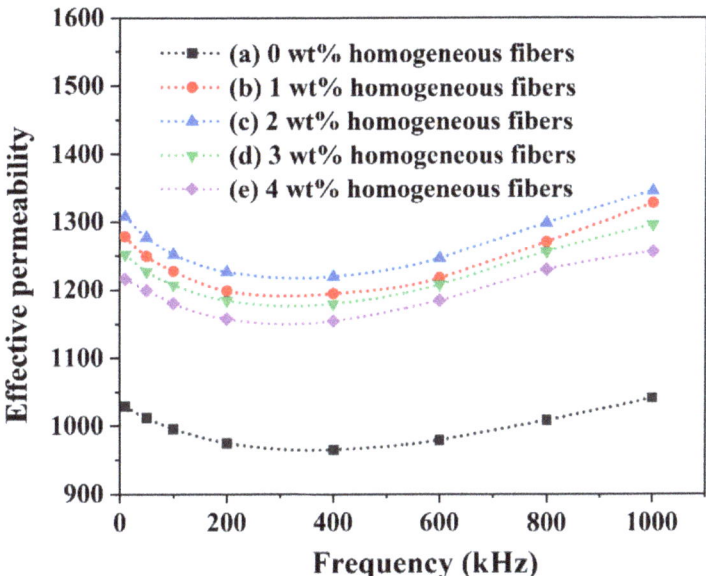

Figure 3. Effective permeability of the homogeneous-fiber-reinforced MnZn ferrite samples versus frequency at room temperature: (a) 0 wt%, (b) 1 wt%, (c) 2 wt%, (d) 3 wt%, and (e) 4 wt%.

It can be known that the internal grain size and grain interface structure of MnZn ferrite materials affected the dynamic balance during the magnetization process [17]. The permeability was directly proportional to the average grain sizes and inversely proportional to the number of grain boundaries and pores [18]. Ordinary MnZn ferrite with 0 wt% homogeneous fibers had more pores and grain boundaries than the homogeneous-fiber-reinforced MnZn ferrites, so the effective permeability was relatively low. As the homogeneous fiber content increased, the crystal grains gradually grew and became uniform, so the pores were gradually eliminated and the effective permeability gradually increased. When the homogeneous fiber content exceeded 2 wt%, owing to the fact that the fibers deposited at grain boundaries and impeded domain wall motion, the permeability decreased.

Figure 4 shows the magnetic loss of the homogeneous-fiber-reinforced MnZn ferrite samples with different fiber contents (0 wt% to 4 wt%) versus frequency at room temperature. It can be found that the magnetic loss of all samples increased gradually with the increase in frequency. At the same frequency, the magnetic loss of the homogeneous-fiber-reinforced MnZn ferrite samples with the fiber content from 1 wt% to 4 wt% was lower than that of the ordinary MnZn ferrite sample with 0 wt% homogeneous fibers. In addition, the magnetic loss of the homogeneous-fiber-reinforced MnZn ferrite samples first decreased and then gradually increased with the rise in the homogeneous fiber contents from 1 wt% to 4 wt%. The minimum magnetic loss was obtained as the fiber content was 2 wt%.

The cause of this phenomenon was that the MnZn ferrite with 2 wt% homogeneous fibers had a uniform and compact crystal structure, which could effectively restrain the interaction between magnetic particles and greatly reduce the magnetic loss. Thus, a certain content of MnZn ferrite homogenous fibers was beneficial to improve the magnetic properties of MnZn ferrite materials. Meanwhile, the homogeneous-fiber-reinforced MnZn ferrite materials exhibited the best magnetic properties as the fiber content was about 2 wt%.

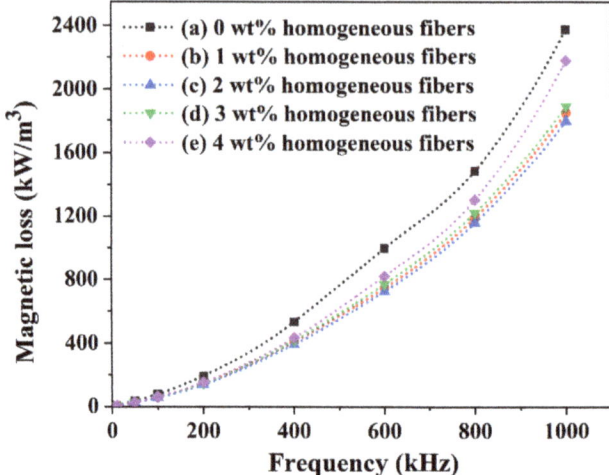

Figure 4. Magnetic loss of the homogeneous-fiber-reinforced MnZn ferrite samples versus frequency at room temperature: (a) 0 wt%, (b) 1 wt%, (c) 2 wt%, (d) 3 wt%, and (e) 4 wt%.

3.4. Mechanical Properties Analysis of the Homogeneous-Fiber-Reinforced MnZn Ferrite

To study the impact of the MnZn ferrite homogeneous fibers on the mechanical properties of the homogeneous-fiber-reinforced MnZn ferrites, the Vickers hardness of the homogeneous-fiber-reinforced MnZn ferrite samples synthesized with different fiber contents (0 wt% to 4 wt%) was measured by the hardness tester. Figure 5 shows the Vickers hardness of the homogeneous-fiber-reinforced MnZn ferrite samples with different fiber contents (0 wt% to 4 wt%). Meanwhile, the Vickers hardness values for seven groups of repetitive tests of the homogeneous-fiber-reinforced MnZn ferrite samples with different fiber contents (0 wt% to 4 wt%) are shown in Table 1. As we can see, the average Vickers hardness values of 0 wt%, 1 wt%, 2 wt%, 3 wt%, and 4 wt% homogeneous-fiber-reinforced MnZn ferrite samples were 530.1, 579.6, 603.3, 585.1, and 540.9 HV, respectively. The corresponding standard deviations of the samples were 3.49, 2.16, 2.06, 3.06, and 2.96 HV. It can be seen that the deviations in Vickers hardness for the samples were within a certain range. The results indicated that the 1 wt%, 2 wt%, 3 wt%, and 4 wt% homogeneous-fiber-reinforced MnZn ferrite samples had a much higher Vickers hardness than the 0 wt% ordinary MnZn ferrite sample. At the same time, the Vickers hardness of the corresponding homogeneous-fiber-reinforced MnZn ferrite samples first increased and then gradually decreased with the increase in homogeneous fiber contents from 1 wt% to 4 wt%. In addition, the maximum Vickers hardness was obtained as the fiber content was 2 wt%, which was consistent with the test results of magnetic properties.

It is known that good crystal structure is the guarantee of materials with high magnetic and mechanical properties. For the homogeneous-fiber-reinforced MnZn ferrite materials, the MnZn ferrite homogeneous fibers were used as the reinforcing phase to construct a co-reinforced structure of homogeneous fibers and particles. In this co-reinforced structure, the homogeneous fibers could effectively weaken the damage caused by stress concentration, so the particles could improve the microregional stress distribution of materials [19,20]. In the crystallization process, the added homogeneous fibers could bridge between the original particles, providing more paths for the diffusion of atoms during sintering, strengthening the reaction and crystallization kinetic conditions. In addition, the addition of fibers had an impact on the movement of dislocations on the matrix. The fibers were obstacles for the dislocation movement, and internal stresses also occurred. Therefore, the hardness of MnZn ferrite materials increased. More importantly, the MnZn ferrite homogeneous fibers had the same chemical composition, perfect lattice matching, and good surface wettability with the MnZn ferrite particles [21,22]. It can completely overcome the problem of unstable interface structures between heterogeneous fibers and particles, due to their different thermal conductivity, thermal expansion coefficient, elastic modulus, and Poisson's ratio. Accordingly, the strength and stability of homogeneous-fiber-reinforced MnZn ferrite materials were enhanced. Thus, doping a certain content of MnZn ferrite homogenous fibers gave an optimized crystal structure of MnZn ferrite materials, improving the strength and stability of materials. Meanwhile, homogeneous-fiber-reinforced MnZn ferrite materials exhibited the best mechanical properties as the fiber content was about 2 wt%.

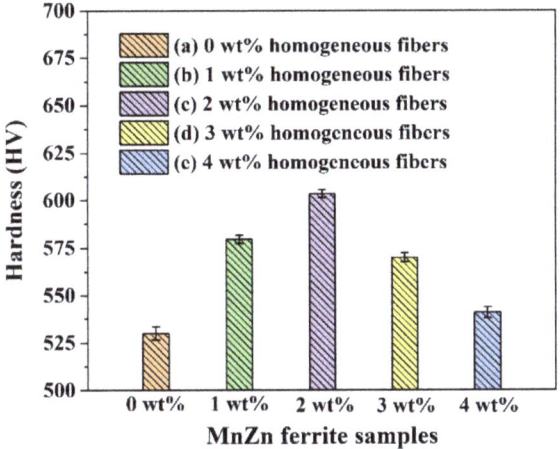

Figure 5. Vickers hardness of homogeneous-fiber-reinforced MnZn ferrite samples with different fiber contents: (a) 0 wt%, (b) 1 wt%, (c) 2 wt%, (d) 3 wt%, and (e) 4 wt%.

Table 1. Vickers hardness values for seven repetitions of homogeneous-fiber-reinforced MnZn ferrite samples with different fiber contents: (a) 0 wt%, (b) 1 wt%, (c) 2 wt%, (d) 3 wt%, and (e) 4 wt%.

Samples	Vickers Hardness (HV)							Average Value (HV)	Standard Deviation (HV)
	1	2	3	4	5	6	7		
(a) 0 wt%	525.2	526.7	529.7	528.2	531.2	534.3	535.3	530.1	3.49
(b) 1 wt%	580.1	579.0	577.8	583.6	578.4	581.3	576.7	579.6	2.16
(c) 2 wt%	606.6	601.6	605.3	601.6	600.4	603.5	604.1	603.3	2.06
(d) 3 wt%	567.6	569.8	573.2	566.5	572.1	571.5	568.1	585.1	3.06
(e) 4 wt%	542.1	537.4	545.2	540.5	541.5	543.6	536.3	540.9	2.96

4. Conclusions

In this paper, we put forward a MnZn ferrite homogeneous-fiber- and particle-co-reinforced structure and revealed the effect of MnZn ferrite homogeneous fibers (0 wt% to 4 wt%) doping on the microstructures as well as magnetic and mechanical characteristics of MnZn ferrite materials. The result showed that MnZn ferrite homogeneous fibers prepared via the solvothermal method exhibited high purity, good crystallinity, and smooth 1D fibrous structures, which were homogeneous with MnZn ferrite materials. In addition, the homogeneous-fiber-reinforced MnZn ferrite materials had a denser crystal structure, less porosity, and fewer grain boundaries, exhibiting excellent properties of high permeability, low magnetic loss, and high Vickers hardness in comparison to 0 wt% ordinary MnZn ferrite materials. Simultaneously, the magnetic and mechanical properties of the homogeneous-fiber-reinforced MnZn ferrites first increased and then gradually decreased with the rise in the homogeneous fiber contents from 1 wt% to 4 wt%. The optimal content of the homogeneous fibers was around 2 wt%.

Author Contributions: Conceptualization, Z.D. and Y.S.; methodology, Y.S. and F.L.; validation, Y.S., F.L. and Z.D.; formal analysis, Y.S.; investigation, Y.S.; resources, Z.D.; data curation, Y.S.; writing—original draft preparation, Y.S.; writing—review and editing, Y.S., F.L. and Z.D.; visualization, Y.S.; supervision, F.L. and Z.D.; project administration, Z.D; funding acquisition, Z.D. All authors have read and agreed to the published version of the manuscript.

Funding: The work was supported by the National Natural Science Foundation of China Youth Fund "Homogeneous fiber and strong magnetic field control the axial crystallization and performance of large aspect ratio MnZn ferrite" (51907186), the STS Regional Key Project of the Chinese Academy of Sciences "Induction heating key magnetic permeability preparation and industrialization technology development" (KFJ-STS-QYZD-142), and the Beijing Municipal Science and Technology Commission Huairou Science City Special Project "Induction heating key permeable magnet preparation and industrial application" (Z201100008420008).

Institutional Review Board Statement: Not applicable.

Informed Consent Statement: Not applicable.

Data Availability Statement: Not applicable.

Conflicts of Interest: The authors declare no conflict of interest.

References

1. Wang, S.-F.; Chiang, Y.-J.; Hsu, Y.-F.; Chen, C.-H. Effects of additives on the loss characteristics of Mn-Zn ferrite. *J. Magn. Magn. Mater.* **2014**, *365*, 119–125. [CrossRef]
2. Wu, G.H.; Yu, Z.; Sun, K.; Guo, R.D.; Zhang, H.Y.; Jiang, X.N.; Wu, C.J.; Lan, Z.W. Ultra-low core losses at high frequencies and temperatures in MnZn ferrites with nano-$BaTiO_3$ additives. *J. Alloy. Compd.* **2020**, *821*, 153573. [CrossRef]
3. Wang, X.Y.; Yi, S.B.; Wu, C.; Bai, G.H.; Yan, M. Correlating the microstructure and magnetic properties of MnZn power ferrites via Co_2O_3 and MoO_3 co-doping for MHz applications. *J. Magn. Magn. Mater.* **2021**, *538*, 168324. [CrossRef]
4. Petrescu, L.-G.; Petrescu, M.-C.; Ioniță, V.; Cazacu, E.; Constantinescu, C.-D. Magnetic properties of manganese-zinc soft ferrite ceramic for high frequency applications. *Materials* **2019**, *12*, 3173. [CrossRef] [PubMed]
5. Yi, S.B.; Bai, G.H.; Wang, X.Y.; Zhang, X.F.; Hussain, A.; Jin, J.Y.; Yan, M. Development of high-temperature high-permeability MnZn power ferrites for MHz application by Nb_2O_5 and TiO_2 co-dopin. *Ceram. Int.* **2020**, *46*, 8935–8941. [CrossRef]
6. Fang, X.G.; Ma, D.Y.; Sun, B.W.; Xu, X.P.; Quan, W.; Xiao, Z.S.; Zhai, Y.Y. A high-performance magnetic shield with MnZn ferrite and Mu-metal film combination for atomic sensors. *Materials* **2022**, *15*, 6680. [CrossRef]
7. Nie, J.H.; Li, H.H.; Feng, Z.K.; He, H.H. The effect of nano-SiO_2 on the magnetic properties of the low power loss manganese-zinc ferrites. *J. Magn. Magn. Mater.* **2003**, *265*, 172–175. [CrossRef]
8. Žnidaršič, A.; Drofenik, M. High-resistivity grain boundaries in CaO-doped MnZn ferrites for high-frequency power application. *J. Am. Ceram. Soc.* **1999**, *82*, 359–365. [CrossRef]
9. Guo, R.D.; Yang, X.F.; Sun, K.; Li, K.W.; Yu, Z.; Wu, G.H.; Zhang, X.F.; Jiang, X.N.; Lan, Z.W. Temperature characteristics of core losses for HfO_2 doped manganese-zinc ferrites. *J. Magn. Magn. Mater.* **2019**, *491*, 165554. [CrossRef]
10. Zaspalis, V.T.; Eleftheriou, E. The effect of TiO_2 on the magnetic power losses and electrical resistivity of polycrystalline MnZn-ferrites. *J. Phys. D Appl. Phys.* **2005**, *38*, 2156. [CrossRef]
11. Huang, A.P.; He, H.H.; Feng, Z.K. Effects of SnO_2 addition on the magnetic properties of manganese zinc ferrites. *J. Magn. Magn. Mater.* **2006**, *301*, 331–335.
12. Sun, K.; Lan, Z.W.; Yu, Z.; Li, L.Z.; Ji, H.N.; Xu, Z.Y. Effects of NiO addition on the structural, microstructural and electromagnetic properties of manganese-zinc ferrite. *Mater. Chem. Phys.* **2009**, *113*, 797–802. [CrossRef]
13. Hussain, A.; Bai, G.H.; Huo, H.X.; Yi, S.B.; Wang, X.Y.; Fan, X.Y.; Yan, M. Co_2O_3 and SnO_2 doped MnZn ferrites for applications at 3–5 MHz frequencies. *Ceram. Int.* **2019**, *45*, 12544–12549. [CrossRef]
14. Chen, S.H.; Chang, S.C.; Tsay, C.Y.; Liu, K.S.; Lin, I.N. Improvement on magnetic power loss of MnZn-ferrite materials by V_2O_5 and Nb_2O_5 co-doping. *J. Eur. Ceram. Soc.* **2001**, *21*, 1931–1935. [CrossRef]
15. Andrei, P.; Caltun, O.F.; Papusoi, C.; Stancu, A.; Feder, M. Losses and magnetic properties of Bi_2O_3 doped MnZn ferrites. *J. Magn. Magn. Mater.* **1999**, *196*, 362–364. [CrossRef]
16. Gu, M.; Liu, G.Q. Effects of MoO_3 and TiO_2 additions on the magnetic properties of manganese-zinc power ferrites. *J. Alloy. Compd.* **2009**, *475*, 356–360. [CrossRef]
17. Matsuo, Y.; Ono, K.; Hashimoto, T.; Nakao, F. Magnetic properties and mechanical strength of MnZn ferrite. *IEEE Trans. Magn.* **2001**, *37*, 2369–2372. [CrossRef]
18. Luo, F.; Duan, Z.X.; Zhang, Y.Y.; Shang, Y.J. Influence of microstructure optimization on magnetic and thermal properties of MnZn ferrite. *J. Mater. Sci. Mater. Electron.* **2021**, *21*, 15633–15642. [CrossRef]
19. Wang, H.L.; Wang, C.A.; Chen, D.L.; Xu, H.L.; Lu, H.X.; Zhang, R.; Feng, L. Preparation and characterization of ZrB_2-SiC ultra-high temperature ceramics by microwave sintering. *Front. Mater. Sci. China* **2010**, *4*, 276–280. [CrossRef]

20. Ma, F.C.; Lu, S.Y.; Liu, P.; Li, W.; Liu, X.K.; Chen, X.H.; Zhang, K.; Pan, D. Evolution of strength and fibers orientation of a short-fibers reinforced Ti-matrix composite after extrusion. *Mater. Des.* **2017**, *126*, 297–304. [CrossRef]
21. Gu, M.; Liu, G.Q.; Wang, W. Study of Mn_3O_4 doping to improve the magnetic properties of MnZn ferrites. *Mater. Sci. Eng. B* **2009**, *158*, 35–39. [CrossRef]
22. Choi, M.S.; Kang, S.-T.; Lee, B.Y.; Koh, K.-T.; Ryu, G.-S. Improvement in predicting the post-cracking tensile behavior of ultra-high performance cementitious composites based on fiber orientation distribution. *Materials* **2016**, *9*, 829. [CrossRef] [PubMed]

Disclaimer/Publisher's Note: The statements, opinions and data contained in all publications are solely those of the individual author(s) and contributor(s) and not of MDPI and/or the editor(s). MDPI and/or the editor(s) disclaim responsibility for any injury to people or property resulting from any ideas, methods, instructions or products referred to in the content.

Article

Analysis and Measurement of Differential-Mode Magnetic Noise in Mn-Zn Soft Ferrite Shield for Ultra-Sensitive Sensors

Danyue Ma [1,2,3], Xiujie Fang [1,3,*], Jixi Lu [2,3], Kun Wang [2], Bowen Sun [2], Yanan Gao [2], Xueping Xu [2,3] and Bangcheng Han [2,3,*]

1 School of Physics, Beihang University, Beijing 100191, China
2 Key Laboratory of Ultra-Weak Magnetic Field Measurement Technology, Ministry of Education, School of Instrumentation and Optoelectronic Engineering, Beihang University, Beijing 100191, China
3 Zhejiang Provincial Key Laboratory of Ultra-Weak Magnetic-Field Space and Applied Technology, Hangzhou Innovation Institute of Beihang University, Hangzhou 310000, China
* Correspondence: fangxiujie@buaa.edu.cn (X.F.); hanbangcheng@buaa.edu.cn (B.H.)

Abstract: The magnetic noise generated by the ferrite magnetic shield affects the performance of ultra-sensitive atomic sensors. Differential measurement can effectively suppress the influence of common-mode (CM) magnetic noise, but the limit of suppression capability is not clear at present. In this paper, a finite element analysis model using power loss to calculate differential-mode (DM) magnetic noise under a ferrite magnetic shield is proposed. The experimental results confirm the feasibility of the model. An ultrahigh-sensitive magnetometer was built, the single channel magnetic noise measured and the differential-mode (DM) magnetic noise are 0.70 fT/Hz$^{1/2}$ and 0.10 fT/Hz$^{1/2}$ @30 Hz. The DM magnetic noise calculated by the proposed model is less than 5% different from the actual measured value. To effectively reduce DM magnetic noise, we analyze and optimize the structure parameters of the shield on the DM magnetic noise. When the outer diameter is fixed, the model is used to analyze the influence of the ratio of ferrite magnetic shielding thickness to outer diameter, the ratio of length to outer diameter, and the air gap between magnetic annuli on DM magnetic noise. The results show that the axial DM magnetic noise and radial DM magnetic noise reach the optimal values when the thickness to outer diameter ratio is 0.08 and 0.1. The ratio of length to outer diameter is negatively correlated with DM magnetic noise, and the air gap (0.1–1 mm) is independent of DM magnetic noise. The axial DM magnetic noise is less than that of radial DM magnetic noise. These results are useful for suppressing magnetic noise and breaking through the sensitivity of the magnetometer.

Keywords: magnetic shield; Mn-Zn ferrite; differential-mode magnetic noise; SERF magnetometer

1. Introduction

Passive magnetic shields can attenuate the magnetic field of the environment and make their internal magnetic field close to zero, which is widely used in biological magnetic measurement [1–3], frontier scientific experiments [4,5], and ultra-high precision physical measurement [6–8].

Mu-metal materials with high permeability (>30,000) are widely used in passive magnetic shields [9–11]. However, the high electrically conducting mu-metal materials also generated high intrinsic magnetic noise owing to Johnson current, which becomes the main factor limiting the sensitivity of many atomic sensors, especially ultra-high sensitive atomic magnetometers [12–14]. MnZn ferrite materials with high resistivity (~1 Ω·m) and high permeability have been increasingly used to provide magnetic shielding performance but have much lower magnetic noise than that of mu-metal materials [15–18]. It has been gradually applied to many atomic sensors, including atomic magnetometer [1–3], atomic gyroscope, low field nuclear magnetic resonance, etc. Ferrite materials were first used in magnetic shields by Kornack et al. and the sensitivity of 0.75 fT/Hz$^{1/2}$ in the ferrite shield was achieved [19]. It is 25 times lower than that of the mu-metal shield in a comparable size. Although the use of ferrite materials significantly reduces the amplitude of magnetic noise, it is still much higher than the quantum noise and probe light noise of the atomic magnetometer, which is the most critical factor limiting the further improvement of the sensitivity, and further improving the sensitivity can discover new natural phenomena and reveal new scientific laws [19]. By subtracting the signals from adjacent sensors, the CM magnetic noise is partially eliminated and the DM noise is obtained, which can improve the sensitivity. It has been used in many kinds of ultra-high sensitivity sensors [20–22].

Many groups have studied differential measurement [23,24]. Differential measurement has been widely used in super conductive quantum interference devices [25,26], and then used in the atomic magnetometer. The magnetic noise gradient and the distance between the measurement points are the most important factors influencing the reduction capability of the CM magnetic noise for differential measurement. After correcting the small difference of absolute sensitivity between two adjacent magnetometer channels, a first-order gradiometer was formed by Kominis et al. [12]. Lee et al. studied the magnetic noise of the magnetic shielding structure, and explained that the ability of differential measurement to reduce the magnetic noise can also be analyzed with fluctuation dissipation theory [15,27,28]. However, there are few studies on the DM magnetic noise of a ferrite magnetic shield.

In this study, by calculating the hysteresis loss of ferrite material, the theoretical analysis model of the DM magnetic noise of a ferrite shield is established by using the finite element method (FEM) with element software (ANSYS Maxwell 3D 16.0). We built a magnetometer experimental setup, measured the differential sensitivity of the device in a ferrite magnetic shield, and separated the DM magnetic noise. The calculated results are in good agreement with the experimental results, which proves the accuracy of the FEM model. Finally, according to the proposed calculation model, the influence of ferrite magnetic shielding thickness to outer diameter ratio, length to outer diameter ratio and the air gap between magnetic annuli on the DM magnetic noise is analyzed using differential measurement.

2. Methods

2.1. Magnetic Field Response and Differential Measurement Model

The ultra-sensitive magnetometer experimental setup based on atomic spin effect usually uses a two-light configuration in which the pump light and probe light is perpendicular to each other. The time evolution process of alkali atom spin dynamics can be solved analytically by using the Bloch equation [21,29,30]

$$\frac{d}{dt}P = \frac{1}{Q}\left[\gamma_e P \times (B_{res} + B) + R_P\left(s\vec{z} - P\right) - R_{rel}P\right] \quad (1)$$

where P is the electron spin polarization vector, B is the magnetic field vector to be measured, B_{res} is the residual interference magnetic field, Q is the slowing-down factor. γ_e is the gyromagnetic ratio, R_P is the pumping rate, and s is the average photon spin. R_{rel} is the spin-relaxation rate.

Spin polarization can be detected by the rotation angle θ of the polarization plane of the linearly polarized light passing through the polarized atom. When the probe light direction is along the x-direction, the rotation angle is

$$\theta = \frac{\pi}{2}lnr_ecP_x^e[-A_{D1}\frac{f_{pr}-f_{D1}}{(f_{pr}-f_{D1})^2-(\Gamma_{D1}/2)^2}+A_{D2}\frac{f_{pr}-f_{D2}}{(f_{pr}-f_{D2})^2-(\Gamma_{D2}/2)^2}] \quad (2)$$

where l is the interaction distance between the probe light and the polarized atom. Γ_{D1} and Γ_{D2} are the pressure broadening of D1 and D2 lines. f_{pr} is the frequency of the probe light. A_{D1} and A_{D2} are the oscillation intensity of probe light D1 and D2 lines, respectively. f_{D1} and f_{D2} are the resonant frequencies of probe light D1 and D2 lines, respectively.

In order to suppress the influence of the light intensity and frequency fluctuation of the probe light on the measurement sensitivity, a photo elastic modulator (PEM) can be used to modulate the optical rotation angle at high frequency. The photoelectric detector is used to convert the optical signal into an electrical signal, and finally the output voltage of the magnetometer is

$$V_{out} = \eta I_0 \alpha e^{-OD(\nu)} \theta \tag{3}$$

where η is the conversion coefficient of the photodetector, I_0 is the incident light intensity of the detection light, α is the modulation angle of the photo elastic modulator, and OD is the optical depth. When the interference magnetic field is compensated, the magnetic field to be measured is small and only in the y-direction, and $\theta \propto B$. In order to further suppress the influence of magnetic noise, the common method is to take the magnetic noise in the central area of the magnetic shield as common mode noise and suppress it by differential measurement [20]. The magnetic noise that cannot be suppressed is named DM magnetic noise. The magnetic field measurement constructed by differential measurement is shown in Figure 1. Two beams are used to detect the magnetic fields at different positions. After being received by two detectors, the voltage values of these two signals are directly subtracted by a differential circuit to eliminate common mode magnetic noise.

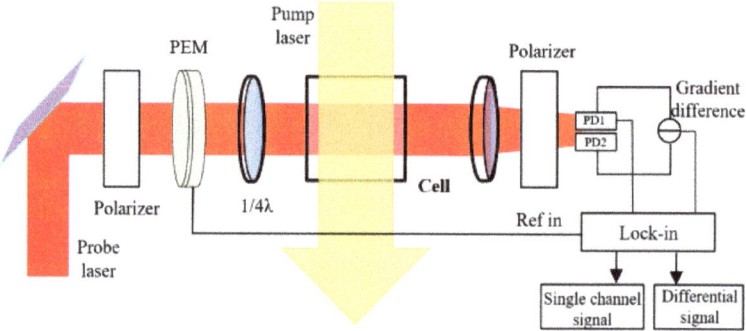

Figure 1. The model of differential measurement.

The reduction capability of differential measurement on CM magnetic noise can be calculated by the frequency response and power spectral density of the magnetometer [20]. In order to facilitate the analysis of the differential measurement method, the output signal in Equation (3) can be written as the property of frequency domain. The output signal $V(f)$ of a channel can be represented by the theoretical frequency response function $K(f)$, the magnetic field $B(f)$, and the voltage noise $N(f)$ independent of the magnetic field:

$$V(f) = K(f)B(f) + N(f) \tag{4}$$

In actual measurement, the frequency response function is also affected by the magnetic field gradient, light field gradient, temperature gradient, and temperature error, so it needs to be recalibrated. It is assumed that the actually calibrated frequency response function is $K_{est}(f)$, the output of the magnetometer will change,

$$\beta(f) = \Delta K(f)B(f) + \frac{N(f)}{K_{est}(f)} \tag{5}$$

where $\beta(f) = V(f)/K_{est}(f)$ and $\Delta K(f) = K(f)/K_{est}(f)$.

After the difference between the two channels, the power spectral density of the differential measurement is

$$P_s(f) = \left|\frac{K_1(f) - K_2(f)}{K^{est}(f)}\right|^2 P[B_{far}(f)] + P[\beta_1^N(f)] + \left|\frac{K_2(f)}{K_1^{est}(f)}\right|^2 P[\beta_2^N(f)] \quad (6)$$

where $P[B_{far}(f)]$ is the power spectral density of far-field magnetic noise. β_1^N and β_2^N are the power spectral density of the output voltage affected by the DM magnetic noise of the first channel and second channel. When the frequency responses of the two channels of the gradiometer are exactly the same, the far-field magnetic noise can be completely eliminated by difference, but the DM magnetic noise cannot be suppressed by this method. In the next section, the theoretical measurement models of space magnetic noise distribution and the DM magnetic noise of a ferrite magnetic shield will be established.

2.2. Analysis of Differential Reduction Capability

The magnetic noise of the magnetic shield was analyzed based on the generalized Nyquist relation. According to this relation, the magnetic noise can be calculated by calculating the power loss P of the magnetic shield induced by the hypothetical excitation coil [19,31,32]

$$\delta B = \frac{\sqrt{4kT}\sqrt{2P}}{\omega AI} \quad (7)$$

where k is the Boltzmann constant, $\omega = 2\pi f$ is the angular frequency, A is the area of the excitation coil, and T is the Kelvin temperature. The hypothetical excitation coil is placed in the hollow cylindrical shield and an oscillating current $I = I_m \sin(\omega t)$ flows in a coil. The power loss mainly contains the eddy current loss $P_e = \int_V 1/2\sigma E^2 dv$ and the hysteresis loss $P_h = \int_V 1/2\omega\mu'' H^2 dv$. H and E are the magnetic field strength and electric field intensity. μ'' represents the imaginary part of the complex permeability $\mu = \mu' - i\mu''$. σ is the electrical conductivity. We cannot measure the loss of the magnetic shield by placing the excitation coil in the magnetic shield by experiment and the actual impedance of the excitation coil cannot be ignored. The loss caused by the current generated by the hypothetical excitation loop is indirectly calculated by the complex permeability and conductivity of the magnetic shielding material. For metal materials, such as silicon steel sheet and mu-metal, the conductivity is high and the eddy current noise is large ($\sigma \sim 10^6\ \Omega^{-1}\ m^{-1}$), resulting in a large magnetic noise, about ~10 fT/Hz$^{1/2}$ [11]. For the ferrite materials, the conductivity is very weak ($\sigma \sim 10\ \Omega^{-1}\ m^{-1}$), so the eddy current loss is very small and hysteresis loss is dominant. Magnetic noise of the ferrite shield is less than ~1 fT/Hz$^{1/2}$ [14]. The radial magnetic noise at the center of the magnetic shield can be rewritten as

$$\delta B^T = \frac{\sqrt{4kT}\sqrt{\int_V \mu'' H^2 dV}}{\sqrt{\omega}AI} \quad (8)$$

where $\int_V H^2 dV$ can be numerically calculated using FEM. During FEM simulation, a single turn excitation coil is placed at the center of the hollow cylindrical shield as shown in Figure 2a, and the volume fraction of the magnetic field intensity of the magnetic shield is calculated using the finite element software (ANSYS Maxwell 3D 16.0). Finally, the magnetic noise is calculated. Ferrite materials have a low permeability and shielding coefficient, so they are usually placed inside the magnetic shield of multi-layer mu-metal to shield the noise of mu-metal, and the noise generated by itself is low. Because the residual magnetic field in the mu-metal shield is small, the complex permeability is a fixed value in the Rayleigh range of ferrite, independent of frequency, which is proved in Reference [14]. The magnetic noise produced by the magnetic shield belongs to far source magnetic noise, which can be suppressed by differential method, and the reduction capability can be simulated by FEM. One excitation coil of the single turn is replaced by the gradient coil.

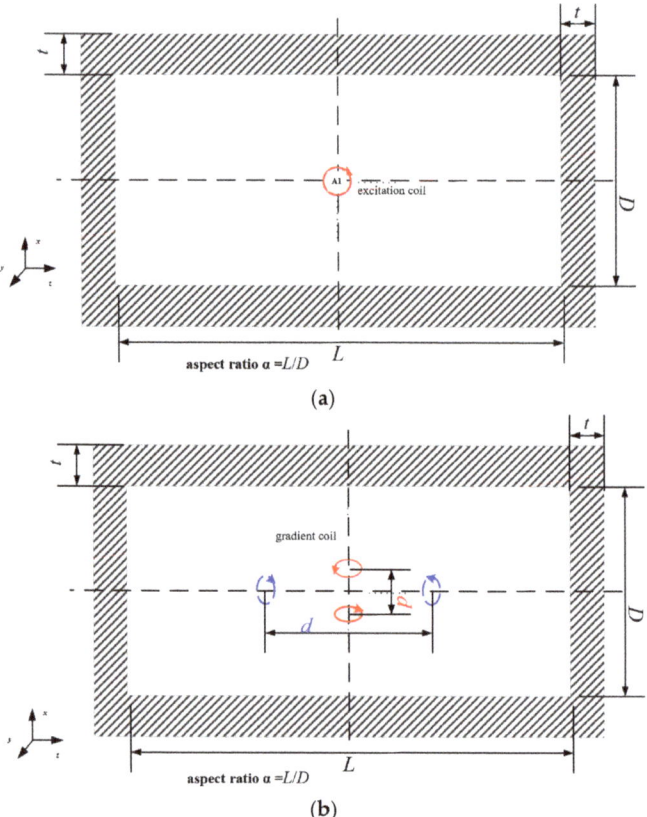

Figure 2. FEM simulation diagram of magnetic noise. (**a**) Simulation diagram of magnetic noise at the center of the magnetic shield, (**b**) simulation diagram of magnetic noise reduction with differential measurement configurations. The red solid line is the excitation gradient coil with axial configuration, and the blue dotted line is the excitation gradient coil with radial configuration.

The FEM simulation diagram is shown in Figure 2b. As mentioned in the previous section, the magnetic noise is still calculated by the power loss P of the magnetic shield caused by the excitation coil. However, in the differential method, the excitation coil is changed from a single coil to a pair of gradient coils, and the central line of the coils is called the baseline length d [14]. Gradient configuration is divided into axial gradient configuration and radial gradient configuration. The baseline of the gradient coil in the axial gradient configuration coincides with the axial center line, and the gradient coil in the radial gradient configuration coincides with the radial center line. The DM magnetic noise can be expressed as

$$\delta B_{\text{diff}} = \frac{\sqrt{4kT}\sqrt{\int_V \mu'' H_{\text{diff}}^2 dV}}{\sqrt{\omega}AI} \tag{9}$$

where $\int_V H_{\text{diff}}^2 dV$ was numerically calculated with the gradient configuration. The reduction capability can be evaluated by the ratio of magnetic noise to DM magnetic noise $\delta B/\delta B_{\text{diff}}$.

To prevent the influence of the parameters set by the simulation on the results, we verified this by changing I from 0.1 to 10 A and the diameter of loop from 1 to 20 mm in the simulation, and the simulation results were approximately the same. In this paper, the global adaptive grid is used to encrypt the grid at the location with large local error (such as the gap), while the grid at other locations is slightly thicker. By setting different grid sizes and comparing the output results, we not only improve the calculation accuracy, but also reduce the simulation time and resource allocation. The suppression effect of the gradient measurement can be further analyzed by the finite element method.

3. Experimental Setup and Results

3.1. Magnetic Field Response and Differential Measurement Model

The main components of MnZn ferrite are Fe_2O_3, Mn_3O_4, and ZnO. The doping of trace elements is used to improve the performance of ferrite, obtain higher mechanical and magnetic properties, improve permeability, and lower hysteresis loss. SiO_2 can improve the temperature coefficient and mechanical strength of soft ferrite, but it will increase the loss. In order to reduce its adverse effects, CaO is added to improve the resistivity and reduce the loss. A small amount of doping of Co^{2+} and Ti^{4+} is also used to improve temperature characteristics and reduce losses. The XRD patterns of $Mn_xZn_{1-x}Fe_{2.06}O_4$ samples are presented in Figure 3. The XRD pattern of Mn-Zn ferrite, which is in good agreement with the standard JCPDS (10-0467), indicates the sample is in the spinel phase structures. The average lattice constant a of the ferrite sample is 8.50 A. The crystallite size is estimated to be 228.6 nm based on the Scherrer's formula, and the main peak occurs at 35°.

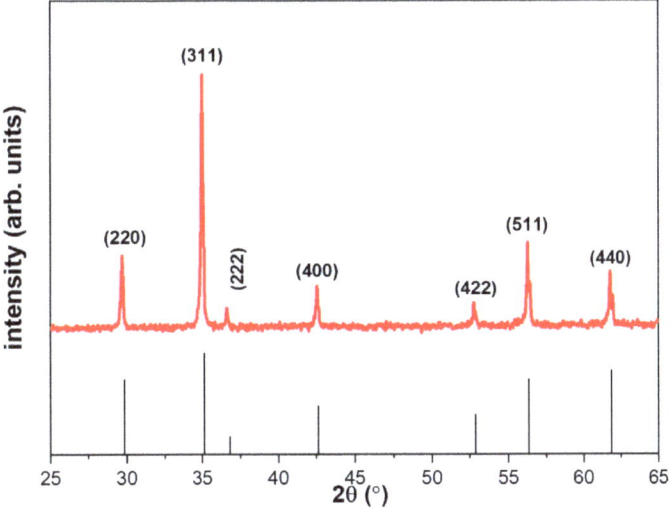

Figure 3. XRD patterns of $Mn_xZn_{1-x}Fe_{2.06}O_4$ samples.

During the ferrite molding process, the molding pressure and temperature were optimized. The final pressure was 130 MPa and the sintering temperature was 1400 °C, and the sintering was carried out under the protection of nitrogen.

In order to accurately calculate the magnetic noise, the complex permeability of ferrite materials is also measured by the experimental impedance analysis method. The real part of the relative complex permeability μ'_0 is 10,500 and the imaginary part of the relative complex permeability μ''_0 is 155.13.

3.2. Magnetic Noise Measurement

The loss generated by the gradient coil is proposed to simulate the differential reduction capability in the ferrite magnetic shield. To verify the accuracy of the calculation, the spin-exchange relaxation-free (SERF) magnetic field measurement device is built to measure the DM magnetic noise. A spherical Pyrex cell (25 mm in diameter) contains a drop of potassium, 50 Torr nitrogen, and 1.3 atm ^{4}He. The cell is heated to 208 °C through an electric heating oven. The magnetic shielding system consists of five layers of mu-metal magnetic shield and one layer of ferrite magnetic shield. At low frequency, the shielding coefficient exceeds 10^6. The ferrite shield is assembled with five annuli and two end caps, as shown in Figure 4. The height of each annulus is 45 mm and the thickness of the end cap is 10 mm. The outer diameter of the ferrite magnetic shield is about 140 mm and the thickness is 13 mm. The air gap between annulus is about 0.1 mm. This is because the ferrite magnetic shield is composed of multiple annuli. The ceramic adhesive is used for bonding between annuli. The thickness of the glue leads to the air gap, which can be measured with a feeler gauge. The width of the air gap can also be controlled by the thickness of the glue. The residual magnetic field inside the magnetic shield is about 0.3 nT and the gradient is about 0.5 nT/cm. The residual magnetic field and magnetic field gradient are compensated in situ by the triaxial coil. The circularly polarized pump beam and linearly polarized probe beam cross orthogonally and illuminate cells at the same time. The probe light is modulated by a PEM, and collected by two detectors at the same time—see Figure 5a. In order to prevent the influence of vibration on the measurement, all optical devices and magnetic shielding systems are placed on the mechanical vibration isolation platform, as shown in Figure 5b.

Figure 4. Ferrite magnetic shield and magnetic annulus. The ferrite magnetic shield has four through holes with a diameter of 25 mm in the transverse direction and two through holes with a diameter of 35 mm in the longitudinal direction.

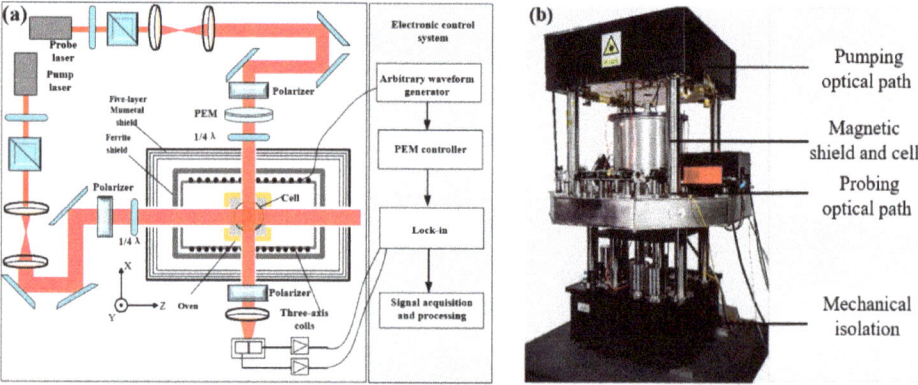

Figure 5. Optical path schematic diagram and actual diagram of magnetic field measuring device. (**a**) Optical path diagram under differential measurement configuration, (**b**) actual diagram of device.

Figure 6 shows the magnetic field sensitivity of the magnetic field measuring device, including single channel sensitivity, differential sensitivity, and various noises. The 30 Hz peak in the figure is the applied 30 Hz, 10 pTrms calibration signal, which is used to calculate the sensitivity. The sensitivity (red solid line) after gradient difference is significantly less than that of single channel sensitivity (purple solid line). The probe noise (blue solid line) is close to the gradient differential sensitivity. We use differential sensitivity δB_{diff} to subtract probe noise δB_{probe} to extract DM magnetic noise (green dotted line). DM magnetic noise is 0.1 fT/Hz$^{1/2}$@30 Hz. According to the comparison between probe noise and DM magnetic noise, it can be seen that in the low-frequency stage (below 30 Hz), probe noise and DM magnetic noise jointly affect the improvement of differential sensitivity, and in the high-frequency stage (above 30 Hz), probe noise is the most important factor. The black solid line is the simulation result of gradient noise using the method proposed in Section 2. The main reason for the difference between the measured noise and the theoretical calculation noise in the low-frequency stage is the influence of low-frequency vibration noise. The DM magnetic noise calculated by the proposed model is less than 5% different from the actual measured value. The calculated results are in good agreement with the actual measured results, which proves the correctness of the proposed model.

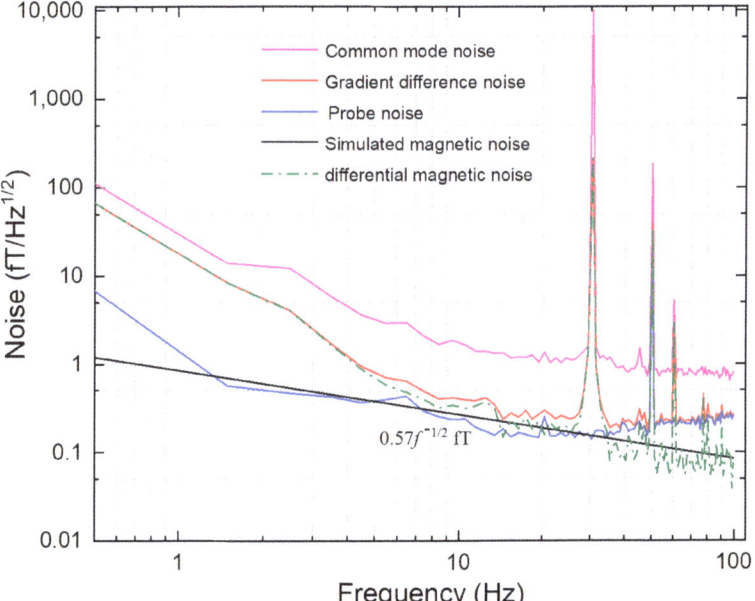

Figure 6. Noise spectrum of magnetic field measuring device. The purple solid line is the single channel sensitivity, about 0.7 fT/Hz$^{1/2}$, above 30 Hz. The red solid line indicates the gradient differential sensitivity. Above 30 Hz, the differential sensitivity is about 0.18 fT/Hz$^{1/2}$. DM magnetic noise is 0.1 fT/Hz$^{1/2}$@30 Hz. The blue solid line indicates probe noise. The green dotted line indicates DM magnetic noise. Black is realized as simulated DM magnetic noise.

4. Noise Reduction Capability of Differential Measurement

After proving the effectiveness of the calculation model, in order to further increase the differential sensitivity in the future, we carry out parameter analysis and optimization to provide guidance. In the study, the real and imaginary parts of the relative complex permeability are actually measured. The resistivity of the ferrite magnetic shield is about 1 Ω·m, and the cut-off frequency is much higher than the measurement bandwidth of the magnetometer. Therefore, within 100 Hz, the imaginary part and real part of complex permeability do not change with frequency. Firstly, in order to evaluate the reduction capability, we first analyzed the magnetic noise of the ferrite magnetic shield under non differential measurement. In different application fields [14,27,32–34], the outer diameters of ferrite magnetic shielding are usually 60 mm, 140 mm, and 200 mm. In this paper, the above dimensions of magnetic shielding are used. When the outer diameter is fixed (D = 60 mm, 140 mm, 200 mm), we analyze the influence of the ratio β of ferrite magnetic shielding thickness to outer diameter on magnetic noise. The curve of magnetic noise of the ferrite magnetic shield with β is shown in Figure 7. The variation law of magnetic noise is the same as that of reference [14], which also proves the accuracy of this model. The optimal value of radial magnetic noise (a) is always β = 0.14, while the optimal value of axial magnetic noise (b) is β = 0.16, and the axial magnetic noise is less than the radial magnetic noise.

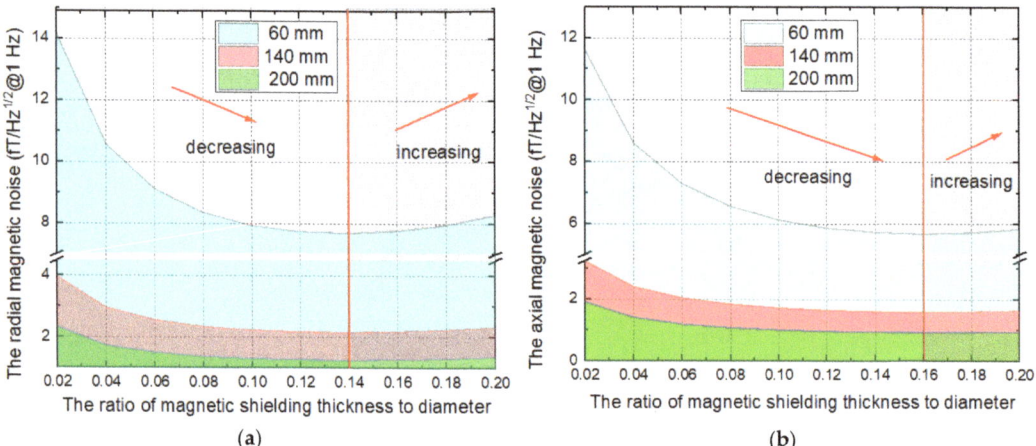

Figure 7. The curve of magnetic noise of ferrite magnetic shielding with β under different diameters, (**a**) radial magnetic noise, (**b**) axial magnetic noise.

In addition, we also explored the relationship between DM magnetic noise and β when the outer diameter is fixed, as shown in Figure 8. Differential measurement can significantly suppress the magnetic noise and the DM magnetic noise increases with the increase of the baseline d. With the increase of β, the radial and axial DM magnetic noise increases initially and decrease afterwards, and there is an optimal value. When β = 0.08, the radial DM magnetic noise (a) reaches the minimum, which is less than the thickness when the radial magnetic noise reaches the minimum, β = 0.14. When the structural parameters of ferrite magnetic shielding are the same, the axial DM magnetic noise (b) is less than that of radial DM magnetic noise. When β = 0.1, the axial DM magnetic noise reaches the minimum. The results of the above analysis are applicable to ferrite magnetic shielding of any outer diameter.

Table 1 summarizes the magnetic noise and DM magnetic noise under optimal β. Under any outer diameter, the axial reduction capability is greater than the reduction capability, and the larger the outer diameter, the greater the difference. The outer diameter increased from 60 mm to 200 mm, and the difference of axial and radial reduction capability increased from 1.03 to 5.02. When the baseline length increased from 0.5 mm to 2 mm, the reduction capability of any outer diameter decreased by 72%. Reducing the magnetic noise of ferrite magnetic shielding can effectively reduce the DM magnetic noise and improve the reduction ability.

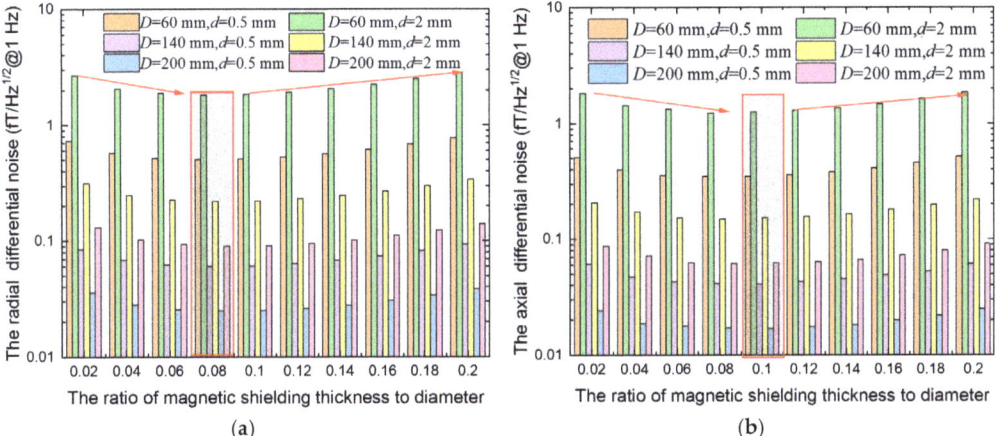

Figure 8. The curve of magnetic noise of ferrite magnetic shielding with β under different diameters and baseline length, (**a**) radial DM magnetic noise, (**b**) axial DM magnetic noise. The optimal DM magnetic noise value is in the red box.

Table 1. Optimal magnetic noise and DM magnetic noise.

Magnetic Noise (fT/Hz$^{1/2}$@1 Hz) Diameter		$D = 60$ mm	$D = 140$ mm	$D = 200$ mm
Radial magnetic noise		7.67	2.16	1.26
Axial magnetic noise		5.66	1.59	0.94
$d = 0.5$ mm	Radial DM noise	0.50	0.06	0.03
	Axial DM noise	0.35	0.04	0.02
	Radial reduction capability	15.24	35.73	50.30
	Axial reduction capability	16.27	38.89	55.32
$d = 1.5$ mm	Radial DM noise	0.96	0.12	0.07
	Axial DM noise	0.66	0.08	0.03
	Radial reduction capability	7.97	18.73	17.83
	Axial reduction capability	8.52	20.23	29.06
$d = 2$ mm	Radial DM noise	1.83	0.22	0.09
	Axial DM noise	1.23	0.15	0.06
	Radial reduction capability	4.19	9.78	13.87
	Axial reduction capability	1.35	10.49	15.11

The ratio of length to outer diameter α of ferrite magnetic shielding is also a main parameter for designing magnetic shielding results. Reference [14] shows that when α is greater than 2.5, the magnetic noise remains unchanged with the change of α. This paper explores the relationship between DM magnetic noise and α, and further improves the design of the ferrite magnetic shielding structure. When α is greater than 3, the change rate of axial DM magnetic noise and radial DM magnetic noise with α is less than 1%, as shown in Figure 9. It is considered that the DM magnetic noise is independent of α. When α is less than 1, the axial DM magnetic noise is greater than the radial DM magnetic noise, and the influence of α on the axial DM magnetic noise is more intense. When α is greater than 1, the axial DM magnetic noise is less than the radial DM magnetic noise.

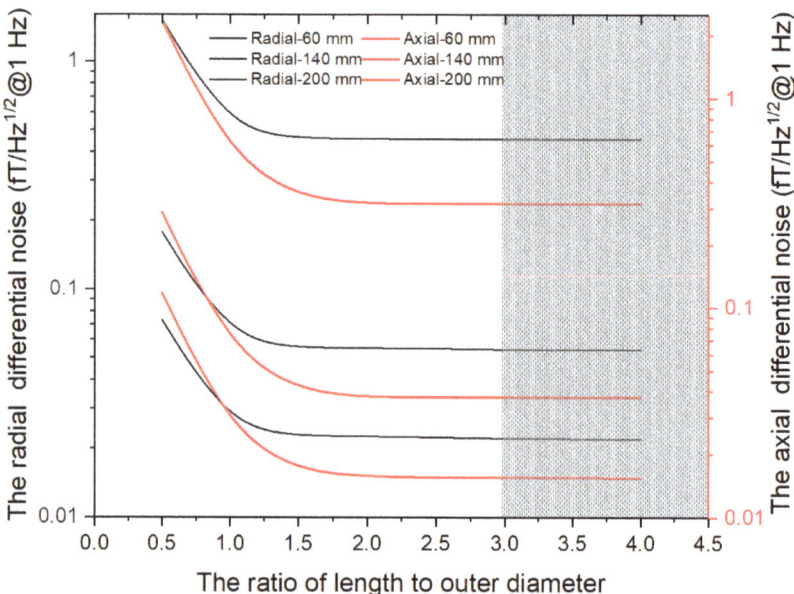

Figure 9. The curve of magnetic noise of ferrite magnetic shielding with α under different diameters with d = 0.5 mm.

We use the same method to discuss the influence of the air gap of the magnetic annuli on the DM magnetic noise. The size of the air gap of the magnetic annuli will not affect the optimal thickness. When α is greater than 3 and consists of five magnetic annuli, the air gap (w = 0.1–1 mm) does not affect the DM magnetic noise at the optimal thickness.

In a word, differential measurement can significantly suppress magnetic noise and the lower baseline indicates the better suppression effect. The outside diameter, thickness, and aspect ratio of ferrite will affect the effect of differential measurement. In order to achieve low gradient DM magnetic noise, the thickness of the ferrite shield should be designed to be 0.008 times the outer diameter, and the aspect ratio should be greater than 3.

5. Conclusions

In this study, a numerical simulation model of the influence of ferrite magnetic shielding on the gradient differential method is established. By calculating the power loss of ferrite caused by the gradient coil, the DM magnetic noise can be calculated equivalently. In order to verify the accuracy of the calculation model, we built a magnetic field measurement device to actually test the gradient magnetic noise. The measured DM magnetic noise is about 0.1 fT/Hz$^{1/2}$, which is in perfect agreement with the calculated value.

Using this model, the influence of structural parameters of ferrite magnetic shielding (outer diameter, thickness, aspect ratio, air gap) on gradient magnetic noise is discussed. The results show that differential measurement can significantly reduce the magnetic noise, and the lower baseline indicates the better suppression effect. When the magnetic shielding diameter increases from 60 mm to 200 mm, the magnetic noise suppression ability increases by four times. When the outer diameter is fixed, the DM magnetic noise decreases first and then increases with the increase of thickness. When the thickness is 0.08 times the outer diameter, the DM magnetic noise reaches the lowest. Under the condition of optimal thickness, the air gap between the magnetic rings does not affect the DM magnetic noise. These results are useful for spatial correlation estimation and common mode magnetic noise suppression. The above results support the design of low-noise ferrite magnetic shielding and lay a foundation for improving the differential sensitivity of magnetic field measurement.

Author Contributions: D.M.: Investigation, Data curation, Writing—original draft. X.F. and J.L.: Conceptualization, Methodology, Funding acquisition, Writing—review and editing. K.W.: Construction of measuring device and data acquisition. B.S. and Y.G.: Simulation analysis, Resources, Visualization. X.X. and B.H.: Supervision, Project administration. All authors have read and agreed to the published version of the manuscript.

Funding: This work was supported by National Natural Science Foundation of China under Grant No. 62203028 and the China Postdoctoral Science Foundation under Grant No. 2022M710321.

Conflicts of Interest: The authors declare no conflict of interest.

References

1. Boto, E.; Holmes, N.; Leggett, J.; Roberts, G.; Shah, V.; Meyer, S.S.; Muñoz, L.D.; Mullinger, K.J.; Tierney, T.M.; Bestmann, S.; et al. Moving magnetoencephalography towards real-world applications with a wearable system. *Nature* **2018**, *555*, 657–661. [CrossRef]
2. Kim, K.; Begus, S.; Xia, H.; Lee, S.; Jazbinsek, V.; Trontelj, Z.; Romalis, M.V. Multi-channel atomic magnetometer for magnetoencephalography: A configuration study. *NeuroImage* **2014**, *89*, 143–151. [CrossRef] [PubMed]
3. He, K.; Wan, S.; Sheng, J.; Liu, D.; Gao, J.H. A high-performance compact magnetic shield for optically pumped magnetometer-based magnetoencephalography. *Rev. Sci. Instrum.* **2019**, *90*, 064102. [CrossRef] [PubMed]
4. Chen, Y.; Zhao, L.; Zhang, N.; Yu, M.; Ma, Y.; Han, X.; Zhao, M.; Lin, Q.; Yang, P.; Jiang, Z. Single beam Cs-Ne SERF atomic magnetometer with the laser power differential method. *Opt. Express* **2022**, *30*, 16541–16552. [CrossRef] [PubMed]
5. Wang, T.; Kimball, D.; Sushkov, A.O.; Aybas, D.; Blanchard, J.W.; Centers, G.; Kelley, S.; Wickenbrock, A.; Fang, J.; Budker, D. Application of spin-exchange relaxation-free magnetometry to the cosmic axion spin precession experiment. *Phys. Dark Universe* **2017**, *19*, 27–35. [CrossRef]
6. Yang, W.; Niu, Y.; Ye, C. Optically pumped magnetometer with dynamic common mode magnetic field compensation. *Sens. Actuator A Phys.* **2021**, *332*, 113195.
7. Satpute, N.; Dhoka, P.; Iwaniec, M.; Jabade, S.; Karande, P. Manufacturing of Pure Iron by Cold Rolling and Investigation for Application in Magnetic Flux Shielding. *Materials* **2022**, *15*, 2630. [CrossRef]
8. Mateos, I.; Patton, B.; Zhivun, E.; Budker, D.; Wurm, D.; Ramos-Castro, J. Noise characterization of an atomic magnetometer at sub-millihertz frequencies. *Sens. Actuator A Phys.* **2015**, *224*, 147–155. [CrossRef]
9. Ma, Y.; Zhao, L.; Chen, Y.; Luo, G.; Yu, M.; Wang, Y.; Guo, J.; Yang, P.; Lin, Q.; Jiang, Z. The micro-fabrication and performance analysis of non-magnetic heating chip for miniaturized SERF atomic magnetometer. *J. Magn. Magn. Mater.* **2022**, *557*, 169495. [CrossRef]
10. Pan, D.; Lin, S.; Li, L.; Li, J.; Jin, Y.; Sun, Z.; Liu, T. Research on the design method of uniform magnetic field coil based on the MSR. *IEEE Trans. Ind. Electron.* **2020**, *67*, 1348–1356. [CrossRef]
11. Fang, X.; Wei, K.; Zhao, T.; Zhai, Y.; Ma, D.; Xing, B.; Liu, Y.; Xiao, Z. High spatial resolution multi-channel optically pumped atomic magnetometer based on a spatial light modulator. *Opt. Express* **2020**, *28*, 26447–26460. [CrossRef] [PubMed]
12. Kominis, I.K.; Kornack, T.W.; Allred, J.C.; Romalis, M.V. A subfemtotesla multichannel atomic magnetometer. *Nature* **2003**, *422*, 596–599. [CrossRef] [PubMed]
13. Dang, H.B.; Maloof, A.C.; Romalis, M.V. Ultrahigh sensitivity magnetic field and magnetization measurements with an atomic magnetometer. *Appl. Phys. Lett.* **2010**, *97*, 151110. [CrossRef]
14. Ma, D.; Lu, J.; Fang, X.; Yang, K.; Wang, K.; Zhang, N.; Han, B.; Ding, M. Parameter modeling analysis of a cylindrical ferrite magnetic shield to reduce magnetic noise. *IEEE Trans. Ind. Electron.* **2022**, *45*, 991–998. [CrossRef]
15. Lee, S.K.; Romalis, M.V. Calculation of magnetic field noise from high-permeability magnetic shields and conducting objects with simple geometry. *J. Appl. Phys.* **2008**, *103*, 084904. [CrossRef]
16. Fan, W.; Quan, W.; Liu, F.; Pang, H.; Xing, L.; Liu, G. Performance of low-noise ferrite shield in a K-Rb-Ne[21] co-magnetometer. *IEEE Sens. J.* **2019**, *20*, 2543–2549. [CrossRef]
17. Ma, D.; Lu, J.; Fang, X.; Dou, Y.; Wang, K.; Gao, Y.; Li, S.; Han, B. A novel low-noise Mu-metal magnetic shield with winding shape. *Sens. Actuator A Phys.* **2022**, *346*, 113884. [CrossRef]
18. Wang, M.; Feng, J.; Shi, Y.; Shen, M. Demagnetization weakening and magnetic field concentration with ferrite core characterization for efficient wireless power transfer. *IEEE Trans. Ind. Electron.* **2019**, *66*, 1842–1851. [CrossRef]
19. Kornack, T.W.; Smullin, S.J.; Lee, S.K.; Romalis, M.V. A low-noise ferrite magnetic shield. *J. Appl. Phys.* **2007**, *90*, 223501. [CrossRef]
20. Savukov, I.; Kim, Y.J. Investigation of magnetic noise from conductive shields in the 10–300 kHz frequency range. *J. Appl. Phys.* **2020**, *128*, 234501. [CrossRef]
21. Ichihara, S.; Mizutani, N.; Ito, Y.; Kobayashi, T. Differential measurement With the Balanced Response of Optically Pumped Atomic Magnetometers. *IEEE Trans. Magn.* **2016**, *52*, 1–9. [CrossRef]
22. Kamada, K.; Ito, Y.; Ichihara, S.; Mizutani, N.; Kobayashi, T. Noise reduction and signal-to-noise ratio improvement of atomic magnetometers with optical gradiometer configurations. *Opt. Express* **2015**, *23*, 6976–6987. [CrossRef] [PubMed]

23. Zhang, G.; Huang, S.; Xu, F.; Hu, Z.; Lin, Q. Multi-channel spin exchange relaxation free magnetometer towards two-dimensional vector, magnetoencephalography. *Opt. Express* **2019**, *27*, 597–607. [CrossRef] [PubMed]
24. Wakai, R.T. The atomic magnetometer: A new era in biomagnetism. *AIP Conf. Proc.* **2014**, *1626*, 46–54.
25. Nenonen, J.; Montonen, J.; Katila, T. Thermal noise in biomagnetic measurements. *Rev. Sci. Instrum.* **1996**, *67*, 2397–2405. [CrossRef]
26. Sandin, H.J.; Volegov, P.L.; Espy, M.A.; Matlashov, A.N.; Savukov, I.M.; Schultz, L.J. Noise modeling from conductive shields using Kirchhoff equations. *IEEE Trans. Appl. Supercon.* **2010**, *21*, 489–492. [CrossRef]
27. Lu, J.; Ma, D.; Yang, K.; Quan, W.; Zhao, J.; Xing, B.; Han, B.; Ding, M. Study of Magnetic Noise of a Multi-Annular Ferrite Shield. *IEEE Access* **2020**, *8*, 40918–40924. [CrossRef]
28. Kubo, R. The fluctuation-dissipation theorem. *Rep. Prog. Phys.* **1996**, *29*, 255. [CrossRef]
29. Ito, Y.; Sato, D.; Kamada, K.; Kobayashi, T. Measurements of Magnetic Field Distributions With an Optically Pumped K-Rb Hybrid Atomic Magnetometer. *IEEE Trans. Magn.* **2014**, *50*, 4006903. [CrossRef]
30. Budker, D.; Romalis, M. Optical Magnetometry. *Nat. Phys.* **2007**, *3*, 227–234. [CrossRef]
31. Smythe, W.R. *Static and Dynamic Electricity*; McGraw-Hill: New York, NY, USA, 1968; pp. 188–190.
32. Ionel, D.M.; Popescu, M.; McGilp, M.I.; Miller, T.J.E.; Dellinger, S.J.; Heideman, R.J. Computation of core losses in electrical machines using improved models for laminated steel. *IEEE Trans. Ind. Electron.* **2007**, *43*, 1554–1564. [CrossRef]
33. Ma, D.; Lu, J.; Xiujie, F.; Wang, K.; Jing, W.; Zhang, N.; Chen, H.; Ding, M.; Han, B. Analysis of coil constant of triaxial uniform coils in Mn–Zn ferrite magnetic shields. *J. Phys. D Appl. Phys.* **2021**, *54*, 275001. [CrossRef]
34. Fan, W.; Quan, W.; Liu, F.; Xing, L.; Liu, G. Suppression of the Bias Error Induced by Magnetic Noise in a Spin-Exchange Relaxation-Free Gyroscope. *IEEE Sens. J.* **2019**, *19*, 9712–9721. [CrossRef]

Article

The Preparation of High Saturation Magnetization and Low Coercivity Feco Soft Magnetic Thin Films via Controlling the Thickness and Deposition Temperature

Wenjie Yang [1,2], Junjie Liu [1,2], Xiangfeng Yu [1,2], Gang Wang [1,2,3], Zhigang Zheng [1,2,3], Jianping Guo [4], Deyang Chen [4], Zhaoguo Qiu [1,2,3,*] and Dechang Zeng [1,2,3]

1. School of Materials Science and Engineering, South China University of Technology, Guangzhou 510640, China
2. Zhongshan R&D Center for Materials Surface and Thin Films Technology of the South China University of Technology, Gent Materials Surface Technology (Guangdong) Co., Ltd., Zhongshan 528437, China
3. Yangjiang Branch, Guangdong Laboratory Materials Science and Technology Yangjiang Advanced Alloys Laboratory, Yangjiang 529500, China
4. Institute for Advanced Materials, South China Academy of Advanced Optoelectronics, South China Normal University, Guangzhou 510006, China
* Correspondence: zgqiu@scut.edu.cn; Tel.: +86-20-8711-1278

Abstract: FeCo thin films with high saturation magnetization (4 πM_s) can be applied in high-frequency electronic devices such as thin film inductors and microwave noise suppressors. However, due to its large magnetocrystalline anisotropy constant and magnetostrictive coefficient of FeCo, the coercivity (H_c) of FeCo films is generally high, which is detrimental to the soft magnetic properties. Meanwhile, the thickness and deposition temperature have significant effects on the coercivity and saturation magnetization of FeCo films. In this paper, FeCo thin films with different thicknesses were prepared by magnetron sputtering at different temperatures. The effects of thickness and deposition temperature on the microstructure and magnetic properties of FeCo thin films were systematically studied. When the film thickness increases from 50 nm to 800 nm, the coercivity would decrease from 309 Oe to 160 Oe. However, the saturation magnetization decreases from 22.1 kG to 15.3 kG. After that, we try to further increase the deposition temperature from room temperature (RT) to 475 °C. It is intriguing to find that the coercivity greatly decreased from 160 Oe to 3 Oe (decreased by 98%), and the saturation magnetization increased from 15.3 kG to 23.5 kG (increased by 53%) for the film with thickness of 800 nm. For the film with thickness of 50 nm, the coercivity also greatly decreased from 309 Oe to 10 Oe (decreased by 96%), but the saturation magnetization did not change significantly. It is contributed to the increase of deposition temperature, which will lead to the increase of grain size and the decrease of the number of grain boundaries. And the coercivity decreases as the number of grain boundaries decreases. Meanwhile, for the thicker films, when increasing the deposition temperature the thermal stress increases, which changes the appearance of (200) texture, and the saturation magnetization increases. Whereas, it has a negligible effect on the orientation of thin films with small thickness (50 nm). This indicates that high-temperature deposition is beneficial to the soft magnetic properties of FeCo thin films, particularly for the films with larger thickness. This FeCo thin film with high saturation magnetization and low coercivity could be an ideal candidate for high-frequency electronic devices.

Keywords: FeCo thin film; saturation magnetization; coercivity; thickness; deposition temperature

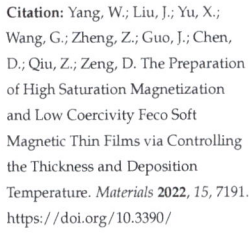

Citation: Yang, W.; Liu, J.; Yu, X.; Wang, G.; Zheng, Z.; Guo, J.; Chen, D.; Qiu, Z.; Zeng, D. The Preparation of High Saturation Magnetization and Low Coercivity Feco Soft Magnetic Thin Films via Controlling the Thickness and Deposition Temperature. *Materials* **2022**, *15*, 7191. https://doi.org/10.3390/ma15207191

Academic Editor: Dominique de Caro

Received: 6 September 2022
Accepted: 12 October 2022
Published: 15 October 2022

Publisher's Note: MDPI stays neutral with regard to jurisdictional claims in published maps and institutional affiliations.

Copyright: © 2022 by the authors. Licensee MDPI, Basel, Switzerland. This article is an open access article distributed under the terms and conditions of the Creative Commons Attribution (CC BY) license (https://creativecommons.org/licenses/by/4.0/).

1. Introduction

With the development of the electronic information industry, the high frequency, miniaturization and integration are required for the electronic devices [1–3]. According to the Snoek's limit, the initial permeability is inversely proportional to the ferromagnetic resonance frequency under a certain saturation magnetization [4]. Therefore, to obtain superior soft magnetic properties at high frequency, magnetic materials should have high saturation magnetization. Among commercial magnetic materials, the saturation magnetization of FeCo is the highest, reaching 24.5 kG when the Fe content is 65% [5]. The study by Najafi et al. also showed that the ratio of Fe and Co had a significant effect on the saturation magnetization, and the saturation magnetization reached the maximum when the Fe content was about 70% [6]. Moreover, according to Acher's limit [7], the film material can exceed the restriction of Snoek's limit, so FeCo films are ideal candidates for high-frequency electronic devices such as thin film inductors [8,9] and microwave noise suppressors [10,11]. However, due to the large magnetocrystalline anisotropy constant ($\sim$10 kJ/m^3) and magnetostrictive coefficient (4~6 $\times$ 10^{-5}), the coercivity of FeCo films is generally high, which is detrimental to the soft magnetic properties and hysteresis loss [12].

In order to reduce the coercivity of FeCo thin films, many studies have been investigated, such as adding under layer [13–16] or adding the third element [17–19] to change the microstructure of the film and reduce the coercivity. However, this method may lead to a decrease in saturation magnetization. Himalay Basumatary et al., studied the effect of deposition temperature on the magnetic properties of FeGa thin films and found that high temperature deposition can greatly reduce the coercivity, since high temperature deposition could be a feasible method to improve the soft magnetic properties of thin films [20]. There exist some studies on the effects of high temperature deposition on the magnetic properties of FeCo-based alloy thin films [21–24]. However, the relationships between structure and magnetic properties have not yet been systematically studied. Moreover, the above studies have only focused on thin films with relatively small thickness. However, in practical applications a certain thickness of the film is required to provide sufficient magnetic signal, and the thickness of the film has a great influence on the magnetic properties [25].

Therefore, in our work, FeCo thin films with different thicknesses were prepared by magnetron sputtering at different temperatures. The effects of thickness and deposition temperature on the microstructure and magnetic properties of FeCo thin films were systematically studied.

2. Experimental

FeCo target is purchased from AT&M Co., Ltd. (Beijing, China), and the composition is Fe: Co = 65:35 (atomic ratio). The Si substrates are purchased from Zhejiang Lijing Silicon Material Co., Ltd. (Quzhou, China) The experimental flow is shown in Figure 1. FeCo thin films with different thicknesses were deposited on Si (100) substrates using the DC magnetron sputtering technique under varying deposition temperatures. The base pressure of the magnetron sputtering system was 5×10^{-4} Pa. Sputtering pressure and power were kept constant at 0.8 Pa and 120 W, respectively, for all the depositions. Thin films with thicknesses of 50 nm, 100 nm, 200 nm, 400 nm, 800 nm and 1200 nm were deposited at room temperature. Additionally, thin films with thicknesses of 50 nm and 800 nm were deposited at different temperatures such as 25 °C, 200 °C, 300 °C, 400 °C, 425 °C, 450 °C, 475 °C, 500 °C, 525 °C and 550 °C. The crystal structure was analyzed by X-ray diffraction (XRD, Philips X'pert pro M, Amsterdam, The Netherlands). Hysteresis loops were detected with the Physical Property Measurement System (PPMS, Quantum Design PPMS-9), and the magnetic field is parallel to the plane of the film in the test. Surface morphology and cross-sectional morphology were characterized by scanning electron microscopy (SEM, SU8220, Hitachi Limited, Tokyo, Japan). Thin film stress is tested by film stress tester (FST, SuPro FST5000, Shenzhen, China).

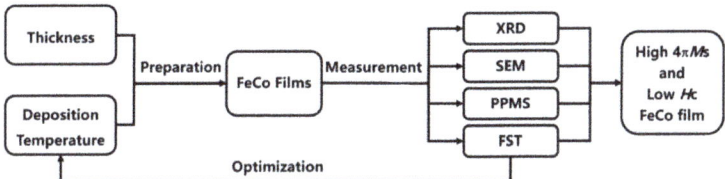

Figure 1. Flow chart for the preparation and testing of FeCo films.

3. Results and Discussion

Representative GIXRD patterns of FeCo thin films with different thicknesses are shown in Figure 2. It can be observed that the (200) diffraction peak gradually disappeared with the increase in thickness. This is due to the film stress decreasing with the increase in film thickness, as shown in Figure 3. Energy minimization is required during grain growth of thin films. When the film thickness is relatively larger, the stress of the film is relatively smaller and the surface energy is dominant. FeCo is body-centered cubic (bcc) structure, and the close-packed plane is (110). Therefore, to minimize the surface energy, the preferred orientation of the film is (110). When the film thickness decreases, the film stress increases. In addition to surface energy, strain energy also needs to be considered. Since the (200) texture of FeCo thin film is insensitive to stress, the surface energy increased by the transformation from (110) texture to (200) texture is less than the decrease in strain energy, resulting in the appearance of (200) diffraction peak [26]. The film stress decreases with increasing the film thickness, because during the film growth process the shrinkage of grain boundaries on the growth surface will lead to a tensile stress. However, with the increase in film thickness the grain size increases, resulting in a decreasing number of grain boundary, so the shrinkage decreases and the stress decreases [25].

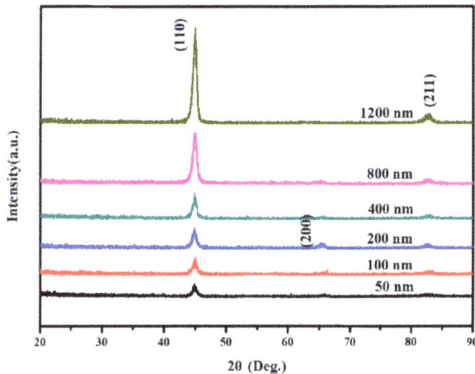

Figure 2. GI-XRD patterns for FeCo thin films with different thickness.

Figure 4 shows that for FeCo films with a thickness of 800 nm, when the deposition temperature increases, the (200) diffraction peak re-emerged. This is due to the thermal expansion coefficients of the film and substrate being different. During thin film deposition, the film and substrate are in a compatible deformation state at a specific temperature. When the film is deposited, both the film and the substrate return to room temperature. The change in temperature leads to an inconsistent shrinkage deformation of the film and the substrate, and the thermal stress will be generated in the film due to the discrepancy of the thermal expansion coefficient. As shown in Figure 3, when the deposition temperature increases from 25 °C to 475 °C the film stress increases from 486.91 MPa to 715.53 MPa. Therefore, to reduce the strain energy (200) texture appears in the film. For FeCo thin films with a thickness of 50 nm, it is different from the film with a thickness of 800 nm, the diffraction peak does not change significantly with the increase in temperature, and the (200) diffraction peak can be obtained when deposited at 25 °C, as shown in Figure 5.

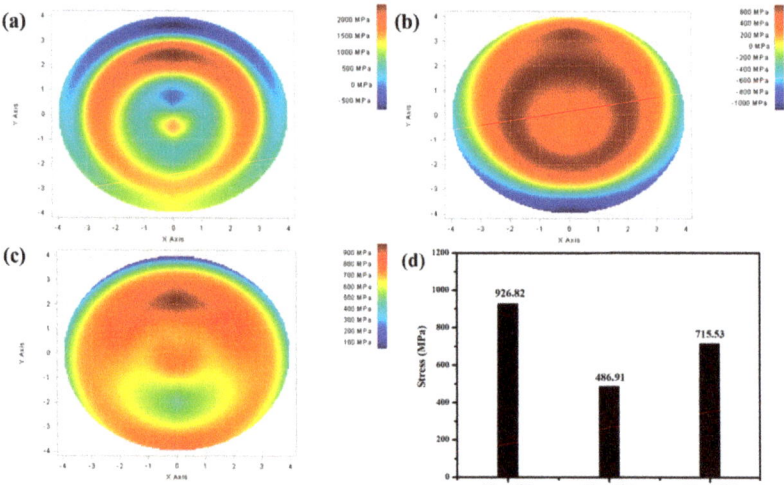

Figure 3. Stress distribution of FeCo thin films with different thicknesses and deposition temperatures: (**a**) 100 nm, 25 °C, (**b**) 800 nm, 25 °C, (**c**) 800 nm, 475 °C; and (**d**) the average stress.

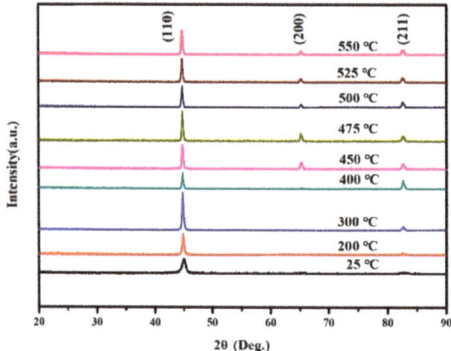

Figure 4. GI-XRD patterns for FeCo thin films with thickness of 800 nm deposited at different temperatures.

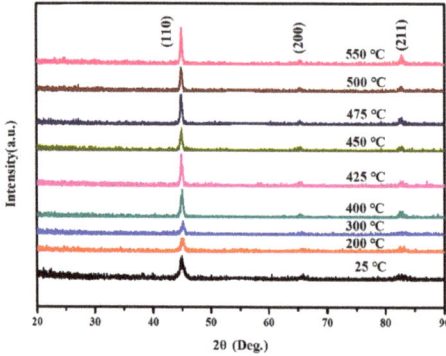

Figure 5. GI-XRD patterns for FeCo thin films with thickness of 50 nm deposited at different temperatures.

In order to evaluate the effects of deposition temperature on the grain size variation, Figure 6 displays the surface morphologies of FeCo thin films with thickness of 800 nm deposited at different temperatures. It shows that as the deposition temperature increases, the grain size increases significantly. Relative works have already demonstrated that an increase in deposition temperature will lead to the increase in grain size of FeCo thin films [22,23,27]. The increase in deposition temperature will lead to the decrease in nucleation rate and the increase in grain growth rate. Under the comprehensive effect of the two factors, the grain size increases with the increase in deposition temperature [20]. Figure 7 shows that when the deposition temperature is lower than 475 °C, the surface roughness increases with enhancing the deposition temperature due to the increase in the grain size, but it is not obvious. However, when the deposition temperature increases from 475 °C to 550 °C, the roughness increases significantly, and the film density obviously decreases. For FeCo films with a thickness of 50 nm, the variation in microstructure with deposition temperatures is similar to that of 800 nm, except for the relatively small grain size, as shown in Figure 8.

Figure 6. Surface morphologies of FeCo thin films with thickness of 800 nm deposited at (**a**) 25 °C, (**b**) 200 °C, (**c**) 400 °C, (**d**) 475 °C and (**e**) 550 °C.

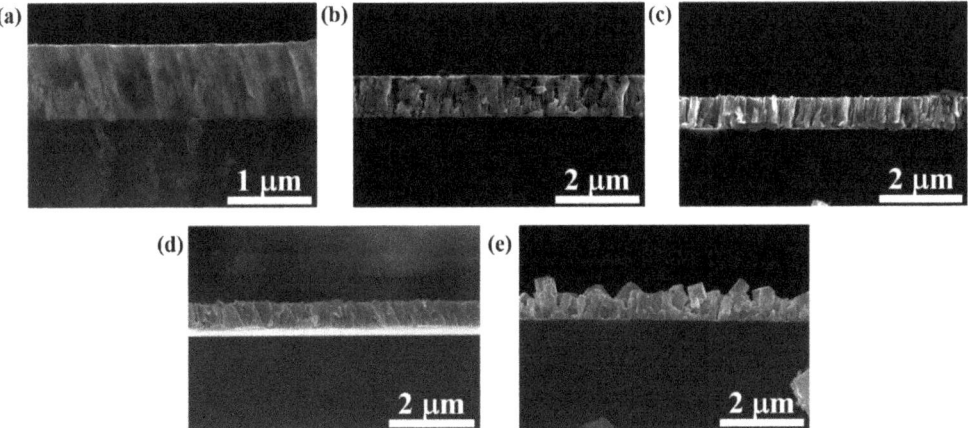

Figure 7. Cross-sectional morphologies of FeCo thin films with thickness of 800 nm deposited at (**a**) 25 °C, (**b**) 200 °C, (**c**) 400 °C, (**d**) 475 °C and (**e**) 550 °C.

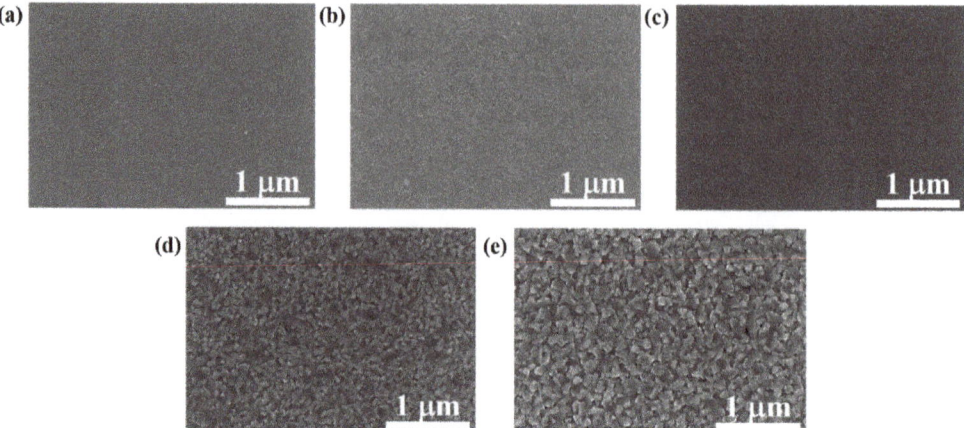

Figure 8. Surface morphologies of FeCo thin films with thickness of 50 nm deposited at (**a**) 25 °C, (**b**) 200 °C, (**c**) 400 °C, (**d**) 475 °C and (**e**) 550 °C.

Figure 9 shows the variation in coercivity and saturation magnetization of FeCo films with different thicknesses. It indicates that the saturation magnetization and coercivity decrease first, and then remain stable with the increase in thickness. When the film thickness increases from 50 nm to 800 nm, the coercivity decreases from 309 Oe to 160 Oe, and the saturation magnetization decreases from 22.1 kG to 15.3 kG. The decrease in saturation magnetization is related to the disappearance of the (200) diffraction peak. Due to the effect of demagnetization field, the effective magnetization of the film should be located in the film plane. However, in practice there is an angle between the direction of effective magnetization and the film plane for FeCo film. Compared with (110) texture, the effective magnetization of (200) texture is closer to the film plane [28]. Therefore, the films with (200) texture would have higher saturation magnetization. Previous studies revealed that reducing film stress could significantly reduce coercivity [29]. This is because the stress in the film may lead to the magnetostrictive effect, resulting in magnetic anisotropy. As the thickness increases, the stress of the film decreases, so the coercivity decreases.

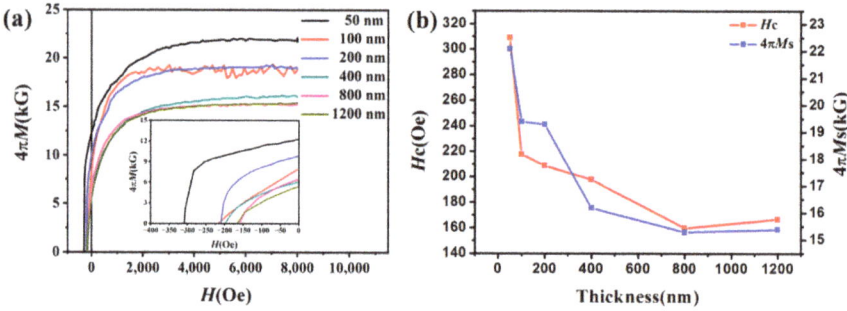

Figure 9. (**a**) Demagnetization curves of FeCo films with different thickness, and (**b**) dependence of the coercivity and saturation magnetization on thicknesses.

In addition, the variation in coercivity and saturation magnetization of FeCo films with thickness of 800 nm and deposited at different temperatures are analyzed in Figure 10. When the deposition temperature is lower than 475 °C, the saturation magnetization increases and the coercivity decreases with gradually increasing the deposition temperature. The reason for the change in saturation magnetization is the same as mentioned above. When the deposition temperature increases, the thermal stress of the film increases, resulting in the existence of (200) texture in the film, and the saturation magnetization of the film increases. The decrease in coercivity is related to the change in grain size. The reverse magnetization process of soft magnetic materials is mainly realized by domain wall displacement. When increasing the deposition temperature, the grain size increases, the number of grain boundaries decreases, the blocking effect of grain boundaries on domain wall displacement is weakened, and the coercivity decreases. The greater roughness in the film gives rise to the surface defects, which can act as pinning centers to hinder domain movement, resulting in a higher coercivity [20]. When the deposition temperature is higher than 475 °C, the roughness of the film becomes even higher. At this condition, the effect of roughness on the coercivity of the film is dominant, so the coercivity begins to increase. Meanwhile, the saturation magnetization starts to decrease because the density of the film decreases.

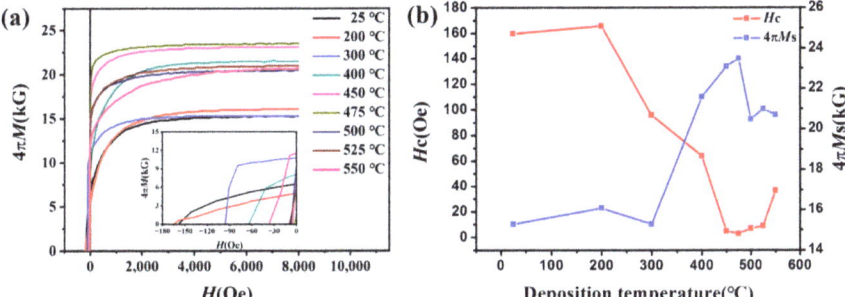

Figure 10. (a) Demagnetization curves of FeCo films with thickness of 800 nm deposited at different temperatures, and (b) dependence of the coercivity and saturation magnetization on deposition temperature.

Figure 11 shows the variation in coercivity and saturation magnetization of FeCo films with thickness of 50 nm deposited at different temperatures. When enhancing the deposition temperature, the variation trend of coercivity of FeCo film with thickness of 50 nm is similar to that of FeCo film with thickness of 800 nm, which decreases first and then increases. The coercivity is also closely related to the grain size and roughness. However, the saturation magnetization changes differently. The deposition temperature has little effect on the saturation magnetization of FeCo films with a thickness of 50 nm. It may be due to the FeCo films with thickness of 50 nm deposited at 25 °C also having (200) texture, and the deposition temperature has negligible effect on the orientation of thin films.

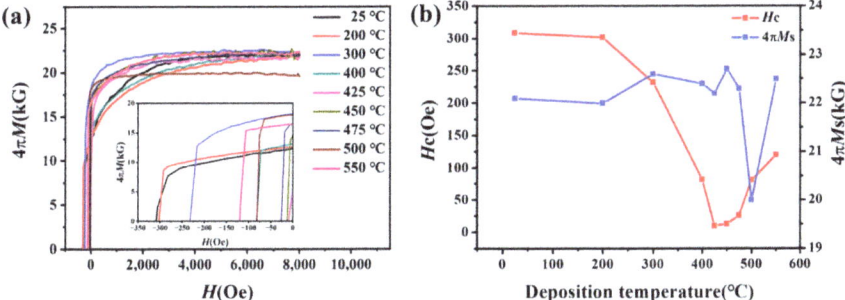

Figure 11. (a) Demagnetization curves of FeCo films with thickness of 50 nm deposited at different temperatures, and (b) dependence of the coercivity and saturation magnetization on deposition temperature.

Comparing the results of Figures 10 and 11, it can be seen that the effect of high temperature deposition on the magnetic properties of FeCo films with different thicknesses is not exactly consistent. For the FeCo film with a thickness of 800 nm, when the deposition temperature increases from 25 °C to 475 °C the coercivity decreases from 160 Oe to 3 Oe (decreased by 98%), and the saturation magnetization increases from 15.3 kG to 23.5 kG (increased by 53%). For FeCo thin films with a thickness of 50 nm, when the deposition temperature increases from 25 °C to 425 °C the coercivity decreases from 309 Oe to 10 Oe (decreased by 96%), but the saturation magnetization does not change significantly. This indicates that high-temperature deposition is beneficial to the soft magnetic properties of FeCo thin films, particularly for the films with larger thickness, and is available for future application in high-frequency electronic devices.

4. Conclusions

In summary, the effects of thickness and deposition temperature on the microstructure and magnetic properties of FeCo thin films were investigated. With the increase in film thickness, the decrease in stress leads to the disappearance of (200) texture, which reduces the saturation magnetization. When the film thickness increases from 50 nm to 800 nm, the saturation magnetization decreases from 22.1 kG to 15.3 kG. However, when the thicker film is deposited at high temperature, the increase in thermal stress leads to the reappearance of (200) texture, and the saturation magnetization of the 800 nm film increases from 15.3 kG to 23.5 kG. Additionally, (200) texture also exists in the thin films with small thickness deposited at room temperature, so the deposition temperature has little effect on the saturation magnetization. When the deposition temperature increases, the grain size obviously increases, resulting in a significant reduction in coercivity. For FeCo films with a thickness of 800 nm, when the deposition temperature increases from 25 °C to 475 °C the coercivity can be reduced by 98% to 3 Oe. This FeCo thin film with high saturation magnetization and low coercivity could be an ideal candidate for high-frequency electronic devices.

Author Contributions: Conceptualization, Z.Q., D.Z.; Investigation, W.Y., J.L., X.Y. and J.G.; Supervision, G.W., Z.Z., D.C., Z.Q. and D.Z.; Writing—original draft, W.Y.; Writing—review and editing, G.W., Z.Z., D.C. All authors have read and agreed to the published version of the manuscript.

Funding: This work was supported by National Natural Science Foundation of China (51901079), Guangdong Basic and Applied Basic Research Foundation (2019A1515010857, 2020A1515010736), Guangzhou Municipal Science and Technology Program (202007020008, 202102020596), Zhongshan Collaborative Innovation Fund (2018C1001), and the Opening Project of National Engineering Research Center for Powder Metallurgy of Titanium & Rare Metals.

Institutional Review Board Statement: Not applicable.

Informed Consent Statement: Not applicable.

Data Availability Statement: Not applicable.

Conflicts of Interest: The authors declare no conflict of interest.

References

1. Zhou, Y.; Kou, X.M.; Mu, M.K.; McLaughlin, B.M.; Chen, X.; Parsons, P.E.; Zhu, H.; Ji, A.; Lee, F.C.; Xiao, J.Q. Loss characterization of Mo-doped FeNi flake for DC-to-DC converter and MHz frequency applications. *J. Appl. Phys.* **2012**, *111*, 07E329. [CrossRef]
2. Nakano, T.; Nepal, B.; Tanaka, Y.; Wu, S.; Abe, K.; Mankey, G.; Mewes, T.; Mewes, C.; Suzuki, T. Soft magnetic and structural properties of (FeCo)-(AlSi) alloy thin films. *J. Magn. Magn. Mater.* **2020**, *507*, 166852. [CrossRef]
3. Li, T.; Liu, X.; Li, J.; Pan, L.; He, A.; Dong, Y. Microstructure and magnetic properties of FeCoHfN thin films deposited by DC reactive sputtering. *J. Magn. Magn. Mater.* **2022**, *547*, 168777. [CrossRef]
4. Snoek, J.L. Dispersion and Absorption in Magnetic Ferrites at Frequencies Above one MC/S. *Physica* **1948**, *14*, 207–217. [CrossRef]
5. Sundar, R.S.; Deevi, S.C. Soft magnetic FeCo alloys: Alloy development, processing, and properties. *Int. Mater. Rev.* **2005**, *50*, 157–192. [CrossRef]
6. Najafi, A.; Nematipour, K. Synthesis and Magnetic Properties Evaluation of Monosized FeCo Alloy Nanoparticles through Microemulsion Method. *J. Supercond. Nov. Magn.* **2017**, *30*, 2647–2653. [CrossRef]
7. Acher, O.; Adenot, A.L. Bounds on the dynamic properties of magnetic materials. *Phys. Rev. B* **2000**, *62*, 11324–11327. [CrossRef]

8. Ludwig, A.; Frommberger, M.; Zanke, C.; Glasmachers, S.; Quandt, E. Magnetoelastic thin films and multilayers for high-frequency applications. In Proceedings of the Smart Structures and Materials 2002 Conference, San Diego, CA, USA, 18–21 March 2002; pp. 542–552.
9. Yuan, B.B.; Liu, X.L.; Wang, L.S.; Peng, D.L.; Wei, J.W.; Wang, J.B. Magnetic properties of FeCo-O/NiZn-Ferrite$_n$ bi-magnetic phase nanocomposite multilayer films. In Proceedings of the 7th IEEE International Nanoelectronics Conference (IEEE INEC), Chengdu, China, 7–11 May 2016.
10. Wang, Y.; Wang, L.; Zhang, H.; Zhong, Z.; Peng, D.; Ye, F.; Bai, F. Investigation of FeCo-Ti-O nanogranular films with tunable permeability spectrum. *J. Alloys Compd.* **2016**, *667*, 229–234. [CrossRef]
11. Hao, G.; Zhang, D.; Jin, L.; Zhang, H.; Tang, X. Nanogranular CoFe-yttrium-doped zirconia films for noise suppressor. In Proceedings of the 2015 IEEE International Magnetics Conference (INTERMAG), Beijing, China, 11–15 May 2015; p. 1. [CrossRef]
12. Bozorth, R.M. *Ferromagnetism*; IEEE: New York, NY, USA, 1993.
13. Wu, Y.P.; Han, G.-C.; Kong, L.B. Microstructure and microwave permeability of FeCo thin films with Co underlayer. *J. Magn. Magn. Mater.* **2010**, *322*, 3223–3226. [CrossRef]
14. Liu, M.; Shi, L.; Lu, M.; Xu, S.; Wang, L.; Li, H. Enhanced (200) Orientation in FeCo/SiO$_2$ Nanocomposite Films by Sol-Gel Spin-Coating on Al Underlayer. *J. Supercond. Nov. Magn.* **2016**, *29*, 835–838. [CrossRef]
15. Akansel, S.; Venugopal, V.A.; Kumar, A.; Gupta, R.; Brucas, R.; George, S.; Neagu, A.; Tai, C.W.; Gubbins, M.; Andersson, G.; et al. Effect of seed layers on dynamic and static magnetic properties of Fe65Co35 thin films. *J. Phys. D-Appl. Phys.* **2018**, *51*, 305001. [CrossRef]
16. Liu, X.; Kanda, H.; Morisako, A. The effect of underlayers on FeCo thin films. In Proceedings of the 2nd International Symposium on Advanced Magnetic Materials and Applications (ISAMMA), Sendai, Japan, 12–16 July 2010.
17. Liu, X.-L.; Wang, L.-S.; Luo, Q.; Xu, L.; Yuan, B.-B.; Peng, D.-L. Preparation and high-frequency soft magnetic property of FeCo-based thin films. *Rare Met.* **2016**, *35*, 742–746. [CrossRef]
18. Xu, F.; Xu, Z.; Yin, Y. Tuning of the Microwave Magnetization Dynamics in Dy-Doped Fe$_{65}$Co$_{35}$-Based Thin Films. *IEEE Trans. Magn.* **2015**, *51*, 2800904. [CrossRef]
19. Liu, M.; Hu, L.; Ma, Y.; Feng, M.; Xu, S.; Li, H. Effect of Pt doping on the preferred orientation enhancement in FeCo/SiO$_2$ nanocomposite films. *Sci. Rep.* **2019**, *9*, 10670. [CrossRef]
20. Basumatary, H.; Chelyane, J.A.; Rao, D.V.S.; Kamat, S.V.; Ranjan, R. Influence of substrate temperature on structure, microstructure and magnetic properties of sputtered Fe-Ga thin films. *J. Magn. Magn. Mater.* **2015**, *384*, 58–63. [CrossRef]
21. Wei, Y.; Brucas, R.; Gunnarsson, K.; Harward, I.; Celinski, Z.; Svedlindh, P. Static and dynamic magnetic properties of bcc Fe$_{49}$Co$_{49}$V$_2$ thin films on Si(100) substrates. *J. Phys. D-Appl. Phys.* **2013**, *46*, 495002. [CrossRef]
22. Umadevi, K.; Chelvane, J.A.; Jayalakshmi, V. Magnetostriction and Magnetic Microscopy Studies in Fe-Co-Si-B Thin Films. *Mater. Res. Express* **2018**, *5*, 036102. [CrossRef]
23. Wang, C.; Xiao, X.; Rong, Y.; Hsu, T.Y. The effect of substrate temperature on the microstructure and tunnelling magnetoresistance of FeCo-Al$_2$O$_3$ nanogranular films. *J. Mater. Sci.* **2006**, *41*, 3873–3879. [CrossRef]
24. Abe, K.; Wu, S.; Tanaka, Y.; Ariake, Y.; Kanada, I.; Mewes, T.; Mankey, G.; Mewes, C.; Suzuki, T. The thickness and growth temperature dependences of soft magnetic properties and an effective damping parameter of (FeCo)-Si alloy thin films. *AIP Adv.* **2019**, *9*, 035139. [CrossRef]
25. Kumari, T.P.; Raja, M.M.; Kumar, A.; Srinath, S.; Kamat, S.V. Effect of thickness on structure, microstructure, residual stress and soft magnetic properties of DC sputtered Fe65Co35 soft magnetic thin films. *J. Magn. Magn. Mater.* **2014**, *365*, 93–99. [CrossRef]
26. Fu, Y.; Cheng, X.; Yang, Z. Magnetically soft and high-saturation-magnetization FeCo films fabricated by co-sputtering. *Phys. Status Solidi A-Appl. Mater. Sci.* **2005**, *202*, 1150–1154. [CrossRef]
27. Platt, C.L.; Berkowitz, A.E.; Smith, D.J.; McCartney, M.R. Correlation of coercivity and microstructure of thin CoFe films. *J. Appl. Phys.* **2000**, *88*, 2058–2062. [CrossRef]
28. Li, Y.; Li, Z.; Liu, X.; Fu, Y.; Wei, F.; Kamzin, A.S.; Wei, D. Investigation of microstructure and soft magnetic properties of Fe65Co35 thin films deposited on different underlayers. *J. Appl. Phys.* **2010**, *107*, 09A325. [CrossRef]
29. Liu, S.Y.; Cao, Q.P.; Mori, S.; Ishigami, K.; Igarashi, K.; Wang, C.; Qian, X.; Wang, X.D.; Zhang, D.X.; Riedmuller, B.; et al. Synthesis and magnetic properties of amorphous Fe-Y-B thin films. *J. Alloys Compd.* **2014**, *606*, 196–203. [CrossRef]

Article

Low Core Losses of Fe-Based Soft Magnetic Composites with an Zn-O-Si Insulating Layer Obtained by Coupling Synergistic Photodecomposition

Siyuan Wang, Jingwu Zheng *, Danni Zheng, Liang Qiao, Yao Ying, Yiping Tang, Wei Cai, Wangchang Li, Jing Yu, Juan Li and Shenglei Che *

Research Center of Magnetic and Electronic Materials, College of Materials Science and Engineering, Zhejiang University of Technology, Hangzhou 310014, China
* Correspondence: zhengjw@zjut.edu.cn (J.Z.); cheshenglei@zjut.edu.cn (S.C.)

Abstract: The major method used to reduce the magnetic loss of soft magnetic composites (SMCs) is to coat the magnetic powder with an insulating layer, but the permeability is usually sacrificed in the process. In order to achieve a better balance between low losses and high permeability, a novel photodecomposition method was used in this study to create a ZnO insulating layer. The effect of the concentration of diethyl zinc on the formation of a ZnO insulating film by photodecomposition was studied. The ZnO film was best formed with a diethyl zinc n-hexane solution at a concentration of around 0.40 mol/L. Combined with conventional coupling treatment processes, a thin and dense insulating layer was coated on the surface of iron powder in situ. Treating the iron powder before coating by photodecomposition led to a synergistic effect, significantly reduced core loss, and the effective permeability only decreased slightly. An iron-based soft magnetic composite with a loss value of 124 kW/m^3 and an effective permeability of 107 was obtained at the frequency of 100 kHz and a magnetic field intensity of 20 mT.

Keywords: Fe-based soft magnetic composite; photodecomposition; insulation layer; coupling treatment; ZnO

1. Introduction

Soft magnetic composites (SMCs) allow the miniaturization and high frequency of electromagnetic devices, exhibit low eddy current losses over a relatively wide frequency range, and fill the application gaps between metal laminates and ferrites. Therefore, SMCs are promising candidates for the development of efficient and lightweight electromagnetic devices [1].

The coating of the insulating layer on a magnetic powder surface is the core process in the preparation of soft magnetic composites, and it can serve to suppress eddy currents in soft magnetic alloys. The materials used for insulation layers are often categorized into three categories: an organic insulation layer [2], an inorganic insulation layer [3–8], and an organic–inorganic composite insulating layer. The inorganic oxide materials commonly used to prepare coatings are mainly Al_2O_3 [3], ZrO_2 [4], MgO [5], Fe_3O_4 [6], and ferrite [7]. There is also a novel Li-Al-O insulating layer [9].

For soft magnetic composites moving towards miniaturization and light weight, in the case of medium-frequency applications, it is necessary to reduce the magnetic loss so as to obtain high permeability. However, during the insulation coating process, it is difficult to achieve a balance between the effective permeability and the core loss, with low losses often decreasing the permeability.

For this reason, the following three requirements are proposed in the preparation of soft magnetic composites. First, the pressure needs to be as high as possible to reduce the presence of pores and to obtain a dense material. Second, the coating process needs to minimize the introduction of non-magnetic phases. Third, the insulating layer should be thin, dense, and uniform to the greatest extent possible, forming a relatively tight bond with the matrix phase so that the insulating layer can firmly adhere to the matrix phase.

A variety of methods for coating, such as the sol–gel method [10], the ball milling method [11], the bonding method [12], and the liquid phase method [13], have been used, all having their own advantages and disadvantages. But in most coating processes, it is very difficult to obtain an ideal coating, i.e., as thin as a mono-atom layer, fully dense, continuous and with a high electric resistance.

In this paper, a novel method with a controllable reaction process is proposed, which can coat a thin, dense, and uniform insulating layer on the surface of soft magnetic powder. It was expected that a sample with low loss and high permeability could be obtained using this method.

2. Experimental

2.1. Procedure

Reduced iron powders (purity > 99.5%; 0.2% Si, 0.003% P, and 0.015% S) provided by Jilin Huaxing Powder Metallurgy Technology Co. Ltd. with an average powder size of 75 µm were used as the matrix material.

The process of coating the insulating layer on the surface of the iron powders was as follows. Firstly, the silane coupling agent (KH550, 5 wt.% of the weight of the magnetic powder) was fully mixed with ethanol solution, then the iron powders were poured into it and stirred, and the surface of the iron powders was coated by ultrasonic treatment for 30 min. Secondly, after coupling treatment, the iron powders were washed with anhydrous ethanol many times and then transferred to a quartz glass bottle. Additionally, diethyl zinc [$Zn(C_2H_5)_2$] solution and n-ethane (C_6H_{14}) solvents were added to the quartz glass bottle. Since diethyl zinc can easily be oxidized and hydrolyzed when encountering air, it was required to add diethyl zinc in a glove box with an oxygen content and water content of less than 5 ppm. Next, the quartz glass bottle was irradiated with an ultraviolet lamp with a wavelength range of 610 nm to 500 nm to achieve the photodecomposition of diethyl zinc (as shown in Figure 1). In the process of photocatalysis, a certain flow of air was regularly introduced into the quartz glass bottle. After a period of photodecomposition, the iron powders were removed by centrifugation and then transferred to a vacuum-drying oven and dried at 80 °C for 3 h.

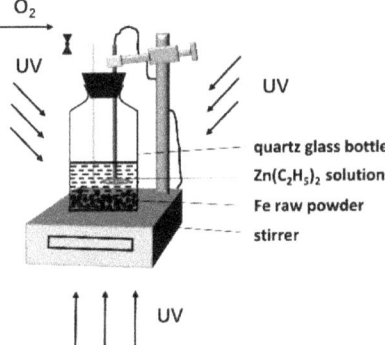

Figure 1. Schematic diagram of iron powders coated with ZnO by photodecomposition.

The preparation process of the Fe-based soft magnetic composites was as follows. Zinc stearate (0.5 wt.%) acted as the lubricant and was mixed with the as-dried powders in the agate mortar. Then, the fully mixed powders were compacted at 1000 MPa into a ring with an outer diameter of 12.7 mm and an inner diameter of 7.5 mm. Finally, the rings were annealed at 773 K for 3 h under Ar atmosphere [14].

2.2. Characterization

The morphology of the coated Fe powders was observed via scanning electron microscopy (SEM, Hitachi SU1510, Tokyo, Japan). The element content of the coating layer was analyzed via X-ray fluorescence (XRF, ZSX PrimusII, Rigaku, Tokyo, Japan). The microscopic structures and elemental distribution of the polished cross-sectional Fe-based soft magnetic composite samples were analyzed via field-emission scanning electron microscopy (FESEM, Tescan mira3 Zeiss sigma 500, Brno, The Czech Republic) coupled with an energy dispersive spectrometer (EDS). Effective permeability was tested using an LCR meter (E4980A, Agilent, Santa Clara, CA, USA) with a frequency from around 5 kHz to 300 kHz at a voltage of 500 mV, based on Equation (1):

$$\mu_e = L \cdot l_e \cdot 10^{-2} / (0.4\pi \cdot A_e \cdot N^2) \tag{1}$$

where L is inductance; l_e and A_e are the effective length and area of the magnetic path, respectively; and N is the number of turns [15]. The core losses of the SMC rings were measured using a B–H curve analyzer (SY-8218, IWATSU, Tokyo, Japan, with less than 0.5% tolerance) in a frequency ambit of 20–100 kHz and with a magnetic excitation level of 20 mT.

The total core loss (P_{cv}) of the sift magnetic materials can be divided into hysteresis loss (P_h), eddy current loss (P_{ed}), and residual loss (P_{exc}) according to the generation mechanism; the relationship can be expressed by the following Equation (2):

$$P_{cv} = P_h + P_{ed} + P_{exc} = K_H B_m^\alpha f + \frac{K_E D^2 B_m^2 f^2}{\rho} + P_{exc} \tag{2}$$

where K_H and K_E are constants; B_m is magnetic flux density (mT); ρ is the resistivity ($\Omega \cdot m$); D is the grain size, which can be regarded as a constant for a certain sample; α corresponds to the Steinmetz coefficient; and f is the frequency [16]. Usually, α = 1–2 for SMCs and α = 2–3 for ferrite materials.

When P_{cv} is divided by f, Equation (2) can be expressed as Equation (3):

$$\frac{P_{cv}}{f} = \frac{P_h}{f} + \frac{P_{ed}}{f} + \frac{P_{exc}}{f} = K_H B_m^\alpha + \frac{K_E D^2 B_m^2 f}{\rho} + \frac{P_{exc}}{f} \tag{3}$$

By measuring the loss at different frequencies, a curve of $\frac{P_{cv}}{f} \sim f$ can be drawn. The intercept of the ordinate is the hysteresis loss ($\frac{P_h}{f}$). Then, the eddy current loss ($\frac{P_{ed}}{f}$) is separated, which is linearly related to the frequency. The rest of the non-linear part is related to the residual loss ($\frac{P_{exc}}{f}$).

3. Results and Discussion

Effect of Surface Treatment Methods on the Morphology and Magnetic Properties of Coated Iron Powders

The quality of the coating on the surface of iron powder depends on the compactness of the coating layer and the adhesion with the iron matrix. For this reason, the effects of different surface treatment methods on the formation of coating layer on the iron powder surface were studied, and the different morphologies are shown in Figure 2. Figure 2a shows the powder coated by ultraviolet light decomposition in the diethyl zinc solution. It can be observed that the surface of the iron powder had many products generated by photodecomposition, and the gullies in the iron powder were obviously filled with coarser particles.

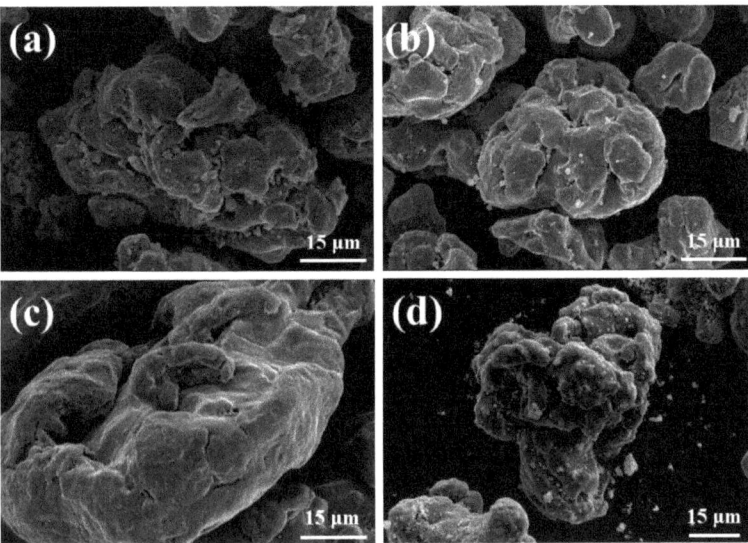

Figure 2. SEM of photolytically coated iron powder with different treatments: (**a**) $Zn(C_2H_5)_2$, (**b**) $Zn(C_2H_5)_2 + C_6H_{14}$, (**c**) 5 wt.% KH550, and (**d**) 5 wt.% KH550 and ($Zn(C_2H_5)_2 + C_6H_{14}$).

Figure 2b shows the powder coated by ultraviolet light decomposition in the same mass diethyl zinc solution diluted to 0.4 mol/L by an n-hexane solvent. The surface insulation products appeared to be slightly fewer compared with those shown in Figure 2a, and the surface was smoother.

Figure 2c shows the powder coated by coupling with 5 wt.% KH550, and it can be seen that there was no significant change on the surface of the iron powder. This is due to the thinness of the organosilane film.

Figure 2d shows the powder coated by coupled with 5 wt.% KH550, and then coated by ultraviolet light decomposition in a mixed solution of 0.4 mol/L diethyl zinc and n-ethane. It can be seen that the coating layer on the powder surface was filmy, and some of the powder surface was scattered with some non-filmed insulating particles.

The iron powder samples obtained from the different solution coating treatments, shown in Figure 2, were analyzed by XRF, and the results are shown in Table 1. Although the accuracy of XRF is not very high, it can be seen that the sample powder treated with diethyl zinc solution only had a higher content of Zn or ZnO in the XRF elements, but as can be seen in Figure 2, the size of the particles produced by photodecomposition was large. However, more non-magnetic phase Zn or ZnO content will lead to the decrease of saturation magnetization (Ms) of magnetic particles (see Figure S1).

Table 1. Elemental effects of different treatments on the prepared powders.

Element (at%)	$Zn(C_2H_5)_2$	$Zn(C_2H_5)_2 + C_6H_{14}$	KH550	KH550 and ($Zn(C_2H_5)_2 + C_6H_{14}$)
Fe	88.27	98.26	99.65	97.57
Zn	11.73	1.74	0	2.17
Si	0	0	0.35	0.26

With respect to diethyl zinc with the same percentage content of iron powder, when the n-hexane solvent was added for dilution, the concentration of diethyl zinc decreased, and the corresponding proportion of zinc or zinc oxide obtained after photodecomposition was also reduced by about one-tenth.

When the iron powder surface was pretreated by coupling with 5 wt.% KH550, the Zn content was elevated relative to that without pretreatment, which indicated that the coupling treatment was beneficial to the formation of a zinc coating layer on the iron powder surface through photocatalytic decomposition. This may be because the coupling modification of the surface by KH550 reduced the agglomeration between the surface particles and increased the specific surface area of the powder, which resulted in more Zn or ZnO deposited on the surface of the sample [17].

Figure 3 shows the core loss and effective permeability of the magnetic ring pressed with iron powders coated using different treatments methods. It can be seen that the highest core loss and the poor high-frequency stability of the permeability were obtained when the iron powders were coated using photolysis with only the diethyl zinc solution. When the iron powders were coated with the product of the photodecomposition of the diethyl zinc diluted with n-hexane solvent, the core loss of the sample was reduced by about two-thirds, and the high-frequency stability of the permeability was better compared to that of the samples prepared without the n-hexane solvent. When the iron powder was pre-coupled with KH550 and then coated with a low concentration of diethyl zinc solution, it was found that the core loss was significantly reduced and the effective permeability not only remained at a high value of about 107, but also had excellent high-frequency stability. Coupling treatment and photodecomposition coating were integrated together and produced good results, which cannot be achieved by using these treatments individually. In order to further explore the influence of different surface treatment methods on the core loss, the measured core loss values of the three samples were separated, respectively.

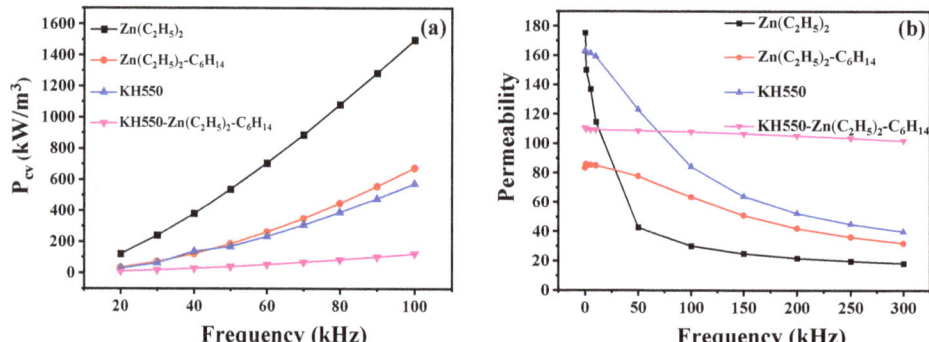

Figure 3. (a) Core loss and (b) effective permeability of the coated Fe-based magnetic ring prepared with different surface treatments.

We referred to the formula according to the Bertotti's loss separation model [18], $P_{cv} = P_{ed} + P_h + P_{exc}$, where P_{cv}, P_{ed}, P_h, and P_{exc} represent total core loss, eddy current loss, hysteresis loss, and residual loss, respectively. Hysteresis loss (P_h) is generally related to the powder particle size, internal stress, and the purity of the entire matrix phase. Eddy current loss (P_{ed}) is related to material resistivity and thickness. Residual loss (P_{exc}) is closely related to the magnetization rate [19]. When the frequency is below 500 KHz, residual loss (P_{exc}) is usually very small for iron-based SMCs; most of the core loss is from hysteresis loss (P_h) and eddy current loss (P_{ed}) [20,21].

The loss separation results for the iron-based soft magnetic composites prepared by different surface treatments are shown in Figure 4 (at $B = 20$ mT, $f = 100$ kHz). As previously discussed, when diethyl zinc was not diluted, the ZnO content produced by photodecomposition was higher. For magnetic particles, the introduction of ZnO increased the impurity mass. Therefore, the magnetic hysteresis loss of the sample coated by photodegradation of undiluted diethyl zinc was higher than that diluted with ethane.

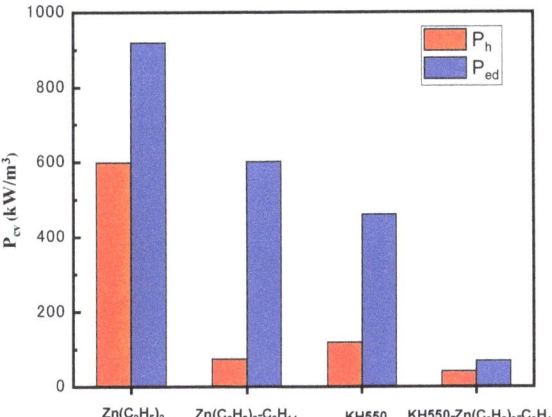

Figure 4. Separation of iron loss for samples prepared by different surface treatments at $f = 100$ kHz and $B = 20$ mT.

The integrity and uniformity of the insulating layer on the surface of iron powder are conducive to improving the resistivity of the composite magnetic ring, thus reducing eddy current loss. The coarse ZnO grains generated by photodecomposition in a high concentration of diethyl zinc solution led to the formation of a thick and discontinuous insulation layer. However, after the concentration of diethyl zinc was diluted, the ZnO grains generated were fine. In addition, after the organic silane film was obtained using the pre-coupling treatment for the iron powder, it was coated with ZnO. The silane film and ZnO played a synergistic role, which increased the adhesion between the coating and the substrate and formed a uniform insulating layer, thereby greatly reducing eddy current loss.

To confirm the formation of a homogeneous insulating layer on the surface of the sample as predicted by the loss separation data, the sample treated using pre-coupling and subsequent photodecomposition was polished on the cross-section, and the element distribution around the iron particles was observed by EDS, as shown in Figure 5. It can be seen from the figure that the iron particles were isolated by Zn, O, and Si elements, and the distribution of these three elements was consistent in the area. This proves that a relatively uniform and dense Zn-O-Si insulation layer was formed. It can also be seen that the thickness of the insulating layer was small, which verifies that the photodecomposition method can indeed deposit a thin, dense, and uniform insulating layer. The thin insulating layer caused a weaker magnetic dilution effect and was able to obtain a higher magnetic permeability. This was also consistent with the high effective permeability shown in Figure 3b.

$$\text{Surface} - \text{OH} + \text{H}_2\text{N}(\text{CH}_2)_3\text{Si}(\text{OCH})_3 + \text{H}_2\text{O} \rightarrow \text{Surface}(\text{O}-\text{Si})_n\text{OH} + \text{CH}_3\text{OH} \quad (4)$$

$$\text{Surface}(\text{O}-\text{Si})_n\text{OH} + \text{C}_2\text{H}_5 - \text{Zn} - \text{C}_2\text{H}_5 \rightarrow \text{Surface}(\text{O}-\text{Si})_n\text{O} - \text{Zn} - \text{C}_2\text{H}_5 + \text{C}_2\text{H}_6 \quad (5)$$

$$\text{Surface}(\text{O}-\text{Si})_n\text{O} - \text{Zn} - \text{C}_2\text{H}_5 + 7\text{O}_2 \xrightarrow{h\nu} \text{Surface}(\text{O}-\text{Si})_n\text{O} - \text{Zn} - \text{O} + \text{C}_2\text{H}_6 \quad (6)$$

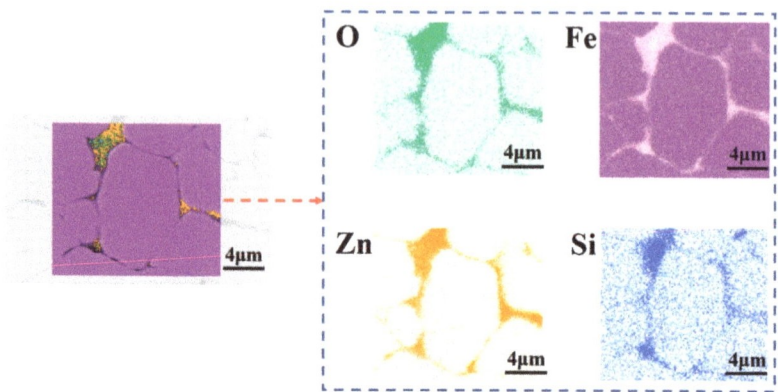

Figure 5. EDS spectra of polished cross-section samples coated with 5 wt.% KH550 + $Zn(C_2H_5)_2$ + C_6H_{14}.

Combining the above results, the simultaneous use of coupling treatment and photodecomposition coating achieved remarkable performance improvements and a good synergistic effect. For this synergistic mechanism, Figure 6 shows our speculations regarding this. During the pre-coupling treatment, the hydroxyl group adsorbed on the surface of iron powder reacted with KH550 for dehydration, producing $(O-Si)_nOH$ organosilane film (see Equation (4)), whereas, after adding diethyl zinc, the diethyl zinc reacted with the hydroxyl group of organosilane film, bonding them to form $(O-Si)_nO-Zn-C_2H_5$ (see Equation (5)) [22]. Under sufficient UV energy and an aerobic environment, the intermediate chemical bond of $-Zn-C_2H_5$ is dissociated and generates ZnO (as in Equation (6)), which is deposited on the surface of iron powder together with organosilane film, thereby achieving a thin and dense insulating layer. IR of iron powder with different treatments are shown in Figure S2. After pre-coupling treatment and photodecomposition of iron powder, the characteristic absorption peaks of Si-O bond and Zn-O bond appear at 1077.58 cm^{-1} and 472.25 cm^{-1} respectively, which may prove the occurrence of Equation (6).

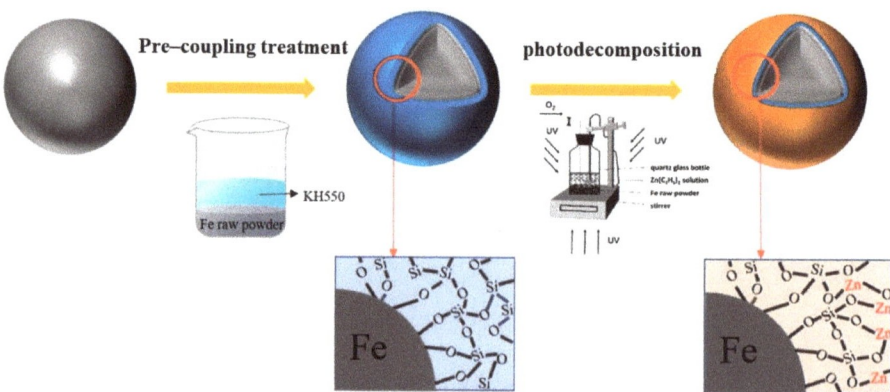

Figure 6. Schematic diagram of Fe@Zn-O-Si prepared using the pre-coupling and photodecomposition treatment.

In order to further investigate the effect of the diethyl zinc concentration on the coating quality of the iron powder surface, we specified the premise of diethyl zinc relative to the mass of the magnetic powder, and diluted the diethyl zinc with an n-ethane solvent to obtain different molar concentrations of diethyl zinc solutions; additionally, photodegradation coating treatment was performed on the iron powder pre-coupled with 5 wt.% KH550. The morphology of the iron powder obtained is shown in Figure 7.

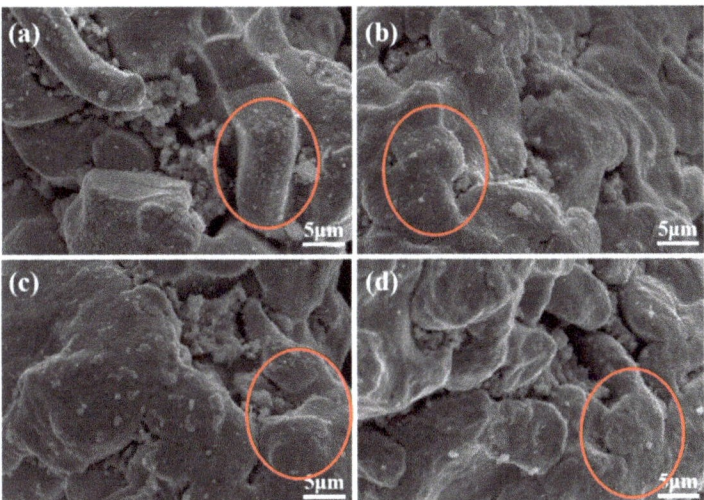

Figure 7. SEM images of iron powder coated using photolysis in diethyl zinc solutions of different concentrations: (**a**) 0.24 mol/L, (**b**) 0.32 mol/L, (**c**) 0.40 mol/L, and (**d**) 0.48 mol/L.

When the concentration of diethyl zinc was low, a cracked orange-peel-shaped insulating layer was formed on the surface of the iron powder, which is shown in the circle of Figure 7a. When the diethyl zinc concentration increased, the cracks on the surface insulation layer gradually started to become smaller, as shown in the circle in Figure 7b. When the concentration rose further to 0.40 mol/L, most of the surface insulating layer had started to form a film. Finally, when the concentration rose to 0.48 mol/L (as shown in Figure 7d), the orange-peel insulating layer appeared again, and the cracks of the surface insulating layer became larger, as shown in the circles.

Figure 8 shows the core loss and effective permeability of the magnetic ring pressed with iron powders coated by photolysis in diethyl zinc solutions of different concentrations. It can be seen that the concentration of the diethyl zinc solution had a large effect on the core loss, and the core loss value was first reduced but then increased again with the increase in the diethyl zinc solution concentration.

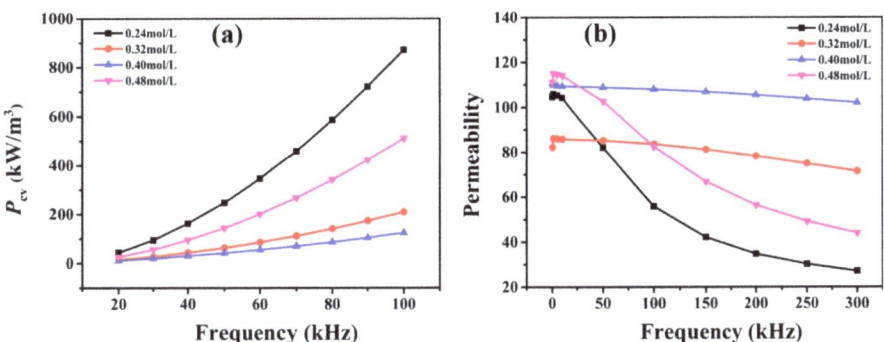

Figure 8. (**a**) Loss and (**b**) effective permeability of SMC rings made by the photodecomposition of coated iron powder with diethyl zinc solutions of different concentrations.

From this, we were able to determine the optimal solution concentration. Combined with the SEM images shown in Figure 7, it was found that the optimal concentration was 0.40 mol/L. At this time, the powder surface was essentially coated with a film similar to a Zn-O-Si compound. When the concentration of the solution was further increased, the core loss was not reduced but was, instead, increased, which corresponds to the rule of insulating film formation quality. In general, the effective permeability decreased with increasing diethyl zinc solution concentrations due to the magnetic dilution effect, but, in this section, the effective permeability first decreased and then increased.

We speculate that the photodecomposition products tended to be produced on the sample surface first, and when the diethyl zinc solution concentration was higher, they tended to be produced in the solution. Thus, the magnetic dilution effect was diminished again, and the effective permeability became higher. When the diethyl zinc solution concentration was 0.40 mol/L, the effective permeability was larger and more stable with regard to the frequency changes. However, when the concentration of KH550 was increased from 2 wt.% to 6 wt.%, the fluctuation of loss and effective permeability of the SMC rings was not as obvious as that of n-ethane, the relatively optimized KH550 concentration was 5 wt.% (see Figures S3 and S4).

In summary, after the pre-coupling treatment with 5 wt.% KH550, the iron powder surface was coated with an excellent insulating film by photolysis in the 0.40 mol/L diethyl zinc solution. The composite magnet prepared with the coated iron powders had an excellent core loss of 124 kW/m^3 and an effective permeability of 107 when the magnetic field strength was 20 mT and the frequency was 100 kHz.

4. Conclusions

In this paper, a thin and dense insulating layer was coated on the surface of iron powder in situ using pre-coupling treatment and the photodecomposition of diethyl zinc so as to obtain samples with low core losses and high effective permeability. The concentration of diethyl zinc was found to be related to the size of the ZnO grains generated by photodecomposition, which also determined the denseness of the insulating film. A high concentration of diethyl zinc resulted in coarse ZnO grains, and the surface insulation products showed the appearance of cracked orange peel, while dilution by the n-hexane solvent resulted in fine zinc oxide grains. The iron powder was pre-coupled and then coated by photodecomposition, and the superposition of the two treatments achieved a synergistic effect, generating a composite insulation layer that was similar to Zn-O-Si. The insulation layer not only improved the adhesion to the matrix, but it was also more dense. The eddy current loss was obviously reduced, and the thin insulation layer weakened the magnetic dilution effect, so the effective permeability was high. At a magnetic field strength of 20 mT and a frequency of 100 kHz, the loss value was about 124 kW/m^3 and the effective permeability was about 107.

Supplementary Materials: The following supporting information can be downloaded at: https://www.mdpi.com/article/10.3390/ma15238660/s1, Figure S1: Magnetisation curves of iron powder and photolytically coated iron powder with different treatments; Figure S2: IR of iron powder; iron powder surface treated with 5 wt.% KH550; iron powder surface treated with 5 wt.% KH550&(Zn(C2H5)2+C6H14) and ZnO obtained by photodecomposition; Figure S3: SEM morphology of iron powders treated with different KH550 ratios (a) 2 wt.%, (b) 3 wt.%, (c) 4 wt.%, (d) 5 wt.% and (e) 6 wt.% after photolytic coating; Figure S4: (a) Loss and (b) effective permeability of the SMC rings obtained after coating with iron powder treated with 2 wt.%, 3 wt.%, 4 wt.%, 5 wt.% and 6 wt.% KH550.

Author Contributions: Conceptualization, S.W and J.Z.; methodology, S.W. and J.Z.; software, W.L. and J.L.; validation, D.Z., L.Q. and W.C.; formal analysis, W.C.; investigation, S.W. and D.Z.; resources, S.C.; data curation, Y.Y. and J.Y.; writing—original draft preparation, S.W.; writing—review and editing, S.W., D.Z., J.Z. and S.C.; visualization, J.Y. and L.Q.; supervision, J.Z. and S.C.; project administration, S.C. and Y.T.; funding acquisition, Y.Y. and Y.T. All authors have read and agreed to the published version of the manuscript.

Funding: This work was financially sponsored by Intergovernmental International Science, Technology and Innovation Cooperation Key Project of the National Key R&D Programme (No. 2022YFE0109800), the National Natural Science Foundation of China (U1809215,52101235), the Basic Public Welfare Research Program of Zhejiang Province (LGG22E010010), and the Key R&D Project of Zhejiang Provincial Department Science and Technology (2021C01172).

Institutional Review Board Statement: Not applicable.

Informed Consent Statement: Not applicable.

Conflicts of Interest: The authors declare no conflict of interest.

References

1. Schoppa, A.; Delarbre, P. Soft Magnetic Powder Composites and Potential Applications in Modern Electric Machines and Devices. *IEEE Trans. Magn.* **2014**, *50*, 1–4. [CrossRef]
2. Wu, S.; Fan, J.L.; Liu, J.X.; Gao, H.X.; Zhang, D.H.; Sun, A.Z. Synthesis and Magnetic Properties of Soft Magnetic Composites Based on Silicone Resin-Coated Iron Powders. *J. Supercond. Nov. Magn.* **2018**, *31*, 587–595. [CrossRef]
3. Azimi-Roeen, G.; Kashani-Bozorg, S.F.; Nosko, M.; Nagy, S.; Matko, I. Formation of Al/(Al$_{13}$Fe$_4$ + Al$_2$O$_3$) Nano-composites via Mechanical Alloying and Friction Stir Processing. *J. Mater. Eng. Perform.* **2018**, *27*, 471–482. [CrossRef]
4. Protsenko, V.S.; Vasil'eva, E.A.; Smenova, I.V.; Baskevich, A.S.; Danilenko, I.A.; Konstantinova, T.E.; Danilov, F.I. Electrodeposition of Fe and Composite Fe/ZrO$_2$ Coatings from a Methanesulfonate Bath. *Surf. Eng. Appl. Electrochem.* **2015**, *51*, 66–75. [CrossRef]
5. Taghvaei, A.H.; Ebrahimi, A.; Gheisari, K.; Janghorban, K. Analysis of the magnetic losses in iron-based soft magnetic composites with MgO insulation produced by sol-gel method. *J. Magn. Magn. Mater.* **2010**, *322*, 3748–3754. [CrossRef]
6. Li, J.X.; Yu, J.; Li, W.C.; Che, S.L.; Zheng, J.W.; Qiao, L.; Ying, Y. The preparation and magnetic performance of the iron-based soft magnetic composites with the Fe@Fe3O4 powder of in situ surface oxidation. *J. Magn. Magn. Mater.* **2018**, *454*, 103–109. [CrossRef]
7. Nie, W.; Yu, T.; Wang, Z.G.; Wei, X.W. High-performance core-shell-type FeSiCr@MnZn soft magnetic composites for high-frequency applications. *J. Alloys Compd.* **2021**, *864*, 158215. [CrossRef]
8. Guo, R.D.; Wang, S.M.; Yu, Z.; Sun, K.; Jiang, X.N.; Wu, G.H.; Wu, C.J.; Lan, Z.W. FeSiCr@NiZn SMCs with ultra-low core losses, high resistivity for high frequency applications. *J. Alloys Compd.* **2020**, *830*, 154736. [CrossRef]
9. Zheng, J.W.; Zheng, H.D.; Lei, J.; Qiao, L.; Ying, Y.; Cai, W.; Li, W.C.; Yu, J.; Liu, Y.H.; Huang, X.L.; et al. Structure and magnetic properties of Fe-based soft magnetic composites with an Li-Al-O insulation layer obtained by hydrothermal synthesis. *J. Alloys Compd.* **2020**, *816*, 152617. [CrossRef]
10. Chen, J.C.; Guo, H.S.; Zhong, X.X.; Fu, J.Y.; Huang, S.J.; Xie, X.H.; Li, L.Z. Microstructural and magnetic properties of core-shell FeSiAl composites with Ni0.4Zn0.45Co0.15Fe2O4 layer by sol-gel method. *J. Magn. Magn. Mater.* **2021**, *529*, 167774. [CrossRef]
11. Luo, Z.; Fan, X.; Feng, B.; Yang, J.; Chen, D.; Jiang, S.; Wang, J.; Wu, Z.; Liu, X.; Li, G.; et al. Highly enhancing electromagnetic properties in Fe-Si/MnO-SiO$_2$ soft magnetic composites by improving coating uniformity. *Adv. Powder Technol.* **2021**, *32*, 4846–4856. [CrossRef]
12. Shi, X.Y.; Chen, X.Y.; Wan, K.; Zhang, B.W.; Duan, P.T.; Zhang, H.; Zeng, X.D.; Liu, W.; Su, H.L.; Zou, Z.Q.; et al. Enhanced magnetic and mechanical properties of gas atomized Fe-Si-Al soft magnetic composites through adhesive insulation. *J. Magn. Magn. Mater.* **2021**, *534*, 168040. [CrossRef]
13. Chan, L.W.; Liu, X.H.; Heng, P.W.S. Liquid phase coating to produce controlled-release alginate microspheres. *J. Microencapsul.* **2005**, *22*, 891–900. [CrossRef]
14. Lei, J.; Zheng, J.W.; Zheng, H.D.; Qiao, L.; Ying, Y.; Cai, W.; Li, W.C.; Yu, J.; Lin, M.; Che, S.L. Effects of heat treatment and lubricant on magnetic properties of iron-based soft magnetic composites with Al$_2$O$_3$ insulating layer by one-pot synthesis method. *J. Magn. Magn. Mater.* **2019**, *472*, 7–13. [CrossRef]
15. Zhou, M.; Peng, X.L. *Fundamentals of Magnetics and Magnetic Materials*; Zhejiang University Press: Hangzhou, China, 2019; pp. 175–176.
16. Li, W.C.; Li, W.; Ying, Y.; Yu, J.; Zheng, J.W.; Qiao, L.; Li, J.; Zhang, L.; Fan, L.; Wakiya, N.; et al. Magnetic and Mechanical Properties of Iron-Based Soft Magnetic Composites Coated with Silane Synergized by Bi$_2$O$_3$. *J. Electron. Mater.* **2021**, *50*, 2425–2435. [CrossRef]
17. Shi, W.X.; Yang, J.; Wang, T.J.; Jin, Y. Surface organic modification of magnetic iron oxide black particles. *Acta Phys.-Chim. Sin.* **2001**, *17*, 507–510. [CrossRef]
18. Berbotti, G. General properties of power losses in Soft ferromagnetic Materials. *IEEE Trans. Magn.* **1988**, *24*, 621. [CrossRef]
19. Zheng, J.W.; Zheng, D.N.; Qiao, L.; Ying, Y.; Tang, Y.P.; Cai, W.; Li, W.C.; Yu, J.; Li, J.; Che, S.L. High permeability and low core loss Fe-based soft magnetic composites with Co-Ba composite ferrite insulation layer obtained by sol-gel method. *J. Alloys Compd.* **2022**, *893*, 162107. [CrossRef]
20. Li, W.C.; Wang, Z.J.; Ying, Y.; Yu, J.; Zheng, J.W.; Qiao, L.; Che, S.L. In-situ formation of Fe$_3$O$_4$ and ZrO$_2$ coated Fe-based soft magnetic composites by hydrothermal method. *Ceram. Int.* **2019**, *45*, 3864–3870. [CrossRef]

21. Li, W.C.; Pu, Y.Y.; Ying, Y.; Kang, Y.; Yu, J.; Zheng, J.W.; Qiao, L.; Li, J.; Che, S.L. Magnetic properties and related mechanisms of iron-based soft magnetic composites with high thermal stability in situ composite-ferrite coating. *J. Alloys Compd.* **2020**, *829*, 154533. [CrossRef]
22. Guziewicz, E.; Godlewski, M.; Wachnicki, L.; Krajewski, T.A.; Luka, G.; Gieraltowska, S.; Jakiela, R.; Stonert, A.; Lisowski, W.; Krawczyk, M.; et al. ALD grown zinc oxide with controllable electrical properties. *Semicond. Sci. Technol.* **2012**, *27*, 074011. [CrossRef]

Article

Study of the Soft Magnetic Properties of FeSiAl Magnetic Powder Cores by Compounding with Different Content of Epoxy Resin

Zhengqu Zhu [1,2], Jiaqi Liu [1], Huan Zhao [2], Jing Pang [2], Pu Wang [1,*] and Jiaquan Zhang [1,*]

1. School of Metallurgical and Ecological Engineering, University of Science and Technology Beijing, Beijing 100083, China
2. Qingdao Yunlu Advanced Materials Technology Co., Ltd., Qingdao 266232, China
* Correspondence: wangpu@ustb.edu.cn (P.W.); jqzhang@metall.ustb.edu.cn (J.Z.); Tel.: +86-139-1117-1237 (J.Z.)

Abstract: FeSiAl is a commonly used soft magnetic material because of its high resistivity, low core loss, and low cost. In order to systematically study the effect of epoxy resin (EP) on the insulated coating and pressing effect of FeSiAl magnetic powders, six groups of composite powders and their corresponding soft magnetic powder cores (SMPCs) were prepared by changing the content of EP, and the soft magnetic properties of the powders and SMPCs were characterized. The results showed that FeSiAl powders exhibited good sphericity and morphology. The M_s of FeSiAl/EP composite powders was between 117.4–124.8 emu·g^{-1} after adding (0.3, 0.5, 0.7, 1, 1.5, and 2 wt. %) EP. The permeability μ_e of SMPCs increased first and then decreased with the increase in EP content. Among them, when the EP content was 1 wt. %, the corresponding SMPCs had the highest μ_e and excellent DC bias performance (63%, 100 Oe). In the whole test frequency range (50~1000 kHz), SMPCs with 1 wt. % EP content had the lowest core loss (1733.9 mW·cm^{-3} at 20 mT and 1000 kHz). After that, the loss separation study in the low-frequency range (50~250 kHz) was conducted, and the hysteresis loss and eddy current loss of SMPCs with 1 wt. % EP content were also the lowest. In addition, SMPCs also exhibited the best overall performance when the EP content was 1 wt. %. The results of this study can guide the design of composite insulation coating schemes and promote the development of soft magnetic materials for medium and high frequency applications.

Keywords: FeSiAl magnetic powder cores; epoxy resin; loss separate; soft magnetic properties

Citation: Zhu, Z.; Liu, J.; Zhao, H.; Pang, J.; Wang, P.; Zhang, J. Study of the Soft Magnetic Properties of FeSiAl Magnetic Powder Cores by Compounding with Different Content of Epoxy Resin. *Materials* 2023, 16, 1270. https://doi.org/10.3390/ma16031270

Academic Editor: Łukasz Hawełek

Received: 9 January 2023
Revised: 20 January 2023
Accepted: 31 January 2023
Published: 2 February 2023

Copyright: © 2023 by the authors. Licensee MDPI, Basel, Switzerland. This article is an open access article distributed under the terms and conditions of the Creative Commons Attribution (CC BY) license (https://creativecommons.org/licenses/by/4.0/).

1. Introduction

Soft magnetic powder cores (SMPCs) are a kind of soft magnetic composite made by pressing magnetic powders coated with insulating materials. They exhibit great potential in downstream fields such as photovoltaics, energy storage, variable frequency air conditioning, new energy vehicles, charging piles, uninterrupted power supplies (UPS), 5G base stations, and servers due to their excellent soft magnetic properties, flexible shape, and easy processing [1–3].

Insulated coating and pressing are the keys to preparing SMPCs. How to minimize the eddy current loss between magnetic powder cores and ensure the pressing effect of composite magnetic powder cores has always been a research hotspot. In recent years, some scholars have deposited inorganic insulation layers such as SiO_2, TiO_2, and Al_2O_3 [4–9] on the surfaces of magnetic powders by the sol–gel method and prepared ferrite insulation layers such as MnZn and NiZn [10–13] by the coprecipitation method to obtain SMPCs with excellent comprehensive performance. However, these two methods have many drawbacks, such as a complex coating process, high production cost, and unstable insulation properties. Additionally, most of the above-mentioned coatings are limited to small-scale research in the laboratory, and it is difficult to promote them to industrial mass production.

Epoxy resin (EP) is an organic insulating agent with both insulation and binding properties. It is widely used in the preparation process of SMPCs because it is easy to adhere to the surfaces of particles to achieve a uniform coating [14]. Based on the sol–gel method, Zhou et al. [15] prepared TiO_2 insulating layers on the surfaces of the magnetic powders by a hydrolysis condensation reaction and added 2 wt. % EP as a binder to improve the compressibility of FeSiBCCr/TiO_2 SMPCs. Chi et al. [16] used FeSiBCP amorphous powders as raw materials and prepared Fe_3O_4 nano-scale insulation layers by hydrothermal reaction in an alkaline environment and then embedded EP into the Fe_3O_4 coating matrix by mechanical mixing, which further improved the structural uniformity of composite coatings and the high-frequency comprehensive performance of SMPCs. Liu et al. [17] prepared FeSiBCCr@phosphate@EP core–shell structured SMPCs with excellent soft magnetic properties by using EP as the second insulation layer after the magnetic powders were passivated by phosphoric acid, which further reduced the core loss (1118.7 $mW·cm^{-3}$ at 20 mT and 1000 kHz). It can be seen that the organic insulating agent EP can effectively enhance the compressibility of SMPCs and reduce the eddy current loss; thus, it is commonly used as the main or secondary insulating material.

The FeSiAl powders produced by gas atomization have high sphericity, high permeability, low coercivity, high resistivity, and good wear resistance. Additionally, the magnetostriction coefficient and magnetic anisotropy constant of the FeSiAl alloy both tend to almost zero, so it is a kind of magnetic powder with excellent soft magnetic properties [18]. However, there are few systematic studies on the preparation of FeSiAl SMPCs only using EP as the insulating agent, and the relationship between the optimum addition amount and the characteristics of FeSiAl powder is still unknown. Therefore, in this paper, FeSiAl@EP SMPCs were prepared by using EP as the insulating agent, and the effects of EP content on the coating effect and soft magnetic properties of SMPCs were studied by various means to investigate the influence mechanism, which could provide the theoretical basis and practical guidance for the development and design of a composite insulated coating scheme.

2. Materials and Methods

2.1. Preparation of FeSiAl Powders

The experimental raw materials used in this study were FeSiAl crystalline powders (Qingdao Yunlu Advanced Materials Technology Co., Ltd, Qingdao, China) prepared by the gas atomization process. The composition was 8.7 wt. % Si, 5.6 wt. % Al, and 85.7 wt. % Fe. Epoxy resin was purchased from Shandong Huijia Magnetoelectricity Technology Co., Ltd. in Dezhou, China.

2.2. Preparation of SMPCs

Firstly, EP solution was prepared with 0.3, 0.5, 0.7, 1, 1.5, and 2 wt. % solute EP and 10 wt. % solvent acetone, respectively. The 2.5 kg FeSiAl powders were poured into the mixer, and the solution was poured after stirring for 30 s at room temperature. The heating temperature was 100 °C, and the stirring was 1~2 h until the powders were uniformly coated. The coated powders were sieved through a standard sieve of −140 mesh and were named P1~P6 according to the EP content. The powders (with the mass of 2 g) were compacted into annular SMPC samples of Φ 14 mm × Φ 8 mm × h 3.1 mm by two-way pressing with a pressure of 1800 MPa at room temperature. Subsequently, the annular SMPCs were annealed in vacuum at 200 °C for 1 h to release the internal stress caused by pressing and to enhance the strength of SMPCs. The specific preparation process is shown in Figure 1. SMPCs corresponding to P1~P6 composite powders were named S1–S6.

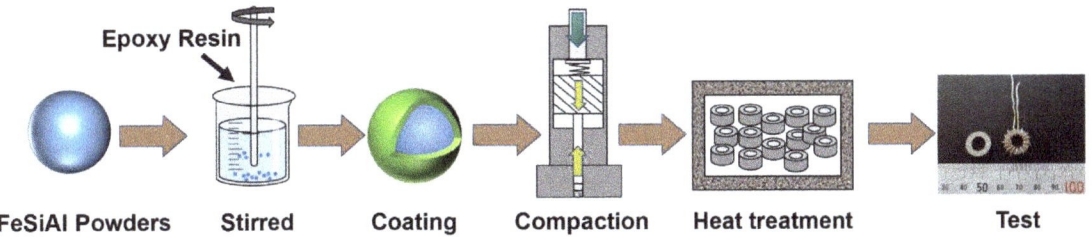

Figure 1. Schematic diagram of fabrication process for FeSiAl SMPCs.

2.3. Characterization

The microscopic morphology of FeSiAl raw powders and coated composite powders was observed by scanning electron microscopy (SEM, Phenom Pro Desktop SEM, Phenom-World BV, Eindhoven, The Netherlands). The particle size of FeSiAl powders was measured by a laser particle size analyzer (BT-9300S, Bettersize Instruments Ltd., Dandong, China). The phase of each sample was analyzed by X-ray diffraction (XRD, D2 PHASER, BRUKER AXS, Karlsruhe, Germany) using Cu Kα radiation with λ = 0.15406 nm. The step size was 0.02°, the scanning range was 30°~90°, the tube voltage was 30 kV, and the tube current was 10 mA. The saturation magnetization of each sample was measured by a vibrating sample magnetometer (VSM, Lake Shore 8604, Lake Shore Cryotronics, Inc., Westerville, OH, USA). The measurement range of saturation magnetization was ± 10,000 Oe, and the step size was 200 Oe. The DC bias performance of each sample at 100 kHz and 1 V and the effective permeability μ_e within the frequency range of 10 kHz~1 MHz were tested using the TH2816B/TH2826 LCR tester. The density of the SMPCs was measured according to the Archimedes drainage method. The coercivity of SMPCs was measured by a TD8220 soft magnetic DC tester. The resistivity of SMPCs was obtained by a four-probe method using an ST2742B automatic powder resistivity tester. The core loss of SMPCs was measured by a B–H analyzer (IWATSU-SY-8219) in the frequency range of 50 to 250 kHz. The maximum magnetic flux densities (B_m) were 0.01 T, 0.03 T, and 0.05 T, respectively. In addition, the core loss P_{cv} of SMPCs was also measured at B_m = 0.02 T and f = 100 kHz~1 MHz.

The effective permeability was calculated by the following equation [17]:

$$\mu_e = \frac{L l_e}{\mu_0 N^2 A_e} \quad (1)$$

where L is the inductance, l_e is the mean flux density path length, N is the number of coil turns, A_e is the cross-section area of the SMPCs, and μ_0 is the permeability of vacuum, $4\pi \times 10^{-7}$ H/m.

3. Results

3.1. Characterization of FeSiAl Powders

Figure 2 is the XRD pattern of FeSiAl raw powder P0 and other composite powders P1~P6. It can be seen from Figure 2 that all the XRD patterns of the powders show three distinct sharp diffraction peaks. According to the Scherrer formula, the average grain size of FeSiAl powder is 22.4 nm. Among them, the peak intensity of P0 is the highest, and the peak intensity of P1~P6 decreases in turn, indicating that the content of the crystal phase in the composite powders gradually decreases with the increase in EP content. The three diffraction peaks are located at 44.73°, 65.21°, and 82.69°, respectively, and Miller indices are (220), (400), and (422), respectively, which are the characteristic peaks of the typical FCC structure $Al_{0.3}Fe_3Si_{0.7}$ crystal phase [19]. EP is a kind of organic polymer compound, so EP is an amorphous material that shows a broad diffuse halo in XRD. However, the EP content added in this study is less, and the corresponding peak is overlapped by the diffraction peak of the $Al_{0.3}Fe_3Si_{0.7}$ crystal phase; thus, the EP is not detected by XRD in the five samples.

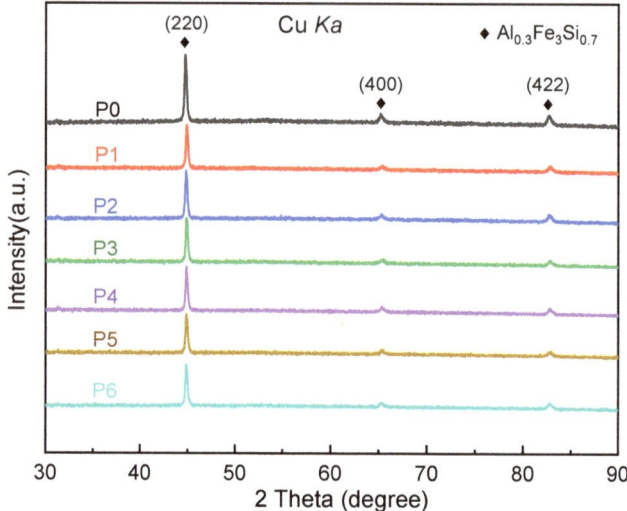

Figure 2. XRD patterns of FeSiAl powders P0~P6.

Figure 3 is the particle size distribution of FeSiAl powders. It can be seen from Figure 3 that the powder particle size deviates from the normal distribution with a d_{50} of 45.86 μm, exhibiting a generally coarse particle size. As can be seen from the SEM image in the inset of Figure 3, the FeSiAl powders show high sphericity and smooth surfaces.

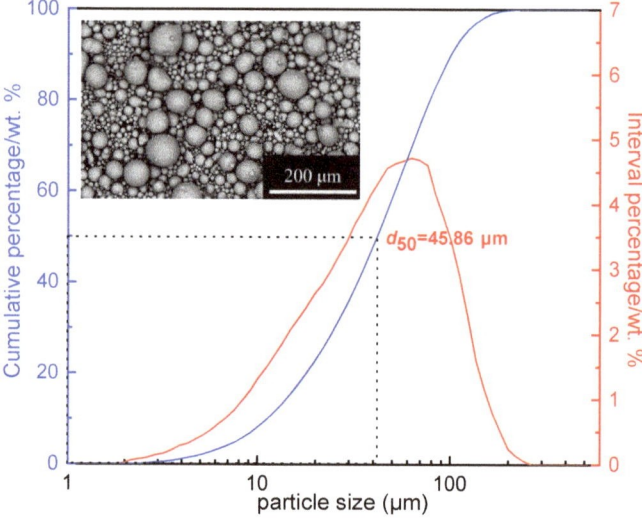

Figure 3. Size distribution of FeSiAl raw powders; the inset is the SEM image of FeSiAl powders.

The SEM images shown in Figure 4 P0–P6 show the evolution of the surface morphology of FeSiAl powders under different EP contents. It can be seen that with the increase of EP content, the surface of the powders is gradually dim, and small particles adhere to the surfaces of large particles. Among them, obvious EP agglomerations on the surface of P5 and P6 can be observed, while there is less for P4. This indicates that when the EP content exceeds 1 wt. %, the powder surface is unevenly coated, resulting in agglomerations of EP. It is worth mentioning that by controlling the amount of EP added, small particles adhere to the surface of large particles, which is beneficial to reduce the air gaps between particles in the subsequent pressing process and improve the compressibility and density of SMPCs.

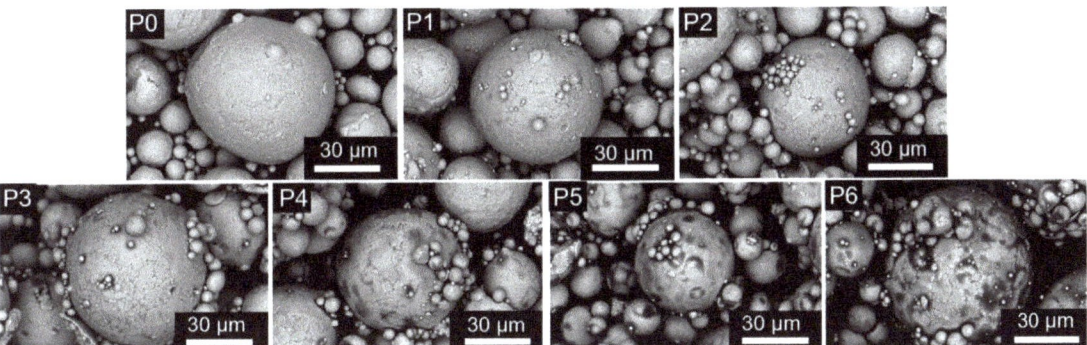

Figure 4. SEM images of FeSiAl powders P0~P6.

Figure 5 shows the hysteresis loops of FeSiAl powders and composite powders P1~P6 measured at room temperature. It can be seen from the enlarged view that the M_s of each sample is not consistent. Among them, the M_s of the raw powders is the highest (127.3 emu·g^{-1}), while the M_s of the coated powders is between 117.4~124.8 emu·g^{-1}, which is lower than that of the raw powders. This decrease indicates that the presence of the non-magnetic insulating layers occupies a certain volume fraction, resulting in a decrease in the M_s of the composite powders under the magnetic dilution effect [20].

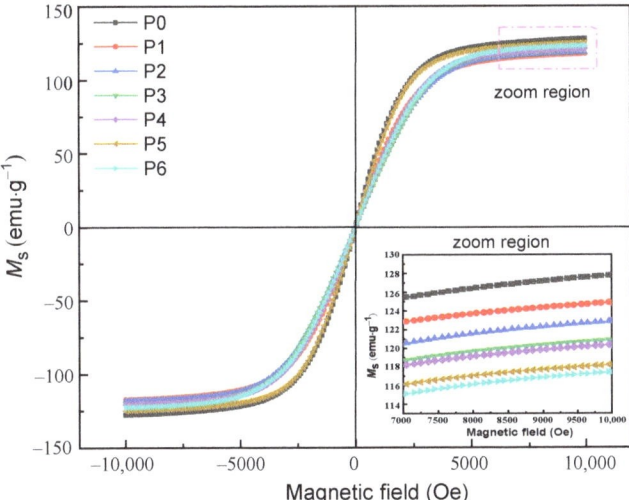

Figure 5. The hysteresis loops of powders P0~P6.

3.2. Performance Characterization of SMPCs

The frequency dependence of the effective permeability μ_e of S1 to S6 is shown in Figure 6a. In the frequency range of 10 kHz to 1 MHz, the μ_e of the 6 SMPCs decreases with increasing frequency. Among them, S1 shows the worst frequency stability because of the thinnest insulating layers. When the EP content increases from 0.5 to 2.0 wt. %, the thickness of the insulating layers increases, and the stability of the permeability of the corresponding SMPCs at high frequency is also improved.

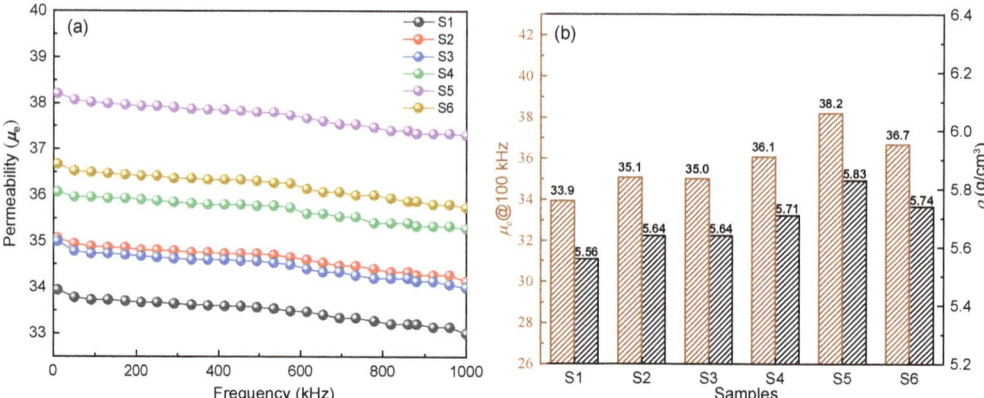

Figure 6. (a) Variations in the effective permeability with frequency for S1~S6; (b) μ_e@100 kHz and density for S1~S6.

It can be seen from Figure 6b that the μ_e of SMPCs increases from 33.9 to 38.2 with the increase of EP content from 0.3 to 1.5 wt. %, but when the EP content continues to increase to 2.0 wt. %, the permeability decreases to 36.7. The permeability of SMPCs increases first and then decreases with the increase of EP content. It is well known that the variation of μ_e is closely related to the density of SMPCs, the air gaps, and the microstructure of the insulating layers. Therefore, in this paper, the Archimedes principle is applied to calculate the density of SMPCs. Firstly, the mass w_1 of the sample in air and the mass w_2 of the sample after being hung on the wire are measured. Then the sample is put into water and the mass w_3 is measured. The volume of the sample is calculated by the equal volume method. Finally, the density is calculated according to Equation (2) as follows:

$$\rho = \frac{w_1 \rho_1}{(w_2 - w_3)} \tag{2}$$

where ρ is the sample density, g·cm^{-3}, and ρ_1 is the density of water used for weighing, which is 1.00 g·cm^{-3}. As shown in Figure 6b, it can be seen that the variation trend of μ_e is consistent with density. It can be speculated that when the EP content increases from 0.3 to 1.5 wt. %, small particles adhere to the surfaces of large particles, and the air gaps between particles decrease, which improves the compressibility of SMPCs and the volume fraction of the magnetic materials, resulting in an increase in μ_e. When the non-magnetic insulating material is excessive, the powder coating is uneven, the air gaps increase, and the demagnetization field in the magnetic powder cores increases, resulting in a decrease in μ_e [21].

The DC bias is defined as the ratio of the permeability of SMPCs in the applied magnetic field to that without the magnetic field. The DC bias of the samples is shown in Figure 7. The DC bias performance of S1~S6 shows an overall attenuation trend with the increase of the applied magnetic field. Especially when the magnetic field intensity is greater than 10 Oe, the percent permeability begins to drop sharply, from 99% at 10 Oe to about 64% at 100 Oe. Theoretically, the permeability of SMPCs is directly related to the volume fraction of air gaps. Although the air gaps in SMPCs play a role in pinning the magnetic domain walls during the magnetization process and hinder the attenuation of the permeability, the density of SMPCs also decreases due to the air gaps, resulting in a decrease in the permeability [22]. Therefore, under the same magnetic field intensity, the lower the permeability of SMPCs, the stronger the ability to resist saturation attenuation, that is, the better the DC bias performance, as can be seen from the inset in Figure 7.

Figure 8a is the core loss versus frequency curve of SMPC samples. It can be found that with the increase in the EP content, the core losses decrease first and then increase.

Among them, the core losses of S4 are the lowest, and the core losses of other samples are relatively close, as shown in Figure 8b. The role of the EP is to increase the resistivity of the SMPCs by uniformly coating the particles with insulation layers, thereby greatly reducing core losses, but the content of EP should not be excessive.

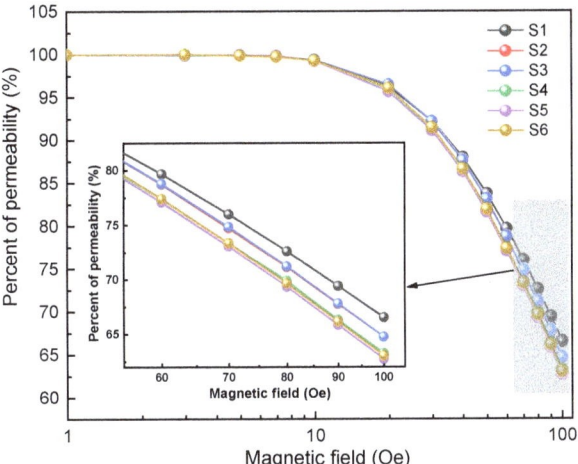

Figure 7. DC bias field dependence for SMPCs S1~S6 at magnetic field of 1~100 Oe.

The core loss of magnetic devices occupies a large proportion of switching between power supply and motor, so it is necessary to study the core losses of SMPCs. Figure 9a shows the relationship between P_{cv} and magnetic induction intensity for 6 samples at $f = 100$ kHz, while Figure 9b shows the relationship between P_{cv} and frequency for 6 samples at $B_m = 50$ mT. The core losses of S1~S6 increase with the increase in magnetic induction intensity and frequency. It can be seen from Figure 9b that, in the range of 50~100 kHz, the P_{cv} of samples is relatively close, but as the frequency continues to increase, the P_{cv} of the 6 samples begins to show a significant difference, indicating that the SMPCs prepared in this study are suitable for medium and high frequencies. In addition, when the EP content is increased from 0.3 to 1 wt. %, the P_{cv} showed a decreasing trend, whereas when the EP content further increased from 1 to 2 wt. %, P_{cv} increased slightly. Among them, S4 shows the lowest core loss of 512 mW·cm^{-3} (at 100 kHz and 50 mT).

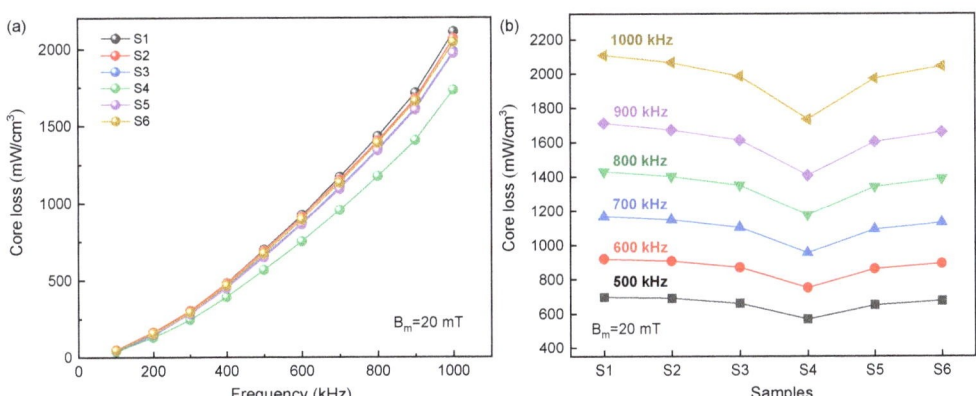

Figure 8. The total core losses of SMPCs for S1~S6: (**a**) under different EP content; (**b**) in the range of 500~1000 kHz.

In order to further explore the influence mechanism of EP content on the core loss of SMPCs, the loss separation of P_{cv} for 6 samples was studied. In theory, the core loss P_{cv} can be expressed as [23]:

$$P_{cv} = P_h + P_e + P_{ex} \tag{3}$$

where P_h, P_e, and P_{ex} are hysteresis loss, eddy current loss, and residual loss, respectively. Among them, the residual loss P_{ex} is mainly caused by the magnetization relaxation process or dispersion. It is a frequency-independent constant under a low-frequency weak magnetic field, and its value is generally small. Compared with P_h and P_e, P_{ex} is negligible. Therefore, P_{ex} is ignored in the loss separation analysis of this paper. Figure 9c,d show the curves of P_h and P_e with frequencies. With the increase in EP content, P_h and P_e decrease first and then increase. P_h refers to SMPCs in an alternating magnetic field due to the repeated magnetization hysteresis phenomenon caused by the consumption of energy, which can be expressed as follows:

$$P_h = C_h \cdot B_m^\alpha \cdot f \tag{4}$$

where C_h is the hysteresis coefficient, B_m is the magnetic induction intensity, f is the frequency, and α is the simulation coefficient. C_h is related to H_c; the lower value of C_h represents a lower P_h. Figure 9e shows the C_h and H_c of FeSiAl@EP SMPCs with different EP content. With the gradual increase in EP content, H_c decreased from 0.801 Oe to 0.664 Oe, and then increased to 1.046 Oe, while C_h decreased from 418 to 334, and then increased to 402.8, which is consistent with the changing trend of P_h in Figure 9c.

The eddy current loss P_e refers to the energy loss caused by the induced current generated by SMPCs in the alternating magnetic field. P_e can be expressed as [24]:

$$P_e = C_e \cdot B_m^2 \cdot f^2 = C \cdot B_m^2 \cdot f^2 \cdot d^2 / \rho \tag{5}$$

where C_e is the eddy current coefficient, C is the proportional constant, d is the diameter of composite magnetic powders, and ρ is resistivity. It can be seen that P_e is proportional to the square of the diameter of the magnetic powders and inversely proportional to the resistivity of SMPCs; that is, the size of the particles should also be taken into account while coating the powders to reduce the eddy current path. It can be seen from Figure 9f that the resistivity of the composite magnetic powders increases continuously with the increase in EP content, which is not consistent with the changing trend of P_e for FeSiAl@EP SMPCs. This is because the agglomeration of the particles occurs due to the excessive amount of EP (1.5~2 wt. %), reducing the local resistivity of SMPCs, as shown in Figure 4f,g. The uneven coating causes the eddy current path of the particles inside the SMPCs to become relatively large, resulting in an abnormal increase in P_e. In addition, the agglomeration of the particles further increases the diameter d of the magnetic powders, resulting in the negative effect of increasing eddy current loss beyond the positive effect of decreasing eddy current loss caused by the increased resistivity, so P_e will increase inversely with the increase in EP. Therefore, the content of the insulating agent EP should be controlled within a proper range. It can be seen from the hysteresis loss in Figure 9c and the eddy current loss in Figure 9d that the core losses are the lowest when the EP content is 1 wt. %, which also shows that the insulation coating effect of the S4 sample in this study is the best.

Based on the discussion of all the above experimental results, considering the properties of saturation magnetization, permeability, DC bias performance, and core losses of the 6 samples, it is considered that the SMPC with the best comprehensive performance in the high-frequency range is sample S4 with 1 wt. % EP. Table 1 summarizes the soft magnetic properties of FeSiAl@EP SMPCs prepared in this study. It can be seen from the table that the samples in this study show high saturation magnetization (117.4~124.8 emu·g^{-1}) and low core losses under different conditions (512.0~630.9 mW·cm^{-3} at 50 mT and 50 kHz; 1733.9~2110.3 mW·cm^{-3} at 20 mT and 1 MHz). The samples also have a high permeability (33.9 to 38.2) and show excellent stability in the high-frequency range. In addition, the excellent DC bias performance (62.7 to 66.5, 100 Oe) of SMPCs meets the requirements of high-end inductors and transformer coils. The improvement of saturation magnetization, permeability, and core losses also expands the application range of SMPCs into the low-frequency, high-frequency, and high current fields. It is worth mentioning that the insulating agent used in this paper is only EP, and its insulation effect is not the best, resulting in relatively high core losses. However, after the improvement of the subsequent coating process, it is expected to further improve the comprehensive performance of SMPCs, especially the core losses.

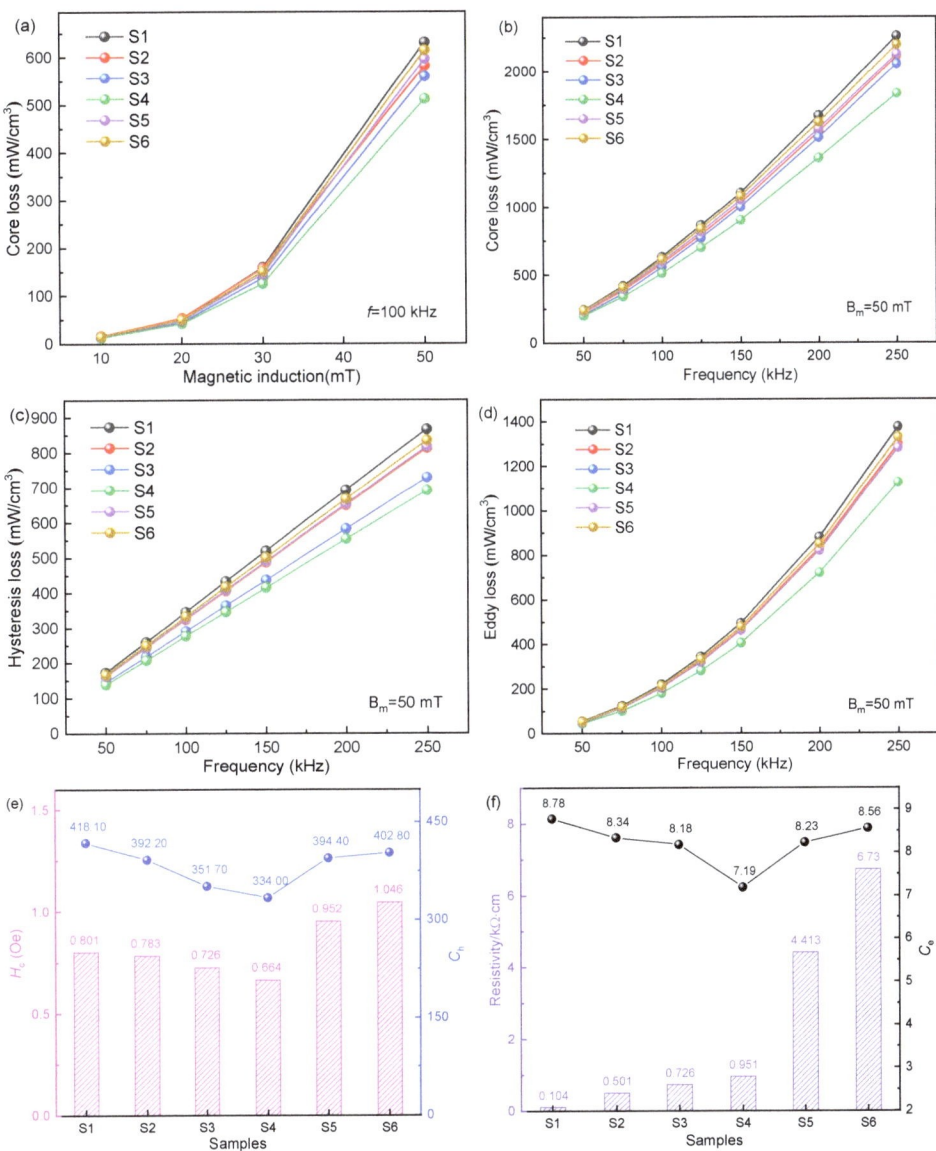

Figure 9. (**a**) Simulation results for frequency dependence of the core losses of S1~S6 under different magnetic induction. P_{cv} (**b**), P_h (**c**), and P_e (**d**) of FeSiAl@EP SMPCs with different frequencies. (**e**) The coercivity and the coefficient of hysteresis loss of the samples. (**f**) The resistivity and the coefficient of eddy current loss of the samples of S1~S6.

Table 1. Comparison of soft magnetic properties of SMPCs prepared in this work.

Sample	M_s	ρ	μ_e	Core Loss, P_{cv} (mW·cm^{-3})			DC-Bias (%)
	emu·g^{-1}	g·cm^{-3}	100 kHz	50 kHz/50 mT	100 kHz/50 mT	1 MHz/20 mT	100 Oe
S1	117.4	5.56	33.9	243.7	630.9	2110.3	66.5
S2	118.2	5.64	35.1	224.9	581.1	2068.1	64.7
S3	120.3	5.64	35.0	210.0	559.3	1985.8	64.8
S4	120.8	5.71	36.1	199.3	512.0	1733.9	63.3
S5	124.8	5.83	38.2	233.9	595.7	1972.5	62.7
S6	122.8	5.74	36.7	239.7	614.4	2044.5	63.1

4. Conclusions

In order to explore the influence mechanism of organic insulating agent EP on the soft magnetic properties of FeSiAl SMPCs, the properties of FeSiAl composite magnetic powders with different EP contents and the corresponding FeSiAl@EP SMPCs were tested and characterized by various means. The main conclusions are as follows:

- With the increase in EP content, the permeability of SMPCs increased from 33.9 to 38.2, and then decreased to 36.7, which showed a trend of increasing first and then decreasing. Six SMPCs had high DC bias performance, between 62~67% (100 Oe).
- The loss separation analysis of SMPCs in the range of 50~250 kHz showed that when the EP content was 1 wt. %, the coating of sample S4 was the most uniform, resulting in the lowest coercivity and hysteresis losses. The smaller magnetic powder particles also curbed the deterioration of eddy current losses and thus obtained the lowest core losses.
- In the high-frequency range, due to uniform EP coating and high density of SMPCs, S4 exhibited the lowest core losses (1733.9 mW·cm^{-3} at 20 mT and 1000 kHz), high permeability (36.1), and excellent DC bias performance (63%, 100 Oe).

The FeSiAl@EP SMPCs with EP content of 1 wt. % prepared in this study have excellent comprehensive soft magnetic properties, and the core losses can be further reduced by the improvement of the subsequent coating process, which is helpful to the development and design of FeSiAl insulated coating schemes and the preparation of high-performance FeSiAl SMPCs.

Author Contributions: Conceptualization, Z.Z. and J.L.; experimental setup design, Z.Z., J.L. and P.W.; validation, P.W. and J.Z.; formal analysis, Z.Z.; investigation, Z.Z., J.L., H.Z. and J.P.; writing—original draft preparation, Z.Z. and J.L.; writing—review and editing, P.W. and J.Z.; visualization, J.L.; project administration, H.Z. and J.Z.; funding acquisition, J.Z. All authors have read and agreed to the published version of the manuscript.

Funding: This research was funded by key research and development project of Shandong province in China, grant number "2022CXGC020308".

Institutional Review Board Statement: Not applicable.

Informed Consent Statement: Not applicable.

Data Availability Statement: The data presented in this study are available on request from the corresponding author.

Acknowledgments: The authors would also like to give thanks for the support from Qingdao Yunlu Advanced Materials Technology Co., Ltd.

Conflicts of Interest: The authors declare no conflict of interest.

References

1. Wu, S.D.; Dong, Y.Q.; Li, X.B.; Gong, M.G.; Zhao, R.L.; Gao, W.; Wu, H.; He, A.N.; Li, J.W.; Wang, X.M.; et al. Microstructure and magnetic properties of FeSiCr soft magnetic powder cores with a MgO insulating layer prepared by the sol-gel method. *Ceram. Int.* **2022**, *48*, 22278–22286. [CrossRef]
2. Sun, H.B.; Wang, C.; Wang, J.H.; Yu, M.G.; Guo, Z.L. Fe-based amorphous powder cores with low core loss and high permeability fabricated using the core-shell structured magnetic flaky powders. *J. Magn. Magn. Mater.* **2020**, *502*, 166548. [CrossRef]
3. Guo, Z.L.; Wang, J.H.; Chen, W.H.; Chen, D.C.; Sun, H.B.; Xue, Z.L.; Wang, C. Crystal-like microstructural Finemet/FeSi compound powder core with excellent soft magnetic properties and its loss separation analysis. *Mater. Des.* **2020**, *192*, 108769. [CrossRef]
4. Wu, Z.Y.; Jiang, Z.; Fan, X.A.; Zhou, L.J.; Wang, W.L.; Xu, K. Facile synthesis of Fe-6.5wt%Si/SiO$_2$ soft magnetic composites as an efficient soft magnetic composite material at medium and high frequencies. *J. Alloys Compd.* **2018**, *742*, 90–98. [CrossRef]
5. Li, W.; Cai, H.; Kang, Y.; Ying, Y.; Yu, J.; Zheng, J.; Qiao, L.; Jiang, Y.; Che, S. High permeability and low loss bioinspired soft magnetic composites with nacre-like structure for high frequency applications. *Acta Mater.* **2019**, *167*, 267–274. [CrossRef]
6. Zhou, B.; Dong, Y.Q.; Liu, L.; Chang, L.; Bi, F.Q.; Wang, X.M. Enhanced soft magnetic properties of the Fe-based amorphous powder cores with novel TiO$_2$ insulation coating layer. *J. Magn. Magn. Mater.* **2019**, *474*, 1–8. [CrossRef]
7. Tang, N.J.; Jiang, H.Y.; Zhong, W.; Wu, X.L.; Zou, W.Q.; Du, Y.W. Synthesis and magnetic properties of Fe/SiO$_2$ nanocomposites prepared by a sol-gel method combined with hydrogen reduction. *J. Alloys Compd.* **2006**, *419*, 145–148. [CrossRef]
8. Peng, Y.D.; Yi, Y.; Li, L.Y.; Yi, J.H.; Nie, J.W.; Bao, C.X. Iron-based soft magnetic composites with Al$_2$O$_3$ insulation coating produced using sol-gel method. *Mater. Des.* **2016**, *109*, 390–395. [CrossRef]
9. Yaghtin, M.; Taghvaei, A.H.; Hashemi, B.; Janghorban, K. Effect of heat treatment on magnetic properties of iron-based soft magnetic composites with Al$_2$O$_3$ insulation coating produced by sol–gel method. *J. Alloys Compd.* **2013**, *581*, 293–297. [CrossRef]
10. Liu, D.; Chen, X.P.; Ying, Y.; Zhang, L.; Li, W.C.; Jiang, L.Q.; Che, S.L. MnZn power ferrite with high Bs and low core loss. *Ceram. Int.* **2016**, *42*, 9152–9156. [CrossRef]
11. Nie, W.; Yu, T.; Wang, Z.G.; Wei, X.W. High-performance core-shell-type FeSiCr@MnZn soft magnetic composites for high-frequency applications. *J. Alloys Compd.* **2021**, *864*, 158215. [CrossRef]
12. Wang, J.H.; Xue, Z.L.; Song, S.Q.; Sun, H.B. Magnetic properties and loss separation mechanism of FeSi soft magnetic composites with in situ NiZn-ferrite coating. *J. Mater. Sci. Mater. Electron.* **2021**, *32*, 20410–20421. [CrossRef]
13. Zhong, Z.Y.; Wang, Q.; Tao, L.X.; Jin, L.C.; Tang, X.L.; Bai, F.M.; Zhang, H.W. Permeability dispersion and magnetic loss of Fe/Ni$_x$Zn$_{1-x}$Fe$_2$O$_4$ soft magnetic composites. *IEEE Trans. Magn.* **2012**, *48*, 3622–3625. [CrossRef]
14. Lu, H.; Dong, Y.; Liu, X.; Liu, Z.; Wu, Y.; Zhang, H.; He, A.; Li, J.; Wang, X. Enhanced magnetic properties of FeSiAl soft magnetic composites prepared by utilizing PSA as resin insulating layer. *Polymers* **2021**, *13*, 1350. [CrossRef]
15. Zhou, B.; Dong, Y.Q.; Liu, L.; Chi, Q.; Zhang, Y.; Chang, L.; Bi, F.Q.; Wang, X.M. The core-shell structured Fe-based amorphous magnetic powder cores with excellent magnetic properties. *Adv. Powder Technol.* **2019**, *30*, 1504–1512. [CrossRef]
16. Chi, Q.; Chang, L.; Dong, Y.Q.; Zhang, Y.Q.; Zhou, B.; Zhang, C.Z.; Pan, Y.; Li, Q.; Li, Q.; Li, J.W.; et al. Enhanced high frequency properties of FeSiBPC amorphous soft magnetic powder cores with novel insulating layer. *Adv. Powder Technol.* **2021**, *32*, 1602–1610. [CrossRef]
17. Liu, J.; Dong, Y.; Zhu, Z.; Zhao, H.; Pang, J.; Wang, P.; Zhang, J. Fe-based amorphous magnetic powder cores with low core loss fabricated by novel gas–water combined atomization powders. *Materials* **2022**, *15*, 6296. [CrossRef]
18. Huang, M.Q.; Wu, C.; Jiang, Y.Z.; Yan, M. Evolution of phosphate coatings during high-temperature annealing and its influence on the Fe and FeSiAl soft magnetic composites. *J. Alloys Compd.* **2015**, *644*, 124–130. [CrossRef]
19. Chen, Z.H.; Liu, X.S.; Kan, X.C.; Wang, Z.; Zhu, R.W.; Yang, W.; Wu, Q.Y.; Shezad, M. Phosphate coatings evolution study and effects of ultrasonic on soft magnetic properties of FeSiAl by aqueous phosphoric acid solution passivation. *J. Alloys Compd.* **2019**, *783*, 434–440. [CrossRef]
20. Lei, J.; Zheng, J.W.; Zheng, H.D.; Qiao, L.; Ying, Y.; Cai, W.; Li, W.C.; Yu, J.; Lin, M.; Che, S.L. Effects of heat treatment and lubricant on magnetic properties of iron-based soft magnetic composites with Al$_2$O$_3$ insulating layer by one-pot synthesis method. *J. Magn. Magn. Mater.* **2019**, *472*, 7–13. [CrossRef]
21. Zhao, R.L.; Huang, J.J.; Yang, Y.; Jiao, L.X.; Dong, Y.Q.; Liu, X.C.; Liu, Z.H.; Wu, S.D.; Li, X.B.; He, A.N.; et al. The influence of FeNi nanoparticles on the microstructures and soft magnetic properties of FeSi soft magnetic composites. *Adv. Powder Technol.* **2023**, *33*, 103663. [CrossRef]
22. Zhang, Y.; Sharma, P.; Makino, A. Fe-rich Fe–Si–B–P–Cu powder cores for high-frequency power electronic applications. *IEEE Trans. Magn.* **2014**, *50*, 2006804. [CrossRef]
23. Zhang, G.D.; Shi, G.Y.; Yuan, W.T.; Liu, Y. Magnetic properties of iron-based soft magnetic composites prepared via phytic acid surface treatment. *Ceram. Int.* **2021**, *47*, 8795–8802. [CrossRef]
24. Hu, F.; Ni, J.L.; Feng, S.J.; Kan, X.C.; Zhu, R.W.; Yang, W.; Yang, Y.J.; Lv, Q.R.; Liu, X.S. Low melting glass as adhesive and insulating agent for soft magnetic composites: Case in FeSi powder core. *J. Magn. Magn. Mater.* **2020**, *501*, 166480. [CrossRef]

Disclaimer/Publisher's Note: The statements, opinions and data contained in all publications are solely those of the individual author(s) and contributor(s) and not of MDPI and/or the editor(s). MDPI and/or the editor(s) disclaim responsibility for any injury to people or property resulting from any ideas, methods, instructions or products referred to in the content.

Article

A Study on an Easy-Plane FeSi$_{3.5}$ Composite with High Permeability and Ultra-Low Loss at the MHz Frequency Band

Peng Wu, Shengyu Yang, Yuandong Huang, Guowu Wang, Jinghao Cui, Liang Qiao *, Tao Wang and Fashen Li *

Key Laboratory for Magnetism and Magnetic Materials of Ministry of Education, Institute of Applied Magnetism, Lanzhou University, Lanzhou 730030, China; wup21@lzu.edu.cn (P.W.); yangshy21@lzu.edu.cn (S.Y.); huangyd20@lzu.edu.cn (Y.H.); wanggw16@lzu.edu.cn (G.W.); cuijh21@lzu.edu.cn (J.C.); wtao@lzu.edu.cn (T.W.)
* Correspondence: qiaoliang@lzu.edu.cn (L.Q.); lifs@lzu.edu.cn (F.L.)

Abstract: An easy-plane FeSi$_{3.5}$ composite with excellent magnetic properties and loss properties at MHz were proposed. The easy-plane FeSi$_{3.5}$ composite has ultra-low loss at 10 MHz and 4 mT, about 372.88 kW/m^3. In order to explore the reason that the P_{cv} of easy-plane FeSi$_{3.5}$ composite is ultra-low, a none easy-plane FeSi$_{3.5}$ composite, without easy-plane processing as a control group, measured the microstructure, and the magnetic and loss properties. We first found that the real reason why magnetic materials do not work properly at MHz due to overheat is dramatical increase of the excess loss and the easy-plane composite can greatly re-duce the excess loss by loss measurement and separation. The total loss of none easy-plane FeSi$_{3.5}$ composite is much higher than that of easy-plane FeSi$_{3.5}$ composite, where the excess loss is a major part in the total loss and even over 80% in the none easy-plane FeSi$_{3.5}$ composite. The easy-plane FeSi$_{3.5}$ composite can greatly reduce the total loss compared to the none easy-plane FeSi$_{3.5}$ composite, from 2785.8 kW/m^3 to 500.42 kW/m^3 (3 MHz, 8 mT), with the main reduction being the excess loss, from 2435.2 kW/m^3 to 204.93 kW/m^3 (3 MHz, 8 mT), reduced by 91.58%. Furthermore, the easy-plane FeSi$_{3.5}$ composite also has excellent magnetic properties, high permeability and ferromagnetic resonance frequencies. This makes the easy-plane FeSi$_{3.5}$ composite become an excellent soft magnetic composite and it is possible for magnetic devices to operate properly at higher frequencies, especially at the MHz band and above.

Keywords: easy-plane soft magnetic composites; MHz frequency band; ultra-low power loss; excess loss

1. Introduction

With the development of electrical and electronic technology, soft magnetic materials are used as core components in daily life. Their magnetic properties and loss properties directly affect the power density and conversion efficiency [1]. In the further development of electrical and electronic devices, magnetic devices (inductor, power converters, etc.) are required to develop in the direction of miniaturization and high efficiency [2–5]. In the past, soft magnetic materials do not need to operate at higher frequencies due to the low operating frequency of power semiconductors. But now, the appearance of the third-generation wide bandgap semiconductors (WBG) has provided a high-frequency scenario to soft magnetic materials [6]. The conversion efficiency P_{th} and size of soft magnetic materials in power devices (for example inductance, transformer, etc.) can be optimized because WBG increases the operating frequency, which can be described:

$$P_{th} = CfB_m A_e W_d \qquad (1)$$

where C is the conversion efficiency coefficient, f is the operating frequency, B_m is the maximum magnetic flux density, A_e is the effective sectional area, and W_d is the number of turns. However, soft magnetic materials have developed a problem, which makes them unable to work properly due to overheating at higher frequencies when WBG increases

the operating frequency above MHz. Especially with the commercialization of SiC and GaN, the problem that soft magnetic materials cannot work properly at higher frequencies becomes more serious. A review, published in 2018 in *Science*, noted that none of the soft magnetic materials available today can cope with this daunting challenge. However, it also pointed out that soft magnetic composites (SMCs) have the potential to solve this problem [6].

Traditional SMCs are often used in electric motors, transformers, inductors and sensors due to high permeability, high-saturation magnetization and high-ferromagnetic resonance frequency [7–10]. When the devices try to work at higher frequencies, SMCs also have the same problem, which makes them unable to work properly due to overheating, resulting in a huge loss [6,11]. This overheating was thought to be caused by the eddy current loss at higher frequencies in the previous research [11–14]. Addressing the sharp increase in the loss at higher frequencies will provide a means of dealing with WBG and making soft magnetic materials work effectively.

The easy-plane soft magnetic materials are materials with a special magnetic configuration, where easy magnetization axes distributed in parallel planes with the same type enables the formation of the easy magnetization plane. According to the physical origins for the formation of the easy magnetization plane, easy-plane soft magnetic materials are divided into the magnetocrystalline-easy-plane soft magnetic materials (the easy magnetization axes are determined by the minimal of the magnetocrystalline anisotropy energy) and the magnetostatic-easy-plane soft magnetic materials (the easy magnetization axes are determined by the minimal of the magnetostatic energy). In the easy magnetization plane, the magnetization of easy-plane soft magnetic material is more easily saturated, remanence and coercivity of easy-plane soft magnetic materials are lower and the easy-plane soft magnetic materials has higher permeability and ferromagnetic resonance frequencies. When prepared as composites, the easy-plane soft magnetic composites maintain the excellent properties of the easy-plane soft magnetic materials, and also have the advantages of the composites. In the field of wave absorption, excellent results have been achieved due to the excellent performance of easy-plane soft magnetic composites [15–18]. Therefore, the easy-plane soft magnetic composites have the potential to address challenges posed by energy problem and WBG.

In the up-to-date published SMCs, measurement frequency of most studies are 1 MHz and below, such as FeSiCr (μ = 37, 560 kW/m^3 at 1 MHz, 20 mT and μ = 47.5, 7086 kW/m^3 at 1 MHz, 50 mT) [19,20], Fe$_{73}$Si$_6$B$_{10}$P$_5$C$_3$Mo$_3$ (μ = 34.63, ~500 kW/m^3 at 1 MHz, 20 mT) [21], flake pure iron (μ = 67, 306.6 kW/m^3 at 1 MHz, 3 mT and μ = 10, 3.285 kW/m^3 at 3.5 MHz, 0.1 mT) [22], etc. There are almost no studies with high frequency and high maximum magnetic flux density. However, in this paper, we measured for the total loss P_{cv} of the easy-plane FeSi$_{3.5}$ composite at 10 MHz and 4 mT, and we found that the easy-plane FeSi$_{3.5}$ composite has an ultra-low loss. The P_{cv} of the easy-plane FeSi$_{3.5}$ composite and none easy-plane FeSi$_{3.5}$ composite was measured and analyzed at 3 MHz and 8 mT for this study; the reason for their ultra-low loss. We first found that the real reason why magnetic materials do not work properly at higher frequencies is due to overheating, which shows a dramatical increase in the excess loss of P_{exc}, and the easy-plane FeSi$_{3.5}$ composite can greatly reduce the total loss of P_{cv}, and the main part of the reduction is the P_{exc}. The easy-plane FeSi$_{3.5}$ composite has high permeability, high-ferromagnetic resonance frequencies and a lower loss compared with the none easy-plane FeSi$_{3.5}$ composite by measurement and separation of loss. It is found that the sharp increase in the excess loss combined with frequency far exceeded the eddy current loss, and the excess loss becomes the most main part of total loss. The easy-plane FeSi$_{3.5}$ composite not only has higher permeability and higher ferromagnetic resonance frequencies, but can also effectively reduce the excess loss. Thus, the easy-plane FeSi$_{3.5}$ composite has the potential to deal with the overheating of soft magnetic materials due to excessive loss at higher frequencies, and is effectively utilized for devices at higher frequencies, especially at MHz and above.

2. Experiment

FeSi$_{3.5}$ are the tradition spherical particles, obtained by purchasing. The tradition spherical FeSi$_{3.5}$ particles are easy-plane processed to obtain the easy-plane FeSi$_{3.5}$ particles, and the tradition spherical FeSi$_{3.5}$ particles are named the none easy-plane FeSi$_{3.5}$ particles. The easy-plane FeSi$_{3.5}$ particles were compounded with polyurethane (PU) to prepare a polyurethane-based composite with 60 vol% (optimal volume fraction obtained after extensive experiments), and oriented in a rotating magnetic field (1 T) for 10 min. After the composite was heated to 90 °C, it was pressed into a ring-shaped composite with an outer diameter of 7 mm, an inner diameter of 3.04 mm and a thickness of 1 mm. And the none easy-plane FeSi$_{3.5}$ composite was prepared using the same method.

The morphology of both particles and the composites were characterized by SEM (Apreo S, Thermo Fisher Scientific, Waltham, MA, USA). The vibrating sample magnetometer (VSM) (Microsence EV9, MicroSense, Lowell, MA, USA) was used to measure the hysteresis loop, and the degree of the plane orientation of two composites were also obtained by VSM data. The complex permeability of the composite was measured by a precision impedance analyzer (Agilent E4991B, Agilent, Santa Clara, CA, USA) and vector network analyzer (Agilent E8363B, Agilent, Santa Clara, CA, USA) at 1 MHz~18 GHz. The power loss was measured by a B-H analyzer (SY-8218, Iwatsu, Tokyo, Japan).

3. Result

The P_{cv} of the easy-plane FeSi$_{3.5}$ composite was measured at 10 MHz and 4 mT. It is found that the easy-plane FeSi$_{3.5}$ composite has an ultra-low loss of about 372.88 kW/m^3. The result is rare in SMC. In order to explore the reason that the P_{cv} of the easy-plane FeSi$_{3.5}$ composite is ultra-low, the none easy-plane FeSi$_{3.5}$ composite, without easy-plane processing as a control group, measured the microstructure, magnetic properties and loss properties.

3.1. Morphology and Microstructure

Figure 1 shows the morphology and microstructures of the none easy-plane FeSi$_{3.5}$ particles (a,b) and the easy-plane FeSi$_{3.5}$ particles (d,e), cross sections of the none easy-plane FeSi$_{3.5}$ (c) and easy-plane FeSi$_{3.5}$ composite (f), respectively. The thickness of the none easy-plane FeSi$_{3.5}$ particles is 44.35 µm, and the thickness of the easy-plane FeSi$_{3.5}$ particles is 1.14 µm, thus the latter is much smaller than the former. The microstructures of the none easy-plane FeSi$_{3.5}$ composite are a homogenous mixture of none easy-plane FeSi$_{3.5}$ particles and PU; the microstructure of the easy-plane composite is an ordered laminar structure because of the orientation process.

3.2. Static Magnetic Performance and Degree of the Plane Orientation

Figure 2 shows the static magnetic properties of FeSi$_{3.5}$ particles (a), hysteresis loops and the degree of plane orientation (DPO) of the none easy-plane FeSi$_{3.5}$ composite (b) and easy-plane FeSi$_{3.5}$ composite (c). The saturation magnetization (M_s) and the coercivity (H_c) of original FeSi$_{3.5}$ is 184 emu/g and 3.32 Oe. The degree of plane orientation was characterized by measuring hysteresis loops of the in-plane and out-of-plane direction. The DPO can be expressed by Equation (2) [23]:

$$DPO(\%) = \frac{M_s - M_{r,\,z-axis}}{M_s} \times 100\% \quad (2)$$

where M_s is the saturation magnetization, and $M_{r,\,z-axis}$ is the remanent magnetism in the z-axis direction. For the none easy-plane FeSi$_{3.5}$ composite, the hysteresis loops of the in-plane and out-of-plane direction coincide. Neither the in-plane direction or out-of-plane direction was magnetized to saturation (about 154 emu/g). For the easy-plane FeSi$_{3.5}$ composite, the hysteresis loops of the in-plane and out-of-plane direction are different. The in-plane direction is more easily magnetized to saturation (about 173.5 emu/g) than the out-of-plane direction (Table 1).

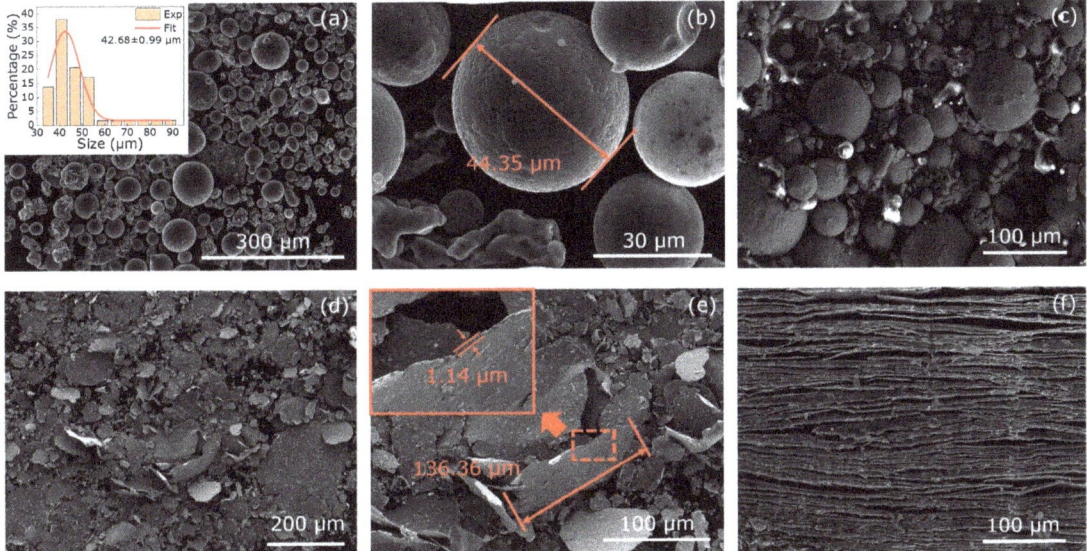

Figure 1. The SEM of the none easy-plane FeSi$_{3.5}$ particles (**a**,**b**) and the cross sections of the composite (**c**), the SEM of the easy-plane FeSi$_{3.5}$ particles (**d**,**e**) and the cross sections of the composite (**f**).

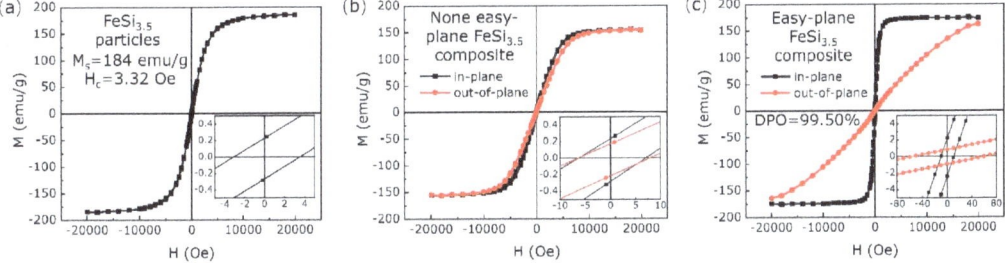

Figure 2. The hysteresis loop of FeSi$_{3.5}$ particles (**a**), hysteresis loops and *DPO* of the none easy-plane FeSi$_{3.5}$ composite (**b**) and easy-plane FeSi$_{3.5}$ composite (**c**).

Table 1. The M_s, M_r and H_c of the in-plane and out-of-plane direction and *DPO* of the none easy-plane FeSi$_{3.5}$ composite and easy-plane FeSi$_{3.5}$ composite.

	In-Plane			Out-of-Plane			*DPO*
	M_s	M_r	H_c	M_s	M_r	H_c	
None easy-plane FeSi$_{3.5}$	154.45	0.24	6.71	153.78	0.18	7.18	
Easy-Plane FeSi$_{3.5}$	173.53	2.28	10.19	163.00	0.87	57.79	99.50%

3.3. Complex Permeability and Magnetic Spectra Simulation

Figure 3 shows the magnetic spectra and spectra simulation of the none easy-plane FeSi$_{3.5}$ and the easy-plane FeSi$_{3.5}$ composite with 60 vol%. For the none easy-plane FeSi$_{3.5}$ composite, the real part of permeability is 30. It has a resonance peak at 20 MHz in the imaginary part of permeability. For the easy-plane FeSi$_{3.5}$ composite, the real part of permeability is 62. Two resonance peaks appear in the imaginary part of the permeability. The domain wall resonance peak is observed in a lower frequency (100 MHz), and the natural resonance is observed in a higher frequency (1.5 GHz). According to the domain wall motion mechanism and spin rotation mechanism, magnetic spectra can be simulated by Equations (3) and (4) [24–28]:

$$\mu' = \mu'_{dw} + \mu'_{spin} = \frac{\omega_{dw}^2 \chi_{dw}\left(\omega_{dw}^2 - \omega^2\right)}{\left(\omega_{dw}^2 - \omega^2\right)^2 + \omega^2 \beta^2} + \frac{\chi_{spin}\omega_{spin}^2\left[\left(\omega_{spin}^2 - \omega^2\right) + \omega^2 \alpha^2\right]}{\left[\omega_{spin}^2 - \omega^2(1+\alpha^2)\right]^2 + 4\omega^2 \omega_{spin}^2 \alpha^2} + 1 \quad (3)$$

$$\mu'' = \mu''_{dw} + \mu''_{spin} = \frac{\omega_{dw}^2 \chi_{dw} \omega \beta}{\left(\omega_{dw}^2 - \omega^2\right)^2 + \omega^2 \beta^2} + \frac{\chi_{spin}\omega_{spin}^2 \omega \alpha \left[\omega_{spin}^2 + \omega^2(1+\alpha^2)\right]}{\left[\omega_{spin}^2 - \omega^2(1+\alpha^2)\right]^2 + 4\omega^2 \omega_{spin}^2 \alpha^2} \quad (4)$$

where μ'_{dw} and μ'_{spin} are the real part of permeability for the domain wall motion mechanism and spin rotation mechanism, and μ''_{dw} and μ''_{spin} are the imaginary part of permeability for the domain wall motion mechanism and spin rotation mechanism, respectively. ω_{dw}, χ_{dw} and β are the domain wall resonance frequency, static susceptibility and damping factor for the domain wall component. ω_{spin}, χ_{spin} and α is the resonance frequency, static susceptibility and damping factor for the spin rotation component. ω is the frequency of the applied field ($\omega = 2\pi f$). According to Equations (3) and (4), the magnetic spectra are simulated and six parameters (ω_{dw}, χ_{dw}, β, ω_{spin}, χ_{spin}, α) are obtained (Figure 3b,c and Table 2).

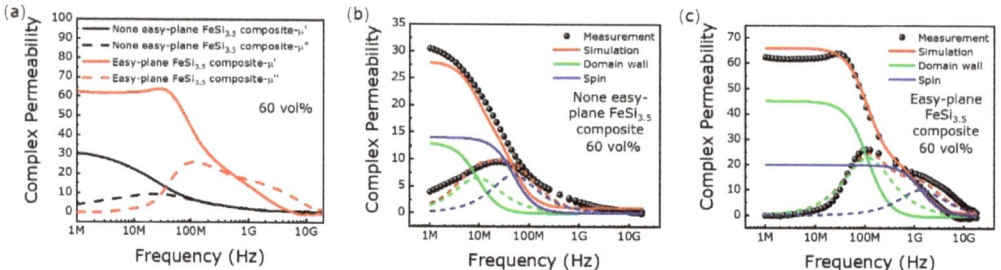

Figure 3. The complex permeability (**a**) and the spectra simulation of the none easy-plane FeSi$_{3.5}$ composite (**b**) and the easy-plane FeSi$_{3.5}$ composite (**c**) with 60 vol%.

Table 2. The six parameters in the magnetic spectra of the none easy-plane FeSi$_{3.5}$ composite and easy-plane FeSi$_{3.5}$ composite.

	μ_i	Domain Wall Motion					Spin Rotation			
		χ_{dw}	f_{dw} (MHz)	$f_{dw}^{\mu'' (max)}$ (MHz)	β	χ_{spin}	f_{spin} (GHz)	$f_{spin}^{\mu'' (max)}$ (GHz)	α
None easy-plane FeSi$_{3.5}$	28	13	120	9	1×10^{10}	14	0.35	0.05	7
Easy-Plane FeSi$_{3.5}$	66	45	840	100	4×10^{10}	20	7	1.5	5

The theoretical domain wall resonance frequency and natural resonance frequency of the none easy-plane FeSi$_{3.5}$ composite are 120 MHz and 350 MHz. The real part of permeability for the domain wall component and spin rotation are 13 and 14. The theoretical domain wall resonance peak and natural resonance peak of the easy-plane FeSi$_{3.5}$ composite are 840 MHz and 7 GHz. The real part of permeability for the domain wall component and spin rotation are 45 and 20.

3.4. Power Loss and Loss Separation

Figure 4 shows the power loss and loss separation of the none easy-plane FeSi$_{3.5}$ composite and the easy-plane FeSi$_{3.5}$ composite. The easy-plane FeSi$_{3.5}$ composite exhibits a desired total loss P_{cv} of 500.42 kW/m^3 under the test conditions of 8 mT and 3 MHz.

In general, the P_{cv} rapidly increases with increasing frequency, and the maximum P_{cv} of the easy-plane FeSi$_{3.5}$ composite (500.42 kW/m^3) is much lower than that of the none easy-plane FeSi$_{3.5}$ composite (2785.8 kW/m^3).

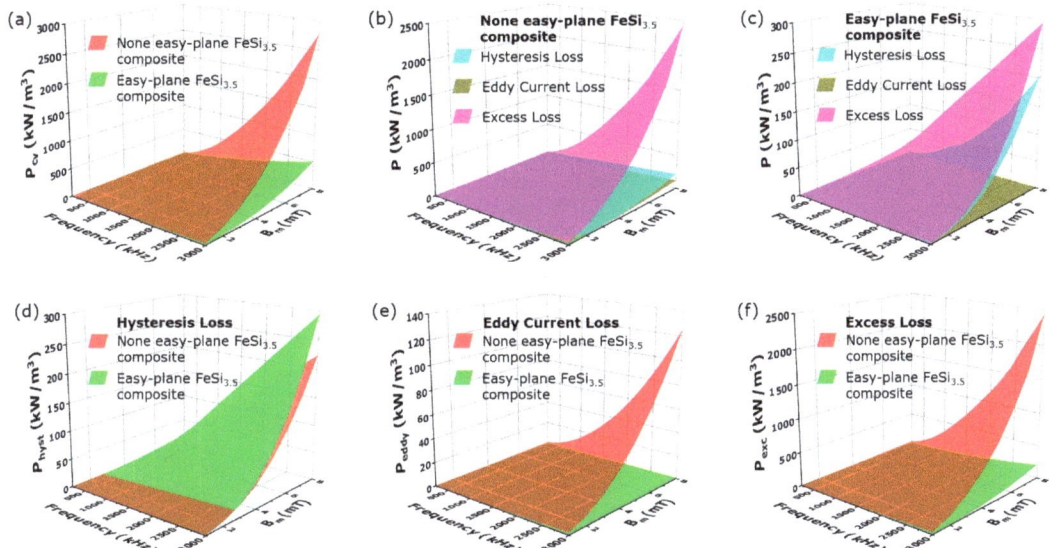

Figure 4. The total loss (**a**), loss separation (**b**,**c**), hysteresis loss (**d**), eddy current loss (**e**) and excess loss (**f**) of the none easy-plane FeSi$_{3.5}$ composite and easy-plane FeSi$_{3.5}$ composite.

According to classic Bertottit's loss separation theory, the total loss of P_{cv} can be separated into three different parts: hysteresis loss P_{hyst}, eddy current loss P_{eddy} and excess loss P_{exc}. The expression can be described [29]:

$$P_{cv} = c_{hyst} B_m^\alpha f + c_{eddy} B_m^2 f^2 + c_{exc} B_m^x f^y \qquad (5)$$

where c_{hyst} and α are the coefficients of hysteresis loss, c_{eddy} is the coefficient of current eddy loss, and c_{exc}, x and y are the coefficients of excess loss. The coefficients in each term are shown in Table 3.

Table 3. Simulated parameters for loss separation of the none easy-plane FeSi$_{3.5}$ composite and easy-plane FeSi$_{3.5}$ composite.

	P_{hyst}		P_{eddy}			P_{exc}	
	c_{hyst}	α	c_{eddy}^{inter}	c_{eddy}^{intra}	c_{exc}	x	y
None easy-plane FeSi$_{3.5}$ composite	9.2463	2.43	5.14×10^{-12}	2.18×10^{-4}	4.02×10^{-4}	2.21	1.76
Easy-Plane FeSi$_{3.5}$ composite	42.897	2.69	8.18×10^{-7}	3.78×10^{-7}	4.16×10^{-6}	2.43	1.97

For the none easy-plane FeSi$_{3.5}$ composite, P_{exc} is the largest proportion of the total loss, about 2435.2 kW/m^3 (8 mT and 3 MHz), and P_{hyst} and P_{eddy} are small percentages of the total loss, about 224.79 kW/m^3 and 125.77 kW/m^3. For the easy-plane FeSi$_{3.5}$ composite, P_{eddy} is the smallest proportion of the total loss, about 0.689 kW/m^3, and P_{hyst} and P_{exc} are main part of total loss, about 294.8 kW/m^3 and 204.93 kW/m^3. Although the P_{hyst} of the none easy-plane FeSi$_{3.5}$ composite is less than that of easy-plane FeSi$_{3.5}$ composite, the difference between them is only 70 kW/m^3 under the test conditions of 8 mT and 3 MHz. For the P_{eddy} and P_{exc}, the advantages of the easy-plane FeSi$_{3.5}$ composite are shown. The P_{eddy} of the easy-plane FeSi$_{3.5}$ composite is $\frac{1}{200}$ for that of the none easy-plane FeSi$_{3.5}$ composite. The P_{exc} of the easy-plane FeSi$_{3.5}$ composite is $\frac{1}{20}$ for that of the none easy-plane FeSi$_{3.5}$ composite.

4. Discussion

The easy-plane FeSi$_{3.5}$ composite has excellent magnetic properties and loss properties at higher frequencies compared with the none easy-plane FeSi$_{3.5}$ composite. The easy-plane FeSi$_{3.5}$ composite has higher permeability and higher ferromagnetic resonance frequencies than the none easy-plane FeSi$_{3.5}$ composite, which is beneficial for reducing the excitation current, turns and achieving device miniaturization. And the P_{cv} of the easy-plane FeSi$_{3.5}$ composite (500.42 kW/m^3) is lower and about $\frac{1}{5}$ for than that of the none easy-plane FeSi$_{3.5}$ composite (2785.8 kW/m^3). It is emphasized that the easy-plane FeSi$_{3.5}$ composite has lower P_{cv} compared with the none easy-plane FeSi$_{3.5}$ composite because of the significant reduction in P_{exc} rather than the decrease of P_{eddy}. The P_{eddy} of the none easy-plane FeSi$_{3.5}$ composite and easy-plane FeSi$_{3.5}$ composite is only a small proportion of total loss (as shown in Figure 4b,c). The P_{eddy} of the none easy-plane FeSi$_{3.5}$ composite and easy-plane FeSi$_{3.5}$ composite, respectively, only accounts for 4.5% and 0.14% of the total loss at 8 mT and 3 MHz (as shown in Figure 5b,e). Both the none easy-plane FeSi$_{3.5}$ composite and easy-plane FeSi$_{3.5}$ composite have a particle size of microns, and are composite materials (high resistivity). Therefore, they have a low intra-particle eddy current loss P_{eddy}^{intra} and inter-particle eddy current loss P_{eddy}^{inter}. And the easy-plane FeSi$_{3.5}$ composite with smaller particle thickness can have a lower P_{eddy} (as shown in Figure 4e) compared with the none the easy-plane FeSi$_{3.5}$ composite. Therefore, the P_{eddy} will not become a relatively large part of the loss and a major issue at higher frequencies of SMCs.

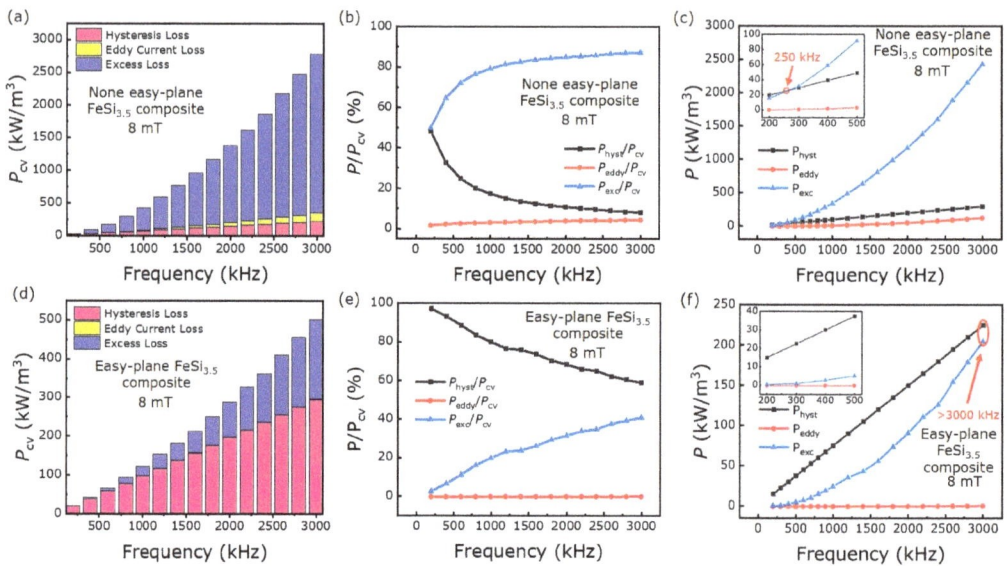

Figure 5. The percentage of loss of each part and loss separation of the none easy-plane FeSi$_{3.5}$ composite (**a**–**c**) and easy-plane FeSi$_{3.5}$ composite (**d**–**f**).

It is found that P_{exc} becomes the main loss when increasing the frequency, according to the experimental results (as shown in Figure 4b,c). The P_{exc} of none the easy-plane FeSi$_{3.5}$ composite and easy-plane composite both occupies the larger portion of the total loss, which accounts for 87.4% (2435.2 kW/m^3:2785.8 kW/m^3) and 40.9% (204.93 kW/m^3: 500.42 kW/m^3) of the total loss at 8 mT and 3 MHz, respectively (as shown in Figure 5b,e).

Whether the none easy-plane FeSi$_{3.5}$ composite or easy-plane FeSi$_{3.5}$ composite, the P_{exc} both dramatically increases with frequency. For the none easy-plane FeSi$_{3.5}$ composite, the P_{exc} sharply increases and becomes the main part of the total loss in the range of 200 kHz to 1200 kHz, accounting for about 80% of the total loss. For the easy-plane FeSi$_{3.5}$ composite, the main loss is P_{hyst} at a low frequency. As the frequency increases, the P_{exc} of the easy-plane FeSi$_{3.5}$ composite rapidly increases and becomes one of the main losses. Therefore, as the frequency increases, the dramatical increase of P_{exc} becomes a serious problem to soft magnetic materials in higher frequency applications, especially after reaching the MHz band.

Both of the none easy-plane FeSi$_{3.5}$ composite and easy-plane FeSi$_{3.5}$ composite can maintain a low loss at a lower frequency, but the easy-plane FeSi$_{3.5}$ composite has an obvious advantage in the high-frequency band. For the none easy-plane FeSi$_{3.5}$ composite, the dramatical increase in P_{exc} alongside frequency leads to a rapid increase in total loss, the P_{exc} is over P_{hyst} at 250 kHz, about 20 kW/m^3; along with the rapid increase in frequency, the P_{exc} accounts for about 80% of the total loss at 1 MHz, about 342.8 kW/m^3; the P_{exc} reaches about 2435.2 kW/m^3 while the frequency continuously increases to 3 MHz (Figure 5c). For the easy-plane FeSi$_{3.5}$ composite, the P_{exc} is about 0.75 kW/m^3 at 250 kHz, 24.61 kW/m^3 at 1 MHz and 204.93 kW/m^3 at 3 MHz (Figure 5f). Compared with the none easy-plane FeSi$_{3.5}$ composite, the P_{exc} in the easy-plane FeSi$_{3.5}$ composite is lower. This phenomenon is possibly related to the domain wall resonance frequency, and we will continue to study this in the future. As shown in Figure 3, the domain wall resonance frequency of the easy-plane FeSi$_{3.5}$ composite and none easy-plane FeSi$_{3.5}$ composite is 100 MHz and 9 MHz, respectively. Due to the domain wall resonance frequency of the none easy-plane FeSi$_{3.5}$ composite being low, the P_{exc} dramatically increases with frequency and the frequency is only 250 kHz when the P_{exc} exceeds the P_{hyst} (Figure 5c). For the easy-plane FeSi$_{3.5}$ composite, P_{exc} still increases with frequency, but the rate of the increase of P_{exc} with frequency is lower, due to the high-domain-wall frequency, and P_{exc} exceeds the P_{hyst} only when the frequency is higher than 3 MHz (Figure 5f). The high-domain-wall resonance frequency can push the frequency point, where the excess loss dramatically increases to a higher frequency, thus the easy-plane material with a higher domain wall resonance frequency can well inhibit the increase in excess loss with frequency.

The performance factor (PF) is a parameter that describes the energy transformation efficiency and can be expressed as the product of B_m and f by Equation (1):

$$PF = B_m f \tag{6}$$

The PF is calculated by simulated parameters in Table 2. It is found that PF and total loss of the easy-plane FeSi$_{3.5}$ composite is much more excellent than that of the none easy-plane FeSi$_{3.5}$ composite. This indicates that the easy-plane FeSi$_{3.5}$ composite has a higher transformation efficiency and lower loss compared with the none easy-plane FeSi$_{3.5}$ composite, and that the easy-plane FeSi$_{3.5}$ composite is more suitable for operating at higher frequencies. In addition, the currently published results of SMCs operating at higher frequencies are compared with those of the easy-plane FeSi$_{3.5}$ composite. The easy-plane FeSi$_{3.5}$ composite still has an advantage in permeability, operating frequency and transformation efficiency (as shown Figure 6b in Table 4).

Table 4. Magnetic properties and loss properties of various up-to-date published SMCs above 1 MHz.

SMCs	Initial Permeability	Power Loss	PF (under 500 kW/m^3)	
The easy-plane FeSi$_{3.5}$ composite	62.5	500.42 kW/m^3 (3 MHz, 8 mT)	23.96	This work
Amorphous FeSiCr + carbonyl iron @ resin	37	560 kW/m^3 (1 MHz, 20 mT)	16	[19]

Table 4. Cont.

SMCs	Initial Permeability	Power Loss	PF (under 500 kW/m³)	
Amorphous $Fe_{73}Si_6B_{10}P_5C_3Mo_3$ @ epoxy resin	34.63	~500 kW/m³ (1 MHz, 20 mT)	20	[21]
Amorphous FeSiCr/phosphating process and polyimide coating	47.5	7086 kW/m³ (1 MHz, 50 mT)	5	[20]
Flake pure iron @ binder composite	67 (at 1 MHz)	306.6 kW/m³ (1 MHz, 3 mT)	~5	[22]
	10 (at 3.5 MHz)	3.285 kW/m³ (3.5 MHz, 0.1 mT)	~15	

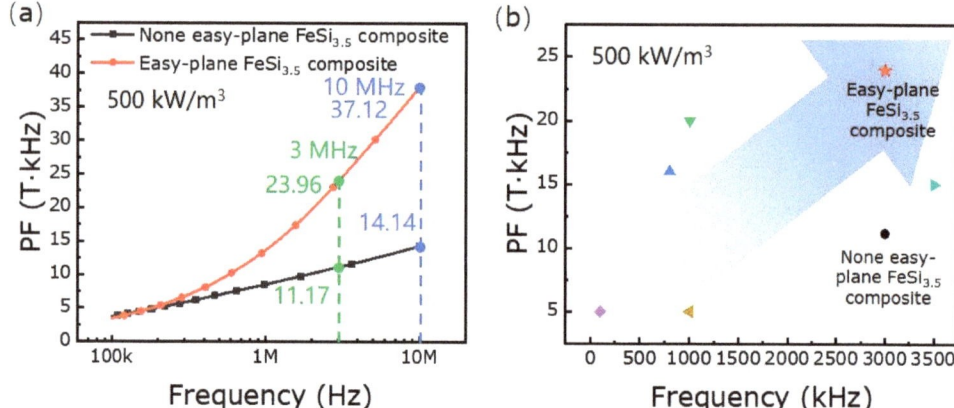

Figure 6. The *PF* calculate curve of the none easy-plane FeSi$_{3.5}$ composite and the easy-plane FeSi$_{3.5}$ composite at 500 kW/m³ (**a**). The *PF* of different SMCs at 500 kW/m³ (**b**) [19–22].

5. Conclusions

The easy-plane FeSi$_{3.5}$ composite has an ultra-low loss at 10 MHz, 4mT (372.88 kW/m³). The reason for the ultra-low loss of the easy-plane FeSi$_{3.5}$ composite at higher frequencies is because of setting up a control group (the none easy-plane FeSi$_{3.5}$ composite). For the first time, we found that the root cause of the failure in soft magnetic composites to work properly at higher frequencies was due to the dramatic increase in the P_{exc}. The rate of increase and the percentage of the P_{exc} is much larger than that of the P_{hyst} and the P_{eddy}. The P_{exc} accounts for a large portion of the P_{cv}; the portion continues to sharply increase with frequency, and conventional loss reduction methods have no effect on the P_{exc}. In addition, we also found that the easy-plane composite can effectively reduce the P_{exc}. Compared with the none easy-plane composite, the easy-plane composite can greatly reduce P_{exc} from 2435.2 kW/m³ to 204.93 kW/m³. This substantial reduction can greatly reduce the P_{cv}. And the easy-plane composite also has excellent magnetic properties, high permeability and ferromagnetic resonance frequencies. Thus, the easy-plane composite has strong potential to be applied for inductor and transformer cores at higher frequencies, especially at the MHz band and above.

Author Contributions: Conceptualization, L.Q., T.W. and F.L.; Methodology, P.W.; Formal analysis, P.W.; Investigation, P.W. and S.Y.; Data curation, P.W.; Writing—original draft, P.W.; Writing—review & editing, P.W., Y.H., G.W. and J.C.; Supervision, L.Q., T.W. and F.L.; Project administration, L.Q., T.W. and F.L.; Funding acquisition, L.Q. All authors have read and agreed to the published version of the manuscript.

Funding: This work was supported by the National Key Research and Development Program of China (Grant No. 2021YFB3501302), the National Natural Science Foundation of China (Grant No. 51731001), and the Fund from the State Key Laboratory of Baiyunobo Rare Earth Resource Researches and Comprehensive Utilization's Key Research and Development Projects.

Institutional Review Board Statement: Not applicable.

Informed Consent Statement: Not applicable.

Data Availability Statement: Not applicable.

Conflicts of Interest: The authors declare no conflict of interest.

References

1. Shokrollahi, H.; Janghorban, K. Soft magnetic composite materials (SMCs). *J. Mater. Process. Technol.* **2007**, *189*, 1–12. [CrossRef]
2. Gilbert, I.P.; Moorthy, V.; Bull, S.J.; Evans, J.T.; Jack, A.G. Development of soft magnetic composites for low-loss applications. *J. Magn. Magn. Mater.* **2002**, *242*, 232–234. [CrossRef]
3. Ozols, A.; Pagnola, M.; García, D.I.; Sirkin, H. Electroless coating of Permalloy powder and DC-resistivity of alloy composites. *Surf. Coat. Technol.* **2006**, *200*, 6821–6825. [CrossRef]
4. Nowicki, M.; Szewczyk, R.; Charubin, T.; Marusenkov, A.; Nosenko, A.; Kyrylchuk, V. Modeling the Hysteresis Loop of Ultra-High Permeability Amorphous Alloy for Space Applications. *Materials* **2018**, *11*, 2079. [CrossRef]
5. Bensalem, R.; Tebib, W.; Alleg, S.; Suñol, J.J.; Bessais, L.; Greneche, J.M. Magnetic properties of nanostructured $Fe_{92}P_8$ powder mixture. *J. Alloys Compd.* **2009**, *471*, 24–27. [CrossRef]
6. Silveyra, J.M.; Ferrara, E.; Huber, D.L.; Monson, T.C. Soft magnetic materials for a sustainable and electrified world. *Science* **2018**, *362*, 418. [CrossRef]
7. Li, W.C.; Wang, W.; Lv, J.J.; Ying, Y.; Yu, J.; Zheng, J.W.; Qiao, L.; Che, S.L. Structure and magnetic properties of iron-based soft magnetic composite with Ni-Cu-Zn ferrite–silicone insulation coating. *J. Magn. Magn. Mater.* **2018**, *456*, 333–340. [CrossRef]
8. Li, J.X.; Yu, J.; Li, W.C.; Che, S.L.; Zheng, J.W.; Qiao, L.; Ying, Y. The preparation and magnetic performance of the iron-based soft magnetic composites with the Fe@Fe_3O_4 powder of in situ surface oxidation. *J. Magn. Magn. Mater.* **2018**, *454*, 103–109. [CrossRef]
9. Wu, C.; Huang, M.Q.; Luo, D.H.; Jiang, Y.Z.; Yan, M. SiO_2 nanoparticles enhanced silicone resin as the matrix for Fe soft magnetic composites with improved magnetic, mechanical and thermal properties. *J. Alloys Compd.* **2018**, *741*, 35–43. [CrossRef]
10. Zhou, B.; Chi, Q.; Dong, Y.Q.; Liu, L.; Zhang, Y.Q.; Chang, L.; Pan, Y.; He, A.; Li, J.W.; Wang, X.M. Effects of annealing on the magnetic properties of Fe-based amorphous powder cores with inorganic-organic hybrid insulating layer. *J. Magn. Magn. Mater.* **2020**, *494*, 165827. [CrossRef]
11. Li, W.C.; Cai, H.W.; Kang, Y.; Ying, Y.; Yu, J.; Zheng, J.W.; Qiao, L.; Jiang, Y.; Che, S.L. High permeability and low loss bioinspired soft magnetic composites with nacre-like structure for high frequency applications. *J. Acta Mater.* **2019**, *167*, 267–274. [CrossRef]
12. Ling, X.; Li, M.; Cheng, W.J.; Fan, H.W.; Liu, G.; Sun, Y.H. Magnetic Properties and Eddy Current Losses in Fe-based Soft Magnetic Composites. *Int. J. Appl. Electromagn. Mech.* **2016**, *52*, 1433–1441.
13. Zhang, C.; Liu, X.S.; Li, M.L.; Liu, C.C.; Li, H.H.; Meng, X.Y. Preparation and soft magnetic properties of γ'-Fe_4N particles. *J. Mater. Sci. Mater. Electron.* **2018**, *29*, 1254–1257. [CrossRef]
14. Wang, W.; Kan, X.C.; Liu, X.S.; Feng, S.J. Effect of Cu on microstructure, magnetic properties of antiperovskite nitrides Cu_xNFe_{4-x}. *J. Mater. Sci. Mater. Electron.* **2019**, *30*, 10383–10390. [CrossRef]
15. Duan, B.F.; Zhang, J.M.; Wang, G.W.; Wang, P.; Wang, D.; Qiao, L.; Wang, T.; Li, F.S. Microwave absorption properties of easy-plane anisotropy Fe-Si powders with surface modification in the frequency range of 0.1–4 GHz. *J. Mater. Sci. Mater. Electron.* **2019**, *30*, 13810–13819. [CrossRef]
16. Gu, X.; Tan, G.; Chen, S.; Man, Q.; Chang, C.; Wang, X.; Li, R.W.; Che, S.; Jiang, L. Microwave absorption properties of planar-anisotropy $Ce_2Fe_{17}N_{3-\delta}$ powders/Silicone composite in X-band. *J. Magn. Magn. Mater.* **2017**, *424*, 39–43. [CrossRef]
17. Zuo, W.L.; Ying, L.; Qiao, L.; Wang, T.; Li, F.S. High frequency magnetic properties of $Pr_2Fe_{17}N_{3-\delta}$ particles with planar anisotropy. *J. Phys. B Condens. Matter.* **2010**, *405*, 4397–4400. [CrossRef]
18. Chi, X.; Yi, H.B.; Zuo, W.L.; Qiao, L.; Wang, T.; Li, F.S. Complex permeability and microwave absorption properties of Y_2Fe_{17} micropowders with planar anisotropy. *J. Phys. D Appl. Phys.* **2011**, *44*, 52–55. [CrossRef]
19. Xia, C.; Peng, Y.; Yi, X.; Yao, Z.; Zhu, Y.; Hu, G. Improved magnetic properties of FeSiCr amorphous soft magnetic composites by adding carbonyl iron powder. *J. Non-Cryst. Solids* **2021**, *559*, 120673. [CrossRef]
20. Long, H.M.; Wu, X.J.; Lu, Y.K.; Zhang, H.; Hao, J. Effect of Polyimide-Phosphating Double Coating and Annealing on the Magnetic Properties of Fe-Si-Cr SMCs. *Materials* **2022**, *15*, 3350. [CrossRef]

21. Kim, H.R.; Jang, M.S.; Nam, Y.G.; Kim, Y.S.; Yang, S.S.; Jeong, J.W. Enhanced Permeability of Fe-Based Amorphous Powder Cores Realized through Selective Incorporation of Carbonyl Iron Powders at Inter-Particle Voids. *Metals* **2021**, *11*, 1220. [CrossRef]
22. Thao, N.G.M.; Fujisaki, K.; That, L.T.; Motozuka, S. Magnetic Comparison between Experimental Flake Powder and Spherical Powder for Inductor Cores at High Frequency. *IEEE Trans. Magn.* **2020**, *57*, 8400407.
23. Wu, P.; Zhang, Y.D.; Hao, H.B.; Qiao, L.; Liu, X.; Wang, T.; Li, F.S. Effects of Nitriding and Ni Doping the Position of Domain Wall Resonance Peak and High-frequency Magnetic Performance of Easy-Plane Y_2Fe_{17}. *J. Magn. Magn. Mater.* **2022**, *549*, 168962. [CrossRef]
24. Wohlfarth, E.P. *Ferromagnetic Materials*; North-Holland: Amsterdam, The Netherlands, 1980; p. 370.
25. Tsutaoka, T. Frequency dispersion of complex permeability in Mn-Zn and Ni-Zn spinel ferrites and their composite materials. *J. Appl. Phys.* **2003**, *93*, 2789. [CrossRef]
26. Nakamura, T. Low-temperature sintering of Ni-Z-Cu ferrite and its permeability spectra. *J. Magn. Magn. Mater.* **1997**, *168*, 285–291. [CrossRef]
27. Han, M.; Liang, D.; Deng, L. Analyses on the dispersion spectra of permeability and permittivity for NiZn spinel ferrites doped with SiO_2. *Appl. Phys. Lett.* **2007**, *90*, 192507. [CrossRef]
28. Liu, X.; Wu, P.; Wang, P.; Wang, T.; Qiao, L.; Li, F.S. High permeability and bimodal resonance structure of flaky soft magnetic composite materials. *Chin. Phys. B* **2020**, *29*, 077506. [CrossRef]
29. Bertotti, G. General properties of power losses in soft ferromagnetic materials. *IEEE Trans. Magn.* **2002**, *24*, 621–630. [CrossRef]

Disclaimer/Publisher's Note: The statements, opinions and data contained in all publications are solely those of the individual author(s) and contributor(s) and not of MDPI and/or the editor(s). MDPI and/or the editor(s) disclaim responsibility for any injury to people or property resulting from any ideas, methods, instructions or products referred to in the content.

Review

Effect of Al and La Doping on the Structure and Magnetostrictive Properties of $Fe_{73}Ga_{27}$ Alloy

Jinchao Du [1], Pei Gong [1], Xiao Li [1,2], Shaoqi Ning [3], Wei Song [1,2], Yuan Wang [1] and Hongbo Hao [2,*]

[1] School of Materials Science and Engineering, Inner Mongolia University of Technology, Hohhot 010051, China
[2] Baotou Rare Earth Research Institute, State Key Laboratory of Baiyunebo Rare Earth Resources Research and Comprehensive Utilization, Baotou 014010, China
[3] Inner Mongolia Autonomous Region Strategic Research Center for Science and Technology, Hohhot 010051, China
* Correspondence: haohongbo@brire.com; Tel.: +86-13484721252

Abstract: The changes of microstructure, magnetostriction properties and hardness of the $Fe_{73}Ga_{27-x}Al_x$ alloy and $(Fe_{73}Ga_{27-x}Al_x)_{99.9}La_{0.1}$ alloy (x = 0, 0.5, 1.5, 2.5, 3.5, 4.5) were studied by doping Al into the $Fe_{73}Ga_{27}$ and $(Fe_{73}Ga_{27})_{99.9}La_{0.1}$ alloy, respectively. The results show that both the $Fe_{73}Ga_{27-x}Al_x$ alloy and $(Fe_{73}Ga_{27-x}Al_x)_{99.9}La_{0.1}$ alloy are dominated by the A2 phase, and the alloy grains are obvious columnar crystals with certain orientations along the water-cooled direction. A proportion of Al atoms replaced Ga atoms, which changed the lattice constant of the alloy, caused lattice distortion, and produced vacancy effects which affected the magnetostriction properties. La atoms were difficult to dissolve in the matrix alloy which made the alloy grains smaller and enhanced the orientation along the (100) direction, resulting in greater magneto-crystalline anisotropy and greater tetragonal distortion, which is conducive to improving the magnetostriction properties. $Fe_{73}Ga_{24.5}Al_{2.5}$ alloy has a saturation magnetostrictive strain of 74 ppm and a hardness value of 268.064 HV, taking into account the advantages of saturated magnetostrictive strain and high hardness. The maximum saturation magnetostrictive strain of the $(Fe_{73}Ga_{24.5}Al_{2.5})_{99.9}La_{0.1}$ alloy is 115 ppm and the hardness is 278.096 HV, indicating that trace La doping can improve the magnetostriction properties and deformation resistance of Fe-Ga alloy, which provides a new design idea for the Fe-Ga alloy, broadening their application in the field of practical production.

Keywords: Al addition; La addition; microstructure; microhardness; magnetostriction

Citation: Du, J.; Gong, P.; Li, X.; Ning, S.; Song, W.; Wang, Y.; Hao, H. Effect of Al and La Doping on the Structure and Magnetostrictive Properties of $Fe_{73}Ga_{27}$ Alloy. *Materials* **2023**, *16*, 12. https://doi.org/10.3390/ma16010012

Academic Editor: Javier Narciso

Received: 25 October 2022
Revised: 13 December 2022
Accepted: 16 December 2022
Published: 20 December 2022

Copyright: © 2022 by the authors. Licensee MDPI, Basel, Switzerland. This article is an open access article distributed under the terms and conditions of the Creative Commons Attribution (CC BY) license (https://creativecommons.org/licenses/by/4.0/).

1. Introduction

Due to the advantages of high energy density, fast response speed, high Curie temperature and excellent magnetostriction properties, the Tb-Dy-Fe alloy was widely used in sensors, brakes, transducers, micro displacement drivers and shock absorbers in the 1970s [1–4], but its large saturation magnetic field and poor mechanical performance limited its extensive application [5]. As a new type of magnetostrictive material developed after Tb-Dy-Fe, the Fe-Ga alloy has the advantages of a low saturated magnetic field, good mechanical properties and low material cost. Although the saturation magnetostrictive strain of a single crystal Fe-Ga alloy along the (100) easy magnetization direction can reach up to 400 ppm [6], its wide utilization is limited by its complex preparation process, poor mechanical properties at room temperature, high cost and size limitation [7]. The saturation magnetostrictive coefficient of the polycrystalline Fe-Ga alloy is low. In order to improve its magnetostriction properties, researchers have carried out extensive research on element doping, alloy preparation processes, heat treatment processes and pre-compression stress treatment. As an effective and simple method of improving the magnetostriction properties, element doping has attracted the attention of many magnetic researchers.

The Al and Ga elements belong to the same main group with similar atomic radii, and the Al atom can replace the Ga atom to solve the Fe (Ga) matrix. Studies have shown that the saturation magnetostrictive strain does not decrease significantly when Al doping is less than 10% in an Fe-Ga alloy, and Al has the advantage of having good ductility, processability and low price. Li Hua [8] studied the $Fe_{81}Ga_{16.5}Al_{3.5}$ alloy and its saturation magnetostrictive strain is 202 ppm. Srisukhumbowornchna [9] studied the directionally grown $Fe_{80}Ga_{15}Al_5$ alloy and found that its saturation magnetostrictive strain reached 234 ppm. Han Zhiyong [10] found that the saturation magnetostrictive strain of the $Fe_{87}Ga_{13}$ polycrystalline alloy reached 286 ppm under a 20 Mpa preload, which was 34.27% higher than the saturation magnetostriction under a zero-preload stress.

By reason of the 4f electron layer and strong spin–orbit effect of rare earth elements, the Fe-Ga alloy A2 phase is preferentially oriented along the (100) direction, contributing to greater magneto-crystalline anisotropy, thus heightening the magnetostriction coefficient. The saturation magnetostrictive value of the $Fe_{83}Ga_{17}Tb_{0.5}$ alloy is 149 ppm, as measured by Yu Quangong [11]. Jiang Liping [12] found that the saturation magnetostrictive value of the $Fe_{83}Ga_{17}Dy_{0.2}$ alloy reached 300 ppm. Meng Chongzheng [13] prepared the $(Fe_{83}Ga_{17})_{99.9}Tb_{0.1}$ alloy by directional solidification and its saturation magnetostrictive value clocked up 255 pm. Yao Zhanquan [14] measured that the saturation magnetostrictive value of the $Fe_{83}Ga_{17}Ce_{0.8}$ alloy was up to 356 ppm and discovered the second phase $CeGa_2$ distributing in a network at the grain boundary of Fe-Ga solid solution phase. All these studies demonstrate that rare earth doping can effectively enhance the magnetostrictive coefficient of the alloy.

As a light rare earth element with the lowest atomic number, La lacks the 4f electron layer and has a price that is more competitive than heavy rare-earth elements. As the Fe-Ga binary phase diagram shows, the saturation magnetostrictive value of the $Fe_{100-x}Ga_x$ alloy presents two peak values at x = 17~19 and x = 27~29 [15]. In this paper, the $Fe_{73}Ga_{27}$ and $(Fe_{73}Ga_{27})_{99.9}La_{0.1}$ alloys are selected as the matrices to be doped with Al element, respectively, and their effects on the saturation magnetostrictive strain and hardness are studied.

2. Materials and Methods

Fe, Ga, Al, La metal elements with purity greater than 99.95% were selected as raw materials and 70 g of raw materials were weighed using an electronic balance. The circulating water cooler was started before smelting. Firstly, the vacuum pressure was kept below 10 Pa by using a mechanical pump. The molecular pump was then turned on, eventually making the vacuum degree under 5×10^{-3} Pa and aerating with high purity argon as the protective gas. During alloy smelting, the arc current was maintained at 50–400 A. After the sample was completely cooled under the action of circulating water, the sample was turned over and smelted repeatedly three times to obtain a sample with uniform composition and good shape.

An AxioImagerAim metallographic microscope was used to observe the structure and grain orientation of alloy grain boundaries and matrices. An X-ray Diffractometer (XRD) of X-PertPro type was used to analyze lattice constants and determine phase compositions of alloys and the Sima500 scanning electron microscope produced by Carl Zeiss of Germany was utilized to observe the microstructure of the alloy and clarify the composition of different areas of the alloy through its own energy spectrometer. The magnetostrictive values of the alloys were measured using the resistance strain gauge method and the microhardness of the alloys were gauged with a MH-5AC microhardness tester.

3. Results

3.1. XRD Analysis

Figure 1 shows the X-ray diffraction patterns of the $Fe_{73}Ga_{27-x}Al_x$ and $Fe_{73}Ga_{27-x}Al_x)_{99.9}La_{0.1}$ alloys. The test angle ranges 20–100° and the step length is 0.02°/min. From Figure 1, the characteristic peak of the alloy remains unchanged after doping with Al and La and it is still composed of three diffraction peaks (110), (200) and (211). The alloy phase is mainly composed of the A2 phase. Compared with the standard diffraction spectrum of bcc Fe, the characteristic peaks of the $Fe_{73}Ga_{27-x}Al_x$ and

$Fe_{73}Ga_{27-x}Al_x)_{99.9}La_{0.1}$ alloys appear to have a small angle deviation, which is related to Al atoms substituting Ga atoms in the Fe (Ga) solid solution [16]. Although Al and Ga belong to the same main group, they dissolved well with each other without a second phase requiring to be generated. The radius of the Al atom (1.43 Å) was larger than the Ga atom, (1.40 Å), when the Al atom replaced the Ga atom in the solid solution lattice resulting in distortion in the alloy lattice, therefore showing a small angle shift of the diffraction peak [17]. Despite no new diffraction peak generating in the alloy diffraction spectrum after doping with La, the intensity of the (200) diffraction peak showed obvious change. On the basis of the following paragraph, when the I_{200}/I_{110} reached a maximum value of 140.77%, the (200) diffraction peak became the main peak, indicating that La doping enhanced the (100) orientation of the alloy.

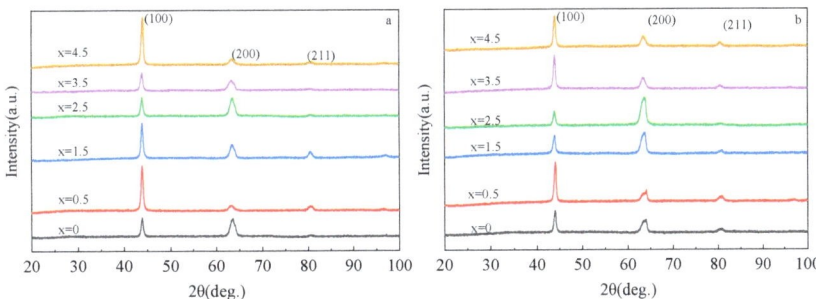

Figure 1. X-ray diffraction patterns of $Fe_{73}Ga_{27-x}Al_x$ and $Fe_{73}Ga_{27-x}Al_x)_{99.9}La_{0.1}$ alloys; (a) $Fe_{73}Ga_{27-x}Al_x$; (b) $Fe_{73}Ga_{27-x}Al_x)_{99.9}La_{0.1}$.

Table 1 is the lattice constant a(Å) and the relative intensity of the diffraction peak (I_{200}/I_{110}) obtained by calculating the XRD patterns of the $Fe_{73}Ga_{27-x}Al_x$ and $(Fe_{73}Ga_{27-x}Al_x)_{99.9}La_{0.1}$ alloys. The second and third columns display the results of lattice constants (a) and I_{200}/I_{110}(%) of the $Fe_{73}Ga_{27-x}Al_x$ alloys which change with Al content. In the $Fe_{73}Ga_{27-x}Al_x$ alloy, the lattice constant first increases and then decreases following the Al content, reaching maximum at x = 2.5% and minimum at x = 1.5% several times. The I_{200}/I_{110}(%) of the alloy also shows a trend of increasing first and then decreasing, showing two inflection points appearing at x = 1.5% and 3.5%, respectively. The Al atoms gradually replace the Ga atoms in the alloy, resulting in the variation of lattice constant and lattice gap generating a vacancy effect [18], which affects the magnetostriction properties. According to the displacement solid solution theory in the basis of material science [19], when atoms with a large radius replace atoms with a small radius to form a solid solution, the lattice constant of the alloy will increase.

Table 1. Lattice constants a(Å) and relative peak intensities (I_{200}/I_{110}) obtained by calculating XRD patterns of $Fe_{73}Ga_{27-x}Al_x$ alloy and $Fe_{73}Ga_{27-x}Al_x)_{99.9}La_{0.1}$ alloy.

Types of Alloys	$Fe_{73}Ga_{27-x}Al_x$		$(Fe_{73}Ga_{27-x}Al_x)_{99.9}La_{0.1}$	
x(%)	a(Å)	I_{200}/I_{110}(%)	a(Å)	I_{200}/I_{110}(%)
0	2.9251	83.8	2.9142	42.8
0.5	2.9263	9.3	2.9155	22.1
1.5	2.9289	33.1	2.9157	89.0
2.5	2.9203	80.4	2.9132	140.77
3.5	2.9215	46.7	2.9184	28.6
4.5	2.9243	10.0	2.9226	26.9

The fourth and fifth columns list the results of lattice constants a(Å) and I_{200}/I_{110}(%) of the $(Fe_{73}Ga_{27-x}Al_x)_{99.9}La_{0.1}$ alloy with varying Al content. The lattice constant generally increases first and then decreases after doping with a small amount of La. However, the lattice constant values are all smaller than those of the $Fe_{73}Ga_{27-x}Al_x$ alloy with the same Al composition, indicating that it is difficult for La to be dissolved in the matrix. The I_{200}/I_{110} (%) of the alloy generally presents a trend of waveform change, attaining the maximum value of 140.77% at x = 2.5%. The increase in the I_{200}/I_{110} ratio reflects that the preferred orientation of the alloy in the (100) direction is enhanced at this time. The (100) direction is the internal stress gradient direction of grain growth and the easy magnetization direction of the Fe-Ga alloy [20], which has a major impact on the magnetostrictive property of the alloy.

3.2. Analysis of Alloy Structure and Morphology

Figure 2 presents a metallographic photo of the $Fe_{73}Ga_{27-x}Al_x$ alloy. It shows that the alloy structure is made up of a gray matrix phase and a black dot which disperses in the matrix and at the grain boundary. The $Fe_{73}Ga_{22.5}Al_{4.5}$ alloy has visibly more black dots than other alloys. The grain shape of the alloy appears columnar and its orientation along the direction of water cooling is very clearly shown. As mentioned in the literature, the gray matrix phase structure is the Fe (Ga) solid solution phase with a bcc structure [21]. There are three reasons for the formation of black dot tissue. Firstly, Al atoms are not completely dissolved in the matrix alloy and the rest of the Al exists in the form of inclusions, growing into the black dots. Secondly, during solidification, a fraction of Ga atoms exists in the form of point precipitates to form a Ga-rich phase. Thirdly, it might be caused by matrix phase shedding in the process of grinding and polishing.

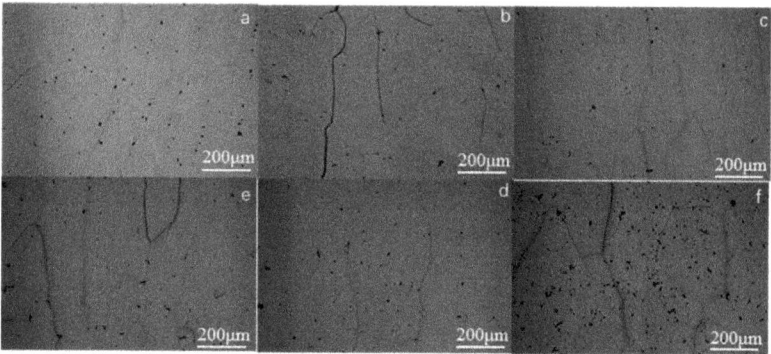

Figure 2. The metallographic photos of $Fe_{73}Ga_{27-x}Al_x$ alloy (**a**) x = 0, (**b**) x = 0.5, (**c**) x = 1.5, (**d**) x = 2.5, (**e**) x = 3.5, (**f**) x = 4.5.

Figure 3 presents the metallographic photo of the $(Fe_{73}Ga_{27-x}Al_x)_{99.9}La_{0.1}$ alloy. It can be seen from Figure 3 that, although the alloy is still dominated by columnar crystals with a certain orientation along the water-cooled direction, the microstructure morphology changes greatly. The black precipitates vary from point to sheet, spreading at the grain boundary. According to the following scanning photos and EDS composition analysis, La elements exist in the form of a precipitated phase, which is consistent with the literature [22]. The atomic radius of La (2.74 Å) is far larger than Ga and Al atoms and the rare earth element is difficult to dissolve in the matrix phase of the Fe-Ga alloy, resulting in the decrease in the lattice constant of the alloy and the role of refining the grains. As the literature references, the finer the metal grains are, the higher the strength and hardness are, and the better the plasticity and toughness [23]. It can be predicted that after adding La the hardness of the alloy will be greater than the alloy without La doping.

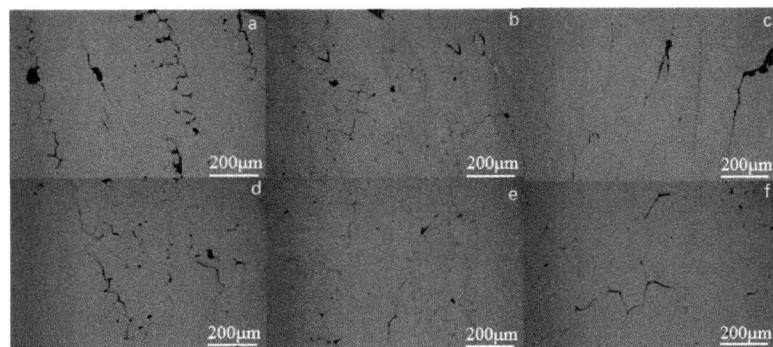

Figure 3. The metallographic photos of $(Fe_{73}Ga_{27-x}Al_x)_{99.9}La_{0.1}$ alloy (**a**) x = 0, (**b**) x = 0.5, (**c**) x = 1.5, (**d**) x = 2.5, (**e**) x = 3.5, (**f**) x = 4.5.

The $Fe_{73}Ga_{27-x}Al_x$ alloy and $(Fe_{73}Ga_{27-x}Al_x)_{99.9}La_{0.1}$ alloy at x = 0 % and x = 2.5% exhibited a preferred orientation in the (100) direction and, as shown in Figure 4, this composition point alloy has excellent magnetostriction properties and hardness. In order to further ascertain the composition of the matrix phase and precipitated phase, spot analysis and surface analysis were carried out using scanning electron microscopy and composition analysis of the selected area was carried out by the EDS. Figure 4a–d are SEM photos of the $Fe_{73}Ga_{27-x}Al_x$ alloy (x = 0% and x = 2.5%), Figure 4e and Figure 4f are SEM photos of the $(Fe_{73}Ga_{27-x}Al_x)_{99.9}La_{0.1}$ alloy (x = 0% and x = 2.5%). The grain boundary phases include A, D, G and J. Matrix phases include B, E, H and K. Precipitated phases include C, F, I and L.

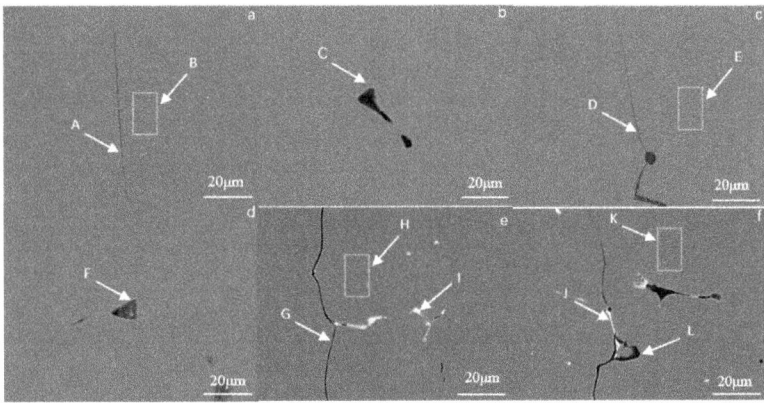

Figure 4. (**a**) x = 0, (**b**) x = 0, (**c**) x = 2.5, (**d**) x = 2.5 are SEM images of $Fe_{73}Ga_{27-x}Al_x$ alloy; (**e**) x = 0, (**f**) x = 2.5 are SEM images of $(Fe_{73}Ga_{27-x}Al_x)_{99.9}La_{0.1}$ alloy.

Table 2 displays the EDS results in different areas of the $Fe_{73}Ga_{27-x}Al_x$ and $(Fe_{73}Ga_{27-x}Al_x)_{99.9}La_{0.1}$ alloys. Zone B is the matrix phase composition of the $Fe_{73}Ga_{27}$ alloy. EDS composition analysis shows that the atomic ratio of Fe to Ga is 73.87:26.13, which is close to the nominal composition, suggesting that the alloy is uniformly melted without component segregation. The H zone and K point are the matrix phase components of $(Fe_{73}Ga_{27})_{99.9}La_{0.1}$ and $(Fe_{73}Ga_{24.5}Al_{2.5})_{99.9}La_{0.1}$, of which the relative content of La elements are 0.06%, which is far lower than the precipitates of I point (15.30%) and L point (23.75%), indicating that it is difficult for the La element to dissolve in the matrix. Zone E and point F are the matrix phase and precipitate phase of the $Fe_{73}Ga_{24.5}Al_{2.5}$ alloy, respectively. The relative content of the Al element at point F (1.42%) is evidently higher than that of the Al element at zone E (0.72%), which evinces a part of the Al atoms in the alloy of this composition entering the matrix to replace Ga atoms. Additionally, because the radius of Al atoms is greater than that of Ga atoms, the lattice constant of the $Fe_{73}Ga_{24.5}Al_{2.5}$ alloy will be reduced when solid solution is formed.

Table 2. EDS analysis results of $Fe_{73}Ga_{27-x}Al_x$ and $(Fe_{73}Ga_{27-x}Al_x)_{99.9}La_{0.1}$ alloys in different regions.

Micro-Zones		$Fe_{73}Ga_{27-x}Al_x$			Micro-Zones		$(Fe_{73}Ga_{27-x}Al_x)_{99.9}La_{0.1}$			
		Fe	Ga	Al			Fe	Ga	Al	La
x = 0	A	69.33	30.67	0	x = 0	G	50.39	39.65	0	9.96
	B	73.87	26.13	0		H	75.25	24.69	0	0.06
	C	77.56	22.44	0		I	36.92	47.78	0	15.30
x = 2.5	D	72.02	27.52	0.46	x = 2.5	J	74.99	23.87	1.10	0.05
	E	72.22	27.06	0.72		K	74.77	23.99	2.19	0.06
	F	80.06	18.52	1.42		L	17.68	58.13	0.44	23.75

In order to determine the composition of the black dots in the $Fe_{73}Ga_{27-x}Al_x$ (x = 0, 0.5, 1.5, 2.5, 3.5, 4.5) alloy, two points of each component were selected for EDS analysis under SEM. Figure 5 presents the SEM images of each component of the $Fe_{73}Ga_{27-x}Al_x$ alloy and Table 3 displays the EDS component analysis results for each alloy.

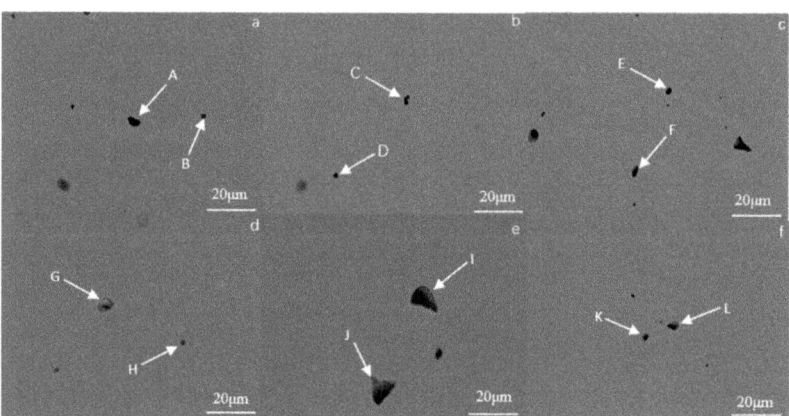

Figure 5. The black dots tissue of $Fe_{73}Ga_{27-x}Al_x$ alloy ((**a**) x = 0, (**b**) x = 0.5, (**c**) x = 1.5, (**d**) x = 2.5, (**e**) x = 3.5, (**f**) x = 4.5) observed by SEM.

Table 3. EDS analysis of black dots tissue of $Fe_{73}Ga_{27-x}Al_x$ alloy (x = 0, 0.5, 1.5, 2.5, 3.5, 4.5) alloy.

Micro-Zones		$Fe_{73}Ga_{27-x}Al_x$			Micro-Zones		$Fe_{73}Ga_{27-x}Al_x$		
		Fe	Ga	Al			Fe	Ga	Al
x = 0	A	83.45	16.55	0	x = 2.5	G	71.75	26.71	1.53
	B	73.87	26.13	0		H	71.94	25.95	2.11
x = 0.5	C	29.81	10.30	59.89	x = 3.5	I	77.77	22.10	0.13
	D	30.39	9.16	60.45		J	78.24	21.76	0
x = 1.5	E	76.82	23.18	0	x = 4.5	K	34.25	10.07	55.68
	F	77.08	22.92	0		L	72.54	24.81	2.65

According to the results in Table 3, the composition of the large proportion of black dots is similar to the composition of their respective matrix. Observing the points C, D and K, the relative contents of Al elements were 59.89, 60.45 and 55.68, respectively, which were higher than those of Fe and Ga, and a Ga-rich region was not observed in the alloy. We can therefore speculate that the black dots tissue in the alloy is more likely to be composed of the deciduous matrix phase and Al element in the form of inclusion.

3.3. Magnetostriction Properties

Figure 6a displays saturation magnetostrictive strain(λs) curve of the $Fe_{73}Ga_{27-x}Al_x$ alloy changing with external magnetic field. As Figure 6a shows, the saturation magnetostrictive strain of the alloy rises with the increase in the external magnetic field. When the magnetic field strength reaches 1500Oe, the λs remains invariable with the external magnetic field, reflecting that it is saturated at this time. Figure 6b presents the curve of the λs for the $Fe_{73}Ga_{27-x}Al_x$ alloy with Al content transformation. In Figure 6b, its λs curve varies within 58–76 ppm. The maximum value of the λs curve reaches 76 ppm at x = 0% and the λs of the alloys descend in the range of 0–1.5%. When the Al content in the alloy is low, Al atoms dissolve into the alloy instead of Ga atoms α-Fe, conducing the augmentation of the lattice constant, the attenuation of the vacancy effect, the increase in electron density around the alloy and the decrease in magnetostriction performance. The λs curve between x = 1.5% and 2.5% ascends slightly. After doping appropriate amounts of the Al element, the adjacent Ga-Ga atom pair in the (100) direction of the Fe-Ga alloy is destroyed [24], contributing to lattice distortion, forming short-range ordered clusters, and reducing the shear modulus of the alloy, which is conducive to improving the magnetostriction properties. As Al content continues to increase, the λs curve decreases within the range of x = 2.5–4.5%. When the addition of Al content becomes higher, the number of valence electrons transferred in the alloy will also rise, leading to an augmentation in the number of electrons on the secondary energy band. Only when the electrons on the secondary energy band are filled, will the remaining valence electrons be transferred to another magnetic energy band, resulting in a diminution in the magnetic moment of the Fe atom and the magnetostriction properties of the alloy [25].

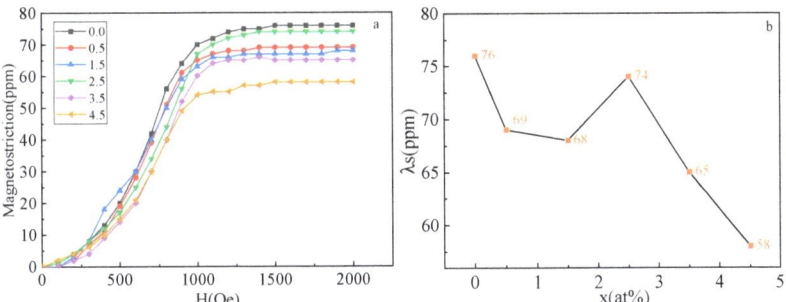

Figure 6. (**a**) Magnetostrictive curves of $Fe_{73}Ga_{27-x}Al_x$ alloy; (**b**) Saturation magnetostrictive strain curve of alloys.

Figure 7a shows the curve of the saturation magnetostriction strain of the $(Fe_{73}Ga_{27-x}Al_x)_{99.9}La_{0.1}$ alloy alteration with the external magnetic field. According to Figures 6 and 7a, the λs curves of the alloy increase with the addition of the external magnetic field. When the external magnetic field strength reaches 1600Oe, the λs curve remains at the saturated state. Figure 6b shows the curve of λs for the alloy changing with the Al content. After adding the La element, the λs of the alloys is kept within 73–115 ppm, which is meliorative in comparison to the $Fe_{73}Ga_{27-x}Al$ alloy. In a range of x = 0–0.5%, the λs of the alloy diminishes due to the increase in the lattice constant. According to literature [26], the lattice constant of the alloy rises, the void in the lattice decreases, and the vacancy effect weakens, which is beneficial to the magnetostriction performance. Within x = 0.5–2.5%, the λs of the alloy rises and reaches the maximum value of 115 ppm at x = 2.5%. After adding the La element in the Fe-Ga alloy, large portions of the La exist in the form of precipitates, but a small amount will still dissolve in the matrix, leading to preferred orientation in the (100) direction, greater magneto-crystalline anisotropy and a greater tetragonal distortion range, which enhances the magnetostrictive property [27]. At 2.5–4.5%, λs reduces continuously, which is the result of the joint action of the increase in the lattice constant and the decrease in I_{200}/I_{110} (%) and is consistent with the previous theory.

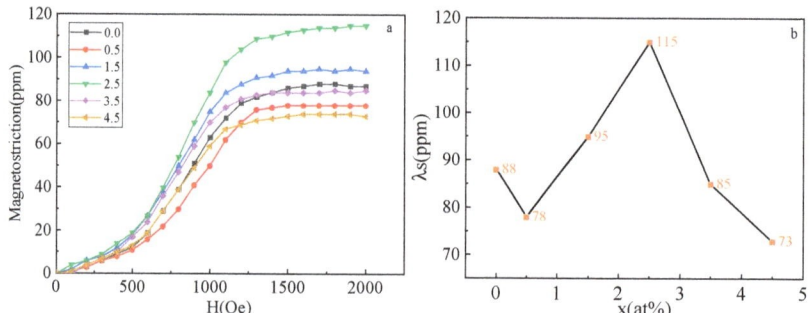

Figure 7. (a) Magnetostrictive curves of $Fe_{73}Ga_{27-x}Al_x)_{99.9}La_{0.1}$ alloy; (b) Saturation magnetostrictive strain curve of alloys.

3.4. Microhardness

Figure 8 presents the curve of microhardness of the $Fe_{73}Ga_{27-x}Al_x$ alloy and the $Fe_{73}Ga_{27-x}Al_x)_{99.9}La_{0.1}$ alloy varying with Al content. Figure 8a shows that the hardness of the $Fe_{73}Ga_{27-x}Al_x$ alloy increases first and then decreases with the change of Al content. In comparison to the $Fe_{73}Ga_{27}$ alloy, after the addition of the Al element, the hardness values of the $Fe_{73}Ga_{27-x}Al_x$ alloy rise, which is related to the increase in ductility and plasticity after adding the Al element to the alloy [28]. The microhardness of the alloy measures up to the maximum value 273.183 HV at 2.5%. Compared with the alloys of undoped Al, although the hardness of the alloy increases by 6.6%, the λs decreases by 10.5%. While the λs of the alloy at x = 3.5% decreases by 2.6%, the hardness value increases by 4.6%. Therefore, by replacing the Ga element with appropriate Al elements in practical production, the Fe-Ga alloy device can obtain more economic benefits and excellent comprehensive properties. In light of Figure 8b, the hardness value of the alloy of x = 2.5% is up to the crest value 278.096 HV, which is higher than any of the $Fe_{73}Ga_{27-x}Al_x$ alloys in accordance with the previous prediction results, suggesting that the appropriate composition of the La element doped into the Fe-Ga alloy can play a role in refining the grain and increasing the hardness of the alloy.

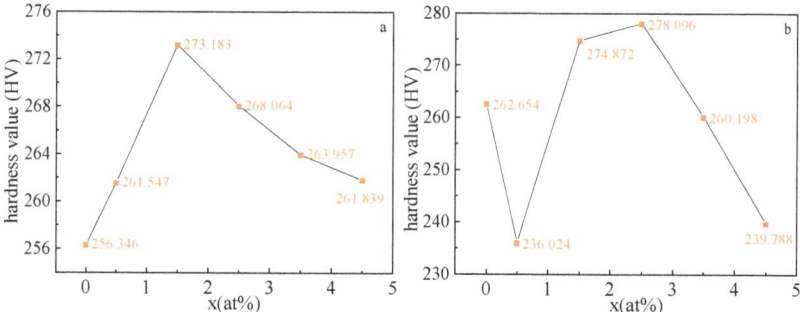

Figure 8. Microhardness (a) $Fe_{73}Ga_{27-x}Al_x$; (b) $Fe_{73}Ga_{27-x}Al_x)_{99.9}La_{0.1}$.

4. Conclusions

The $Fe_{73}Ga_{27-x}Al_x$ alloy and $(Fe_{73}Ga_{27-x}Al_x)_{99.9}La_{0.1}$ alloys are still dominated by the A2 phase, but its diffraction peak position has a small angle deviation in comparison to the standard diffraction spectrum of α-Fe. After Al and La doping, the lattice constant and I_{200}/I_{110} ratio of the alloy change, resulting in lattice distortion, further affecting the magnetostriction properties of the alloy.

The grains of the $Fe_{73}Ga_{27-x}Al_x$ and $(Fe_{73}Ga_{27-x}Al_x)_{99.9}La_{0.1}$ alloys appear as distinct columnar crystals with certain orientations along the water-cooled direction. Moreover, the grains of the $(Fe_{73}Ga_{27-x}Al_x)_{99.9}La_{0.1}$ alloy are finer, exhibiting higher hardness values. It can be further speculated that the other mechanical properties of the alloys will be more excellent. EDS analysis shows that the Al element partially dissolves in the Fe-Ga solid solution, while the La element hardly gets inside the Fe-Ga alloy matrix, with masses of them existing in the form of the precipitation phase.

The saturation magnetostrictive strains of the $Fe_{73}Ga_{27-x}Al_x$ alloy are between 58–76 ppm. Doping Al with proper composition can not only reduce the production cost, but also improve the hardness and magnetostriction properties. The saturation magnetostrictive strains of the $(Fe_{73}Ga_{27-x}Al_x)_{99.9}La_{0.1}$ alloy is in the range of 74–115 ppm, which proves that a small amount of La doping can wholly enhance the saturation magnetostrictive strain of Fe-Ga-Al. The hardness (278.096 HV) and the maximum saturation magnetostrictive strain (115 ppm) of the alloy at x = 2.5% are higher than those of other alloys, indicating that La doping can effectively meliorate the saturation magnetostrictive coefficient of the alloy and augment the hardness of the alloy.

Author Contributions: The work of conceptualization and methodology was provided by X.L.; the work of metallographic polishing and preparing raw materials was provided by W.S. and Y.W.; part of the investigation was provided by S.N.; the work of writing—review and supervision was provided by P.G.; the work of funding acquisition and project administration was provided by H.H.; the work of writing—original draft preparation and editing was provided by J.D. All authors have read and agreed to the published version of the manuscript.

Funding: This research was funded by Key Project of Inner Mongolia Natural Science Foundation, No.2018ZD10.

Conflicts of Interest: There are no conflict of interest in the study.

References

1. Clark, A.E. *Ferromagnetic Materials*; North-Holland: Amsterdam, The Netherlands, 1980.
2. Liu, H.F.; Lim, C.W.; Gao, S.Z.; Zhao, J. Effects analysis of bias and excitation conditions on power output of an environmental vibration energy harvesting device using Fe-Ga slice. *Mechatronics* **2019**, *57*, 20. [CrossRef]
3. Flatau, A.B.; Stadler, B.J.H.; Park, J.; Reddy, K.S.M.; Chaitany, P.R.D.; Mudivarthi, C.; van Ordera, M. *Magnetostrictive Fe-Ga Nanowires for Actuation and Sensing Applications*; Woodhead Publishing: Sawston, UK, 2020; p. 737.
4. Qiao, R.H.; Gou, J.M.; Yang, T.Z.; Zhang, Y.; Liu, F.; Hu, S.; Ma, T. Enhanced damping capacity of ferromagnetic Fe-Ga alloys by introducing structural defects. *J. Mater. Sci. Technol.* **2021**, *84*, 173. [CrossRef]
5. Wang, N.J.; Liu, Y.; Zhang, H.W.; Chen, X.; Li, Y. Fabrication, magnetostriction properties and applications of Tb-Dy-Fe alloys: A review. *China Foundry* **2016**, *13*, 75. [CrossRef]
6. Clark, A.E.; Wun-Fogle, M. Restorff, J.B.; Dennis, K.W.; Lograsso, T.A.; McCallum, R.W. Temperature dependence of the magnetic anisotropy and magnetostriction of Fe100−xGax (x = 8.6, 16.6, 28.5). *J. Appl. Phys.* **2005**, *97*, 10M316. [CrossRef]
7. Xu, S.F. *Study on the Physical Properties of Fe-Ga Magnetostrictive Alloys*; Jilin University: Changchun, China, 2008.
8. Li, H. *Effect of Ce and Tb Doping on Magnetostrictive Properties of Fe-Ga-Al Alloy*; Inner Mongol University of Technology: Hohhot, China, 2021.
9. Srisukhumbowornchai, N.; Guruswamy, S. Large magnetostriction in directionally solidified Fe-Ga and Fe-Ga-A1 alloys. *J. Appl. Phys.* **2001**, *90*, 5680–5688. [CrossRef]
10. Han, Z.Y.; Gao, X.S.; Zhang, M.C.; Zhou, S. <110>Pressure effect of magnetostrictive strain in axially oriented polycrystalline Fe-Ga alloy. In Proceedings of the 2002 China Materials Conference, Beijing, China, 1 October 2002; pp. 375–379.
11. Yu, Q.G.; Jiang, L.P.; Zhang, G.R.; Hao, H.B.; Wu, S.X.; Zhao, Z.Q. Effect of Tb on the magnetostrictive properties of $Fe_{83}Ga_{17}$ alloy. *Rare Earth* **2010**, *31*, 21–24.
12. Jiang, L.; Zhang, G.; Yang, J.; Hao, H.; Wu, S.; Zhao, Z. Research on microstructure and magnetostriction of $Fe_{83}Ga_{17}Dy_x$ alloys. *J. Rare Earths* **2010**, *28*, 409–412. [CrossRef]

13. Meng, C.Z.; Wu, Y.; Jiang, C.B. Design of high ductility FeGa magnetostrictive alloys: Tb doping and directional solidification. *Mater. Des.* **2017**, *130*, 183–189. [CrossRef]
14. Yao, Z.Q.; Zhao, Z.Q.; Jiang, L.P.; Hao, H.B.; Wu, S.X.; Zhang, G.R.; Yang, J.D. Effect of rare earth Ce addition on microstructure and magnetostrictive properties of Fe83Ga17 alloy. *Acta Metall. Sin.* **2013**, *49*, 87–91. [CrossRef]
15. Clark, A.E.; Hathaway, K.B.; Wun-Fogle, M.; Restorff, J.B.; Lograsso, T.A.; Keppens, V.; Petculescu, G.; Taylor, R.A. Extraordinary magnetoelasticity and lattice softening in bcc Fe-Ga alloys. *J. Appl. Phys.* **2003**, *93*, 8621–8623. [CrossRef]
16. Ning, S.Q. *Effect of Al on Microstructure and Magnetostrictive Strain of $Fe_{81}Ga_{19}$ Alloy*; Inner Mongol University of Technology: Hohhot, China, 2015.
17. Yao, Z.; Tian, X.; Jiang, L.; Hao, H.; Zhang, G.; Wu, S.; Zhao, Z.; Gerile, N. Influences of rare earth element Ce-doping and melt-spinning on microstructure and magnetostriction of $Fe_{83}Ga_{17}$ alloy. *J. Alloy. Compd.* **2015**, *42*, 431–435. [CrossRef]
18. Li, J.X.; Gao, X.S.; Zhu, J. Effect of atomic vacancies on magnetostrictive properties of directionally solidified Fe-Ga alloys. *Chin. J. Rare Met.* **2017**, *41*, 155–162.
19. Xu, H.J. *Fundamentals of Materials Science*, 1st ed.; Beijing University of Technology Press: Beijing, China, 2001; p. 56.
20. Cheng, S.; Das, B.; Wun-Fogle, M.; Lubitz, P.; Clark, A. Structure of melt-spun Fe-Ga-based magnetostrictive alloys. *IEEE Trans. Magn.* **2002**, *38*, 2838–2840. [CrossRef]
21. Gong, P.; Jiang, L.P.; Yan, W.J.; Zhao, Z.Q. Effect of Y on microstructure and magnetostrictive properties of as-cast $Fe_{81}Ga_{19}$ alloy. *Chin. Rare Earths* **2016**, *37*, 91–95.
22. Zhang, M.C.; Jiang, H.L.; Gao, X.X.; Zhu, J.; Zhou, S.Z. Magnetostriction and microstructure of rolled $Fe_{83}Ga_{17}Er_{0.4}$ alloy. *Chin. J. Rare Met.* **2016**, *37*, 75–79.
23. Cui, Z.X. *Metallogy and Heat Treatment*, 1st ed.; China Machine Press: Beijing, China, 1989; p. 54.
24. Clark, A.E.; Wun-Fogle, M.; Restorff, J.B.; Lograsso, T.A. Effect of quenching on the magnetostriction of $Fe_{1-x}Ga_x$. *IEEE Trans. Magn.* **2001**, *37*, 2678–2680. [CrossRef]
25. Liu, G.Z. *Effect of Adding the Third Group of Elements (C, B, Al, Cu) on Phase Structure and Magnetostrictive Properties of Fe-Ga Alloy*; Lanzhou University of Technology: Lanzhou, China, 2010.
26. Yao, T. *Study on the Influence of Adding Co and Tb Elements on the Structure and Magnetic Properties of Fe-Ga Alloy*; Inner Mongol University of Technology: Hohhot, China, 2021.
27. Wang, R.; Tian, X.; Yao, Z.Q.; Zhao, X.; Hao, H.B.; Tian, R.N.; Ou, Z.Q.; Zhao, X.T. Effects of trace rare earth elements Tb and La doping on the structure and magnetostrictive properties of Fe-Al alloy. *Chin. Rare Earths* **2020**, *41*, 24–31.
28. Yan, S.W.; Mou, X.; Qi, Y.; Xu, L.H.; Zhang, H.P. Research progress on mechanical properties strengthening of Fe-Ga magnetostrictive alloy. *Iron Steel* **2022**, *57*, 79–90.

Disclaimer/Publisher's Note: The statements, opinions and data contained in all publications are solely those of the individual author(s) and contributor(s) and not of MDPI and/or the editor(s). MDPI and/or the editor(s) disclaim responsibility for any injury to people or property resulting from any ideas, methods, instructions or products referred to in the content.

Article

A Method to Measure Permeability of Permalloy in Extremely Weak Magnetic Field Based on Rayleigh Model

Jinji Sun [1,2], Yan Lu [1,2], Lu Zhang [1,3,*], Yun Le [1,2,*] and Xiuqi Zhao [1,2]

1. School of Instrumentation and Optoelectronic Engineering, Beihang University, Beijing 100191, China
2. Ningbo Institute of Technology, Beihang University, Ningbo 315800, China
3. Hangzhou Extremely Weak Magnetic Field Major Science and Technology Infrastructure Research Institute, Hangzhou 310000, China
* Correspondence: zhanglu723@hust.edu.cn (L.Z.); leyun@buaa.edu.cn (Y.L.)

Abstract: In order to solve the problem that the relative permeability of the permalloy is missing and difficult to measure accurately in an extremely weak magnetic field (EWMF, <1 nT), a method to measure the permeability in EWMF based on the Rayleigh model is proposed in this paper. In this method, the Rayleigh model for the magnetic material was first introduced. Then, the test system for measuring the permeability of permalloy for the standard ring specimen was set up. Based on the test data and the Rayleigh model, the functional expression applied to obtain the permeability in EWMF is achieved. Finally, the feasibility and accuracy of the method are verified by the permeability measurement of the custom large-size ring specimen in EWMF (<1 nT) and residual field measurement based on the four-layer shielding cylinder. This method can obtain the relative permeability in any EWMF and avoid test errors caused by extremely weak magnetization signals.

Keywords: permalloy; extremely weak magnetic field; relative permeability; Rayleigh model

Citation: Sun, J.; Lu, Y.; Zhang, L.; Le, Y.; Zhao, X. A Method to Measure Permeability of Permalloy in Extremely Weak Magnetic Field Based on Rayleigh Model. *Materials* **2022**, *15*, 7353. https://doi.org/10.3390/ma15207353

Academic Editor: Emil Babić

Received: 25 August 2022
Accepted: 2 October 2022
Published: 20 October 2022

Publisher's Note: MDPI stays neutral with regard to jurisdictional claims in published maps and institutional affiliations.

Copyright: © 2022 by the authors. Licensee MDPI, Basel, Switzerland. This article is an open access article distributed under the terms and conditions of the Creative Commons Attribution (CC BY) license (https://creativecommons.org/licenses/by/4.0/).

1. Introduction

The magnetically shielded room (MSR) used high permeable magnetic materials and high conductivity materials, such as permalloy and aluminum, to prevent the external magnetic field from entering its interior. Based on the principle of flux shunt [1–3] and eddy current loss [4,5], it can be used to provide an extremely weak magnetic field (EWMF) environment. In this environment, many frontier research studies for different kinds of fields can be carried out: for example, the measurement of weak signals such as magnetic heart and brain [6–8], the measurement of geophysical research samples [9], the measurement of electric dipole moments [10], the study of high-precision magnetic measuring instruments [11,12] and so on.

The MSR was first built by Bob J. Patton and John L. Fitch, who mainly used it for geophysical research [13]. Since then, many researchers around the world have successively developed the MSR, whose types include cuboid [14–18], cylinder [19] and spheroid-like [20,21]. The common feature of the existing MSR is the nested structure with multi-layer permalloy and aluminum, with the purpose of obtaining an EWMF environment (<1 nT). One of the most typical representatives is the Berlin Magnetically Shielded Room 2 (BMSR-2) built by the Physikalisch-Technische Bundesanstalt (PTB) [16]. BMSR-2 is composed of seven layers of permalloy and one layer of aluminum. With this design structure, the innermost layer of permalloy needs to work in the extremely low residual field: about 1 nT or even lower. Because the permeability of the permalloy varies nonlinearly with the applied magnetic field, it is necessary to obtain the permeability in EWMF. However, the permeability is generally measured at an applied magnetic field about 100 nT, which is considered to be the initial permeability. Using these data in the development of MSR could lead to a considerable overestimation of the shielding factor [22].

In order to solve this problem, Li et al. built an EWMF test system with a specialized large-sized ring specimen placed in a magnetic shielding cylinder, and the capability of the test system for the applied static magnetization field can reach as low as 4.5×10^{-5} A/m ($\approx$0.057 nT) [22]. This method can directly obtain the desired permeability under EWMF, but the problems of the test errors and additional cost of time and money also exist due to the magnetic flux drift and the specialized large size ring and more turns of secondary windings (>400). Yamazaki et al. proposed a method of shielding effects (SEs) measurement for the multi-layer mu-metal cylinders and finite element analysis for the nonlinear model to obtain the incremental permeability of mu-metal in low magnetic fields [23]. In this method, a much more high-resolution magnetic field measurement sensor, such as a superconducting quantum interference device magnetometer and atomic magnetometer, is needed to achieve permeability in low magnetic fields, which limits its applied range due to the high-cost and special working environment of the sensors. British physicist Lord John Rayleigh [24] proposed that the parabolic dependence between the magnetic flux density of ferromagnetic materials and the magnetization field can be used to describe the hysteresis phenomenon of iron and steel under the low magnetizing fields (called the Rayleigh model). Since then, some scholars used the Rayleigh model to obtain the initial permeability of magnetic materials [25,26]. They established a shielding factor test system with Helmholtz coils applied for generating a 5 Hz and low-amplitude homogeneous magnetic field and finite element analysis applied for identifying the two important parameters in the Rayleigh model. These methods can be used to test the amplitude permeability of mu-metal thin foils less than 0.2 mm thick at 5 Hz, but the relative initial permeability at static magnetic fields cannot be obtained.

To obtain the data at an extremely weak static magnetic field, a method to measure the permeability of permalloy in EWMF based on the Rayleigh model is proposed. The method is divided into two processes: testing and identification. A ring specimen with standard size was used to obtain the magnetic characteristics of permalloy in the conventional magnetic field range. The identification process uses the Rayleigh model, fitting the test data in the Rayleigh region to obtain the initial permeability and the Rayleigh constant. The feasibility and accuracy of the proposed method were verified by building a permeability test system with an applied magnetic field less than 1 nT and a residual field test system with a four-layer permalloy cylinder. The proposed method can accurately obtain the relative permeability of permalloy in EWMF without the custom larger-size ring specimen, low magnetic field test environment and high-resolution sensor, and even the initial permeability when the magnetic field intensity tends to zero.

The arrangement of the remaining contents of this paper is as follows. In Section 2, the Rayleigh model for the magnetic material is first introduced. Then, the principle of the permeability measurement is presented. Test results and discussion are presented in Section 3. In Section 4, brief conclusions are summarized.

2. Principle of Permeability Measurement

2.1. Rayleigh Model for Magnetic Material

The model of the magnetization curve in a low magnetic field (Rayleigh region) proposed by Lord Rayleigh is shown in Equation (1), which is used to describe the nonlinear relationship between the magnetic flux density (B) and magnetic field strength (H)

$$B = \mu_0 \mu_i H + aH^2, \tag{1}$$

where $\mu_0 = 4\pi \times 10^{-7}$ T·m/A is the permeability of vacuum; μ_i is the relative initial permeability of the magnetic material; $\mu_i = \lim(B/H/\mu_0)(H\to 0)$, and a is the Rayleigh constant.

According to Equation (1) and the function between the magnetic flux density (B) and the magnetic field strength (H), $B = \mu\mu_0 H$, the relative permeability of the permalloy, μ, can be expressed as

$$\mu = \frac{B}{H\mu_0} = \mu_i + \frac{a}{\mu_0} H. \quad (2)$$

Equation (2) shows that the relative permeability of the material increases linearly with H in the Rayleigh region, and the intercept is the initial permeability.

The two parameters, μ_i and a, are especially significant for the magnetization process of the Rayleigh region caused by the shift of the magnetic domain walls. The initial permeability μ_i is related to the reversible motion, which represents the linear increase in the magnetic flux density within the volume of the material under the influence of an increasing magnetizing field. The Rayleigh constant a is used to sketch nonlinear effects from the irreversible magnetization shift, which gradually covers the linear dependence with the increasing magnetizing field. The two parameters need to be identified by the experimental test and function fitting. Then, the B-H curves and the relative permeability curves can be obtained based on Equations (1) and (2), respectively.

2.2. The Measurement for Magnetic Material

In this paper, the simulated impact method was presented to test the B-H curves. During the test, the closed ring specimen with a primary winding and a secondary winding is used. The primary winding with direct-current (I_1) injected is used to generate a magnetic field strength H, with a minimum value to be 0.01 mA. Based on the Ampere circuital theorem, the magnetic field strength (H) can be expressed as

$$H = \frac{N_1 I_1}{L_e}, \quad (3)$$

where N_1 is the turn number of the primary winding, and L_e is the effective magnetic circuit length. The secondary winding is used to detect flux linkage (ϕ) in the ring with a minimum value to be 0.1 μWb. According to the law of electromagnetic induction, the magnetic flux density (B) can be expressed as

$$B = \frac{\phi}{N_2 S_e}, \quad (4)$$

where N_2 is the turn number of the secondary winding, and S_e is the effective cross-sectional area. Before the experiment, L_e and S_e of the specimen can be calculated by using the size of the ring according to the SJ/T10281 standard, which can be expressed as

$$L_e = \frac{C_1^2}{C_2}, \quad (5)$$

$$S_e = \frac{C_1}{C_2}, \quad (6)$$

where C_1 and C_2 are the constants of the ring specimen and can be expressed as

$$C_1 = \frac{2\pi}{d} \ln\left(\frac{R_1}{R_2}\right)^{-1}, \quad (7)$$

$$C_2 = \frac{4\pi\left(\frac{1}{R_2} - \frac{1}{R_1}\right)}{d^2} \ln\left(\frac{R_1}{R_2}\right)^{-3}, \quad (8)$$

where R_1, R_2 and d are the outer diameter, inner diameter and thickness of the ring specimen, as shown in Figure 1, respectively. As the parameters are determined, the B-H curves can be calculated according to the applied current (I_1) and the tested flux linkage (ϕ).

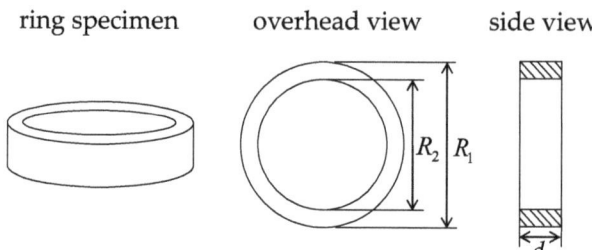

Figure 1. The overhead view and side view of the ring specimen.

Equations (3)–(8) clearly show that the measurement range of the permeability is determined by the size of the ring specimen and the turns of the primary and secondary windings. To measure the B-H curve in the EWMF, we can need to increase the effective magnetic circuit length and reduce the turn of the primary winding to apply desired low magnetic field, and we also need to increase the effective cross-sectional area and the turns of the secondary winding to improve the detection signal of the magnetic flux.

3. Test Results and Discussion
3.1. The Test Method in EWMF Based on the Rayleigh Model

In order to obtain the data of an EWMF accurately and simply, two parts including the measurement process and identification process were performed. The measurement process provides relevant data for the identification process.

In the measurement process, the test system, mainly used to measure the B-H curve of permalloy, is first set up, as shown in Figure 2. The test system consists of a ring specimen with a primary winding and a secondary winding, test equipment with a direct-current power supply and a flux meter, and a software system. The test steps are performed as follows:

1. The dimensions of the ring specimen (outer diameter A, inner diameter B, and height C) were measured using vernier calipers.
2. The effective magnetic circuit length (L_e) and cross-sectional area (S_e) of the ring specimen were calculated according to the SJ/T10281 standard, and the turns of the primary and secondary windings were obtained according to the measured magnetic field range and Equations (3)–(8).
3. The primary and secondary windings are wound around the ring specimen. Notably, the high-temperature resistant tape should be used for insulation between windings and ring specimen, and the secondary winding should be wound before the primary winding.
4. The primary and secondary windings are connected to the power supply and the flux meter, respectively. The size of the measured ring, the turns of two windings, the range of magnetic field strength (H is set from 0.08 to 800 A/m) and other parameters are accurately set in the software. The value of the measured initial permeability corresponds to the one at the magnetic field strength of 0.08 A/m.
5. Based on Equations (3) and (4), the B-H curve, the permeability curve and the parameters including initial permeability (μ_i) (H is 0.08 A/m) can be determined.

Then, the identification process is performed as follows. First, based on the measured B-H data in the Rayleigh region, the two key parameters, the initial permeability and the Rayleigh constant, can be identified by the function fitting. After that, extrapolating the measured data of the B-H curve according to the Rayleigh model with determined parameters, the B-H curve at an EWMF can be determined.

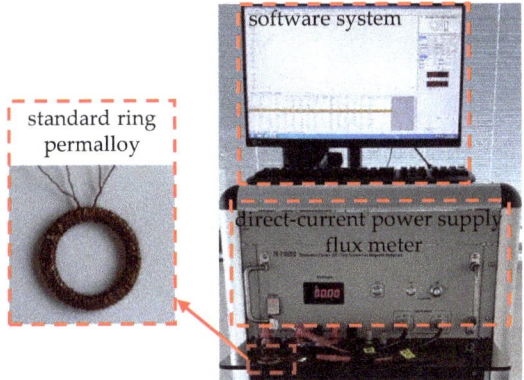

Figure 2. The outline of the common test system.

3.2. The Test Results and Discussion

(a) Test and parameter identification results for the Rayleigh model

In this work, 1J85, one of the permalloys with high permeability, provided by Beijing Beiye Functional Materials Corporation, was chosen. The parameters of the ring specimen are shown in Table 1, which are the commonly used parameters for the magnetic material test. The result for the B-H curve with H ranging from 0.008 to 800 A/m is shown in Figure 3a. Notably, the lower limit value of H is one order of magnitude lower than the set value, which is determined by the working principle of the test equipment. In addition, the curve related to the permeability and H was obtained by the first-order derivative of the B-H curve, which is shown in Figure 3b.

Table 1. Parameters of the tested standard ring specimen.

Symbol	Quantity	Standard Ring
A	outer diameter	40 mm
B	inner diameter	32 mm
C	Thickness	2 mm
N_1	turns of the primary winding	60
N_2	turns of the secondary winding	20

In order to determine Rayleigh constant a and the relative initial permeability μ_i, the identification process related to the function fitting was performed based on the measurement data and Equation (1). The Rayleigh region should be lower than the coercivity (H_c) for permalloy with high permeability. The fitting data were chosen to be in the range of 0.01 A/m to (0.5 × H_c) A/m according to the previous studies for other soft magnetic materials in [27,28]. The fitting results are shown by the blue solid line in Figure 3, demonstrating a good agreement between the fitting curve and the test data. The fitting function is $B = 0.3001H^2 + 0.0906H$. The Rayleigh constant is 0.3001, and the relative initial permeability is 72097, which is significantly lower than the measured relative permeability, being 91072 at 0.08 A/m. It can intuitively illustrate that the shielding effectiveness of the MSR can be overestimated using the permeability at 0.08 A/m.

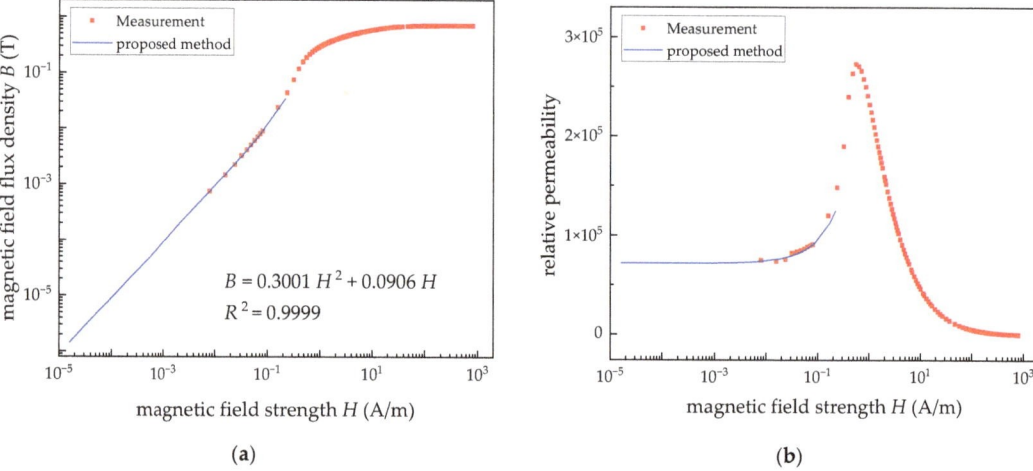

Figure 3. The comparison of the measurement and fitting curves. (**a**) The *B-H* curves. (**b**) The relative permeability curves.

(b) Verification for the proposed method

Firstly, in order to verify the feasibility of the proposed method, an EWMF test system was built to measure the permeability of the permalloy in a low magnetic field, as shown in Figure 4. Different from the common test system, a custom large-size ring specimen and a four-layer magnetic shielding cylinder were introduced in the EWMF test system. The magnetic shielding cylinder is utilized to eliminate the interference of magnetostatic signals and extremely low-frequency magnetic fields in the geomagnetic environment. The custom ring specimen is used to measure the permeability in the EWMF, and the parameters are shown in Table 2. Based on the parameters of the custom ring specimen and Equations (3)–(8), the minimum of the magnetization field is calculated to be about 4.59×10^{-5} A/m, less than 1 nT, and the minimum of the measurable magnetic flux density is about 1×10^{-8} T, which is larger than the magnetic flux measurement resolution of the magnetic material testing instrument. Considering the accuracy and resolution of the test system, the permeability in the magnetic field strength from 8×10^{-5} A/m to 25 A/m was chosen. In addition, due to the testing magnetic flux signal being so small as an EWMF applied in permalloy that it is easily affected by geomagnetic field and other interfering magnetic fields, the tested ring specimen was put in the magnetic shielding cylinder with the residual field less than 1 nT as the measurement is performed. The procedure of the measurement in the EWMF is the same as that of the common test system except for the range of magnetic field strength (here, *H* is from 8×10^{-5} A/m to 25 A/m).

Table 2. Parameters of the tested custom ring specimen.

Symbol	Quantity	Custom Ring
A	outer diameter	100 mm
B	inner diameter	50 mm
C	Thickness	1 mm
N_1	turns of the primary winding	1
N_2	turns of the secondary winding	450

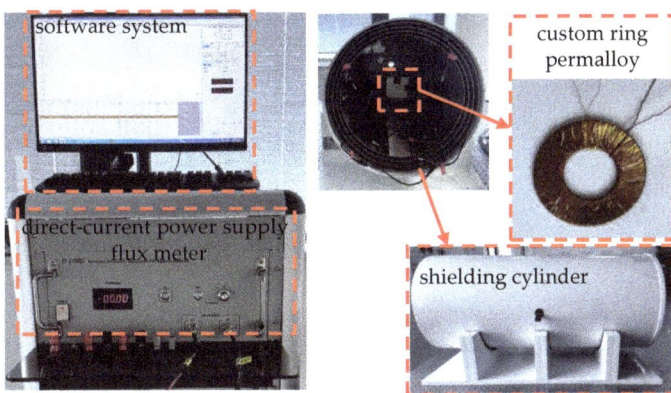

Figure 4. The outline of an extremely weak magnetic field (EWMF) test system.

The measured results for the *B-H* curve and permeability curve of a custom large-size ring specimen in the EWMF are shown as the green dot line in Figure 5, which is consistent with the fitting curve of the Rayleigh model, as the blue solid line in Figure 5 shows. The measured minimum permeability at the *H* being 8×10^{-5} A/m (0.1 nT) is 71,876, illustrating a 0.33% difference from the permeability (72,116) obtained by the fitting function of the Rayleigh model.

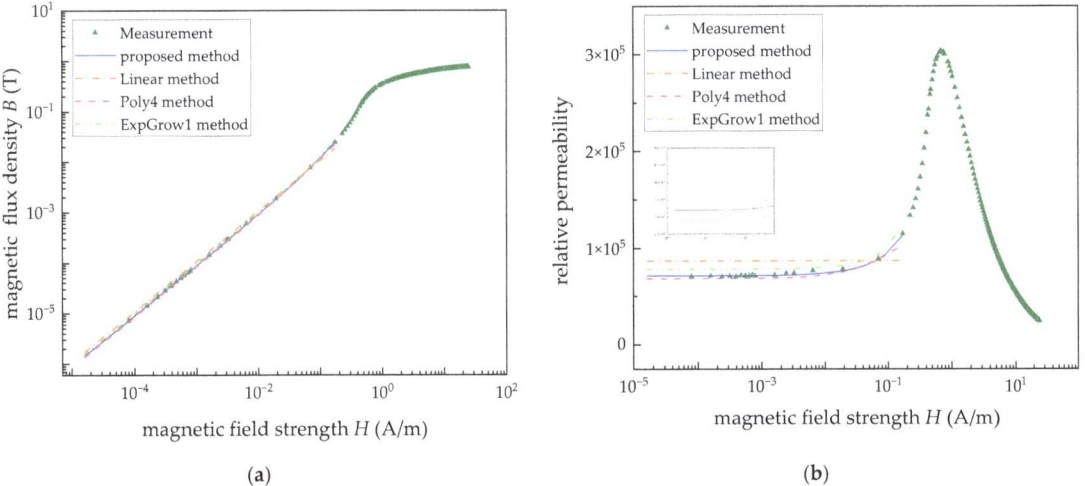

Figure 5. The comparison of four methods and the measurement data. (**a**) The *B-H* curves. (**b**) The relative permeability curves.

Secondly, to verify the optimal applicability of the Rayleigh model for the description of the relationship between the permeability and extremely weak magnetic field, the other three function models, including linear fitting (Linear method), 4th order polynomial fitting (Poly4 method) and one-phase exponential growth function (ExpGrow1 method), are selected for comparison. The comparison results are shown in Table 3 and Figure 5. As can be seen from the data in Table 3, compared with the other three models, the error between the fitted permeability and the measured value of the proposed method is the smallest. In addition, the curve obtained by the proposed method in Figure 5 is more consistent with the variation trend of the test data in the EWMF.

Table 3. Comparison of four methods and the measurement data.

Symbol	Fitting Function	μ_i	$\mu_{8 \times 10^{-5} \text{ A/m}}$ *	Error Ratio *
Measurement	-	-	71,876	-
proposed method	$B = 0.3001H^2 + 0.0906H$	72,097	72,116	0.33%
Linear method	$B = 0.1095H$	87,137	87,137	21.23%
Poly4 method	$B = 3.5652H^4 - 1.9207H^3 + 0.4775H^2 + 0.0865H$	68,834	68,864	4.19%
ExpGrow1 method	$B = -0.020188 + 0.01559e^{(H+0.05273)/0.20399}$	78,757	78,787	8.77%

* The penultimate column in the table represents the relative permeability when H is 8×10^{-5} A/m. The last column in the table represents the error ratio between the measurement data and the permeability when H is 8×10^{-5} A/m.

Finally, using the available four-layer magnetic shielding cylinder in our laboratory, the accuracy and validity of the Rayleigh model in estimating the shielding effect under EWMF were further analyzed and verified. The shielding cylinder shown in Figure 4 is composed of four layers of permalloy and one layer of aluminum, and its inner layer diameter and length are 36 cm and 90 cm, respectively. The thickness and layer spacing of the four-layer permalloy are the same as 1.6 mm and 1 cm, respectively. In this paper, finite element analysis was utilized to calculate the residual field in the center of the shielding cylinder. As the background magnetic field and B-H curve obtained by the Rayleigh model were set in the finite element model, the residual field in the center of the shielding cylinder was calculated to be 0.938 nT. While at the same position, the residual field was measured to be 0.987 nT after the shielding cylinder was degaussed, which has only 5% difference from the simulation. There is a good agreement between the simulation analysis and the actual measurement.

4. Conclusions

This paper proposes a method to measure the permeability of permalloy in EWMF based on the Rayleigh model. In this method, the magnetization and permeability curves were first obtained by measuring the ring permalloy of standard size. Then, two important parameters of the Rayleigh model, the relative initial permeability and the Rayleigh constant, were identified by function fitting the test data in the Rayleigh region. With the functional expression of the Rayleigh model, the measured data for the magnetization and permeability curves can be extrapolated to any desired weak magnetic fields or even zero magnetic fields.

The feasibility of the proposed method was verified by establishing an EWMF test system. The relative permeability of 8×10^{-5} A/m, the lowest magnetic field that can be measured by EWMF test system, is only 0.33% different from that obtained by the proposed method. Furthermore, the accuracy of the proposed method was also verified by comparing the results between analysis and measurement with the existing four-layer magnetic shielding cylinder in the laboratory. Using the relative permeability obtained by the proposed method in the analysis, the central remanences obtained by the analysis and measurement have a 5% difference, showing a good agreement.

The method proposed in this paper for obtaining the permeability in EWMF is accurate and easy to carry out, without considering the custom large-size ring specimen, extremely weak magnetic test environment and high-resolution magnetic field measurement sensor. It also can be applied for the permeability measurement in EWMF of other soft magnetic materials. The method proposed mainly studies the magnetization curve and relative permeability curve of permalloy when the amplitude of the low magnetic field changes while its magnetic characteristics under varying magnetic fields are not involved in this paper.

Author Contributions: Conceptualization, J.S. and Y.L. (Yan Lu); methodology, J.S. and Y.L. (Yan Lu); software, Y.L. (Yan Lu); validation, J.S., Y.L. (Yan Lu), L.Z., Y.L. (Yun Le) and X.Z.; formal analysis, Y.L. (Yan Lu); investigation, Y.L. (Yan Lu) and L.Z.; resources, J.S. and Y.L. (Yun Le); data curation,

Y.L. (Yan Lu) and X.Z.; writing—original draft preparation, Y.L. (Yan Lu); writing—review and editing, J.S., Y.L. (Yan Lu), L.Z., Y.L. (Yun Le) and X.Z.; supervision, Y.L. (Yan Lu) and X.Z.; project administration, J.S., L.Z. and Y.L. (Yun Le); funding acquisition, J.S. All authors have read and agreed to the published version of the manuscript.

Funding: This research was funded by the National Natural Science Foundation of China, grant number 52075017.

Institutional Review Board Statement: Not applicable.

Informed Consent Statement: Not applicable.

Data Availability Statement: Not applicable.

Conflicts of Interest: The authors declare no conflict of interest.

References

1. Li, J.; Quan, W.; Han, B.; Wang, Z.; Fang, J. Design and Optimization of Multilayer Cylindrical Magnetic Shield for SERF Atomic Magnetometer Application. *IEEE Sens. J.* **2020**, *20*, 1793–1800. [CrossRef]
2. Kvitkovic, J.; Patel, S.; Pamidi, S. Magnetic Shielding Characteristics of Hybrid High-Temperature Superconductor/Ferromagnetic Material Multilayer Shields. *IEEE Trans. Appl. Supercond.* **2017**, *27*, 1–5. [CrossRef]
3. Packer, M.; Hobson, P.J.; Davis, A.; Holmes, N.; Leggett, J.; Glover, P.; Hardwicke, N.L.; Brookes, M.J.; Bowtell, R.; Fromhold, T.M. Magnetic Field Design in a Cylindrical High-Permeability Shield: The Combination of Simple Building Blocks and a Genetic Algorithm. *J. Appl. Phys.* **2022**, *131*, 093902. [CrossRef]
4. Ates, K.; Carlak, H.F.; Ozen, S. Dosimetry Analysis of the Magnetic Field of Underground Power Cables and Magnetic Field Mitigation Using an Electromagnetic Shielding Technique. *Int. J. Occup. Saf. Ergon.* **2022**, *28*, 1672–1682. [CrossRef]
5. Matsuzawa, S.; Kojima, T.; Mizuno, K.; Kagawa, K.; Wakamatsu, A. Electromagnetic Simulation of Low-Frequency Magnetic Shielding of a Welded Steel Plate. *IEEE Trans. Electromagn. Compat.* **2021**, *63*, 1896–1903. [CrossRef]
6. Holmes, N.; Leggett, J.; Boto, E.; Roberts, G.; Hill, R.M.; Tierney, T.M.; Shah, V.; Barnes, G.R.; Brookes, M.J.; Bowtell, R. A Bi-Planar Coil System for Nulling Background Magnetic Fields in Scalp Mounted Magnetoencephalography. *Neuroimage* **2018**, *181*, 760–774. [CrossRef]
7. Yang, J.; Zhang, X.; Han, B.; Wang, J.; Wang, L. Design of Biplanar Coils for Degrading Residual Field in Magnetic Shielding Room. *IEEE Trans. Instrum. Meas.* **2021**, *70*, 1–10. [CrossRef]
8. Yang, K.; Wu, D.; Gao, W.; Ni, T.; Zhang, Q.; Zhang, H.; Huang, D. Calibration of SQUID Magnetometers in Multichannel MCG System Based on Bi-Planar Coil. *IEEE Trans. Instrum. Meas.* **2022**, *71*, 1–9. [CrossRef]
9. Afach, S.; Bison, G.; Bodek, K.; Burri, F.; Chowdhuri, Z.; Daum, M.; Fertl, M.; Franke, B.; Grujic, Z.; Helaine, V.; et al. Dynamic Stabilization of the Magnetic Field Surrounding the Neutron Electric Dipole Moment Spectrometer at the Paul Scherrer Institute. *J. Appl. Phys.* **2014**, *116*, 084510. [CrossRef]
10. Kuchler, F.; Babcock, E.; Burghoff, M.; Chupp, T.; Degenkolb, S.; Fan, I.; Fierlinger, P.; Gong, F.; Kraegeloh, E.; Kilian, W.; et al. A New Search for the Atomic EDM of 129Xe at FRM-II. *Hyperfine Interact.* **2016**, *237*, 95. [CrossRef]
11. Tashiro, K.; Wakiwaka, H.; Matsumura, K.; Okano, K. Desktop Magnetic Shielding System for the Calibration of High-Sensitivity Magnetometers. *IEEE Trans. Magn.* **2011**, *47*, 4270–4273. [CrossRef]
12. Zhao, J.; Ding, M.; Lu, J.; Yang, K.; Ma, D.; Yao, H.; Han, B.; Liu, G. Determination of Spin Polarization in Spin-Exchange Relaxation-Free Atomic Magnetometer Using Transient Response. *IEEE Trans. Instrum. Meas.* **2020**, *69*, 845–852. [CrossRef]
13. Patton, B.J.; Fitch, J.L. Design of a Room-Size Magnetic Shield. *J. Geophys. Res.* **1962**, *67*, 1117–1121. [CrossRef]
14. Mager, A. Grossräumige magnetische abschirmungen. *J. Magn. Magn. Mater.* **1976**, *2*, 245–255. [CrossRef]
15. Mager, A. The Berlin Magnetically Shielded Room (BMSR), Section A: Design and Construction. In *Biomagnetism*; Erné, S.N., Hahlbohm, H.-D., Lübbig, H., Eds.; De Gruyter: Berlin, Germany; Boston, MA, USA, 1981; pp. 51–78.
16. Bork, J. The 8-Layered Magnetically Shielded Room of the PTB: Design and Construction. In Proceedings of the Biomag 2000 12th International Conference on Biomagnetism, Espoo, Finland, 13–17 August 2001; pp. 970–973.
17. Jin, S.; Kuwahata, A.; Chikaki, S.; Hatano, M.; Sekino, M. Passive Shimming of Magnetically Shielded Room Using Ferromagnetic Plates. *IEEE Trans. Magn.* **2022**, *58*, 1–5. [CrossRef]
18. Zhao, Y.; Sun, Z.; Pan, D.; Lin, S.; Jin, Y.; Li, L. A New Approach to Calculate the Shielding Factor of Magnetic Shields Comprising Nonlinear Ferromagnetic Materials under Arbitrary Disturbances. *Energies* **2019**, *12*, 2048. [CrossRef]
19. Li, J.; Quan, W.; Han, B.; Liu, F.; Xing, L.; Liu, G. Multilayer Cylindrical Magnetic Shield for SERF Atomic Co-Magnetometer Application. *IEEE Sens. J.* **2019**, *19*, 2916–2923. [CrossRef]
20. Harakawa, K.; Kajiwara, G.; Kazami, K.; Ogata, H.; Kado, H. Evaluation of a High-Performance Magnetically Shielded Room for Biomagnetic Measurement. *IEEE Trans. Magn.* **1996**, *32*, 5256–5260. [CrossRef]
21. Cohen, D. Large-Volume Conventional Magnetic Shields. *Rev. Phys. Appl.* **1970**, *5*, 53–58. [CrossRef]
22. Liyi, L.; Zhiyin, S.; Donghua, P.; Jiaxi, L.; Feng, Y. An Approach to Analyzing Magnetically Shielded Room Comprising Discontinuous Layers Considering Permeability in Low Magnetic Field. *IEEE Trans. Magn.* **2014**, *50*, 1–4. [CrossRef]

23. Yamazaki, K.; Kato, K.; Muramatsu, K.; Haga, A.; Kobayashi, K.; Kamata, K.; Fujiwara, K.; Yamaguchi, T. Incremental Permeability of Mu-Metal in Low Magnetic Fields for the Design of Multilayer-Type Magnetically Shielded Rooms. *IEEE Trans. Magn.* **2005**, *41*, 4087–4089. [CrossRef]
24. Rayleigh, L. XXV Notes on Electricity and Magnetism. —III. On the Behavior of Iron and Steel under the Operation of Feeble Magnetic Forces. *Philos. Mag. J. Seri.* **1887**, *23*, 225–245. [CrossRef]
25. Canova, A.; Freschi, F.; Giaccone, L.; Repetto, M. Numerical Modeling and Material Characterization for Multilayer Magnetically Shielded Room Design. *IEEE Trans. Magn.* **2018**, *54*, 1–4. [CrossRef]
26. Canova, A.; Freschi, F.; Giaccone, L.; Repetto, M.; Solimene, L. Identification of Material Properties and Optimal Design of Magnetically Shielded Rooms. *Magnetochemistry* **2021**, *7*, 23. [CrossRef]
27. Kachniarz, M.; Szewczyk, R. Study on the Rayleigh Hysteresis Model and Its Applicability in Modeling Magnetic Hysteresis Phenomenon in Ferromagnetic Materials. *Acta Phys. Pol. A* **2017**, *131*, 1244–1250. [CrossRef]
28. Kachniarz, M.; Kołakowska, K.; Salach, J.; Szewczyk, R.; Nowak, P. Magnetoelastic Villari Effect in Structural Steel Magnetized in the Rayleigh Region. *Acta Phys. Pol. A* **2018**, *133*, 660–662. [CrossRef]

Article

Testing and Analysis Method of Low Remanence Materials for Magnetic Shielding Device

Yuan Cheng [1,2], Yaozhi Luo [1], Ruihong Shen [1,2,*], Deyu Kong [3,*] and Weiyong Zhou [4]

1. College of Civil Engineering and Architecture, Zhejiang University, Hangzhou 310027, China
2. China United Engineering Co., Ltd., Hangzhou 310022, China
3. College of Civil Engineering, Zhejiang University of Technology, Hangzhou 310023, China
4. School of Instrumentation Science and Optoelectronics Engineering, Beihang University, Beijing 100191, China
* Correspondence: srh_1023@163.com (R.S.); kongdeyu@zjut.edu.cn (D.K.)

Abstract: Magnetic shielding devices with a grid structure of multiple layers of highly magnetically permeable materials (such as permalloy) can achieve remanent magnetic fields at the nanotesla (nT) level or even lower. The remanence of the material inside the magnetic shield, such as the building materials used in the support structure, can cause serious damage to the internal remanence of the magnetic shield. Therefore, it is of great significance to detect the remanence of the materials used inside the magnetic shielding device. The existing test methods do not limit the test environment, the test process is vulnerable to additional magnetic field interference and did not consider the real results of the material in the weak magnetic environment. In this paper, a novel method of measuring the remanence of materials in a magnetic shielding cylinder is proposed, which prevents the interference of the earth's magnetic field and reduces the measurement error. This method is used to test concrete components, composite materials and metal materials commonly applied in magnetic shielding devices and determine the materials that can be used for magnetic shielding devices with 1 nT, 10 nT and 100 nT as residual magnetic field targets.

Keywords: magnetic shielding device; low remanence material; extremely weak magnetic field

1. Introduction

Large-scale magnetic shielding devices, used high permeable magnetic materials and high conductivity materials, such as permalloy and aluminum, to prevent the external magnetic field from entering its interior, based on the flux shunt effect [1–3] and eddy current effect [4,5], are used to provide extremely weak magnetic field environment. In the weak magnetic environment, many scholars have carried out frontier work, such as the measurement of biological signals such as magnetocardiography and magnetoencephalography [6–8], the measurement of geophysical research samples [9], the measurement of electric dipole moments [10], research on high-precision magnetic measuring instruments, such as serf atomic gyroscopes and magnetometers [11,12] and so on. Since the internal magnetic field of the magnetic shielding device is very small, the remanence of internal supporting structures (such as platforms for placing instruments) and scientific instruments equipment (such as cardio brain magnetic equipment) may have a greater impact on the internal magnetic field, so the remanence of the magnetic shielding device should be analyzed when selecting the supporting material of the magnetic shielding device.

Borradaile et al. [13] put the tested sample into a square container and used Molspin "BigSpin" magnetometer to test the remanence of the material. Feinberg et al. [14] used DC-SQUID sensors to design a three-axis remanence to test the low-temperature remanence and magnetism of magnetic minerals and provide explanations for geophysical phenomena such as core dynamics and paleoclimate. Bilardello et al. [15] tested the hysteresis loop of rock samples in geology through vibrating sample magnetometer (VSM), and analyzed the remanence and other magnetic properties. He et al. [16] improved the low-temperature calibration system of the Hall probe and established a low-temperature remanence measurement system for bulk permanent magnets, which can accurately test the remanence performance of bulk permanent magnets. Biedermann et al. [17] used a vibrating sample magnetometer to test the hysteresis and thermomagic of rock samples. It is used to determine the saturation magnetization, coercivity and remanence of rock samples to analyze the influence of seismic slip-on rock magnetism. Robustelli et al. [18] measured the Thermomagnetic susceptibility curves of the sample by heating the sample to 700 °C, and then cooling it to room temperature to measure the magnetic signature along interpolate sheet zones. Lund et al. [19] tested the hysteresis loop of deep-sea sediment samples through the MicroMag cycle. The remanence, saturation magnetization and coercivity were calculated to estimate the particle size distribution and percentage of ferromagnetic minerals and titanomagnetite minerals. Monaico et al. [20] studied the magnetization curve of coaxial core–shell magnetic nanostructures with vibrating sample magnetometer, analyzed the influence of different Ni content on remanence, squareness ratio, and coefficient. Wei et al. [21] tested the remanence in the iron core of the power transformer through the LCR tester and predicted the remanence through the relationship between it and the measured magnetization inductance.

For previous studies, the main purpose is to infer geological changes based on the magnetic analysis of rocks, there are mainly two methods to test the remanence of materials: one is to obtain the material's remanence properties by testing the material's hysteresis loop, which tests the magnitude of remanence in relation to the magnetization magnitude and cannot accurately reflect the material's remanence during application. Additionally, the other is to test the magnetic field near the material directly by a magnetometer to obtain the material's remanence value, but the existing methods for measuring the remanence of materials do not limit the testing environment, and they are directly tested in the earth's magnetic field. Because the earth's magnetic field has hundreds of nT fluctuations, when measuring the remanence of materials less than hundreds of nT, it is impossible to accurately calculate the remanence of materials according to the change of the fluxgate value, which leads to large test errors. Furthermore, the previous test results of material remanence did not consider the real results of the material in the weak magnetic environment, but the purpose of this study is to analyze the remanence of building and support structure materials used in the magnetic shielding device, so it is not applicable. In order to select the building materials and support materials that can be used in large-scale extremely weak magnetic field magnetic shielding devices, this paper uses a fluxgate magnetometer to test the material samples. The test process is carried out in a magnetic shielding cylinder to eliminate the interference of the external magnetic field. Firstly, test the internal environmental magnetic field of the magnetic shielding cylinder, and then test the magnetic field around the material. The true remanence of the material is the magnetic field around the material minus the internal environmental magnetic field. The method proposed in this paper avoids the interference of the earth's magnetic field and other magnetic fields, improves the measurement accuracy, and can directly test the remanence of materials in a weak magnetic environment, it is significant to analyze the materials of building and support structure inside the magnetic shielding device.

The rest of this paper is arranged as follows. In Section 2, the principles of testing and calculation of material remanence are first introduced, followed by the testing procedures. Section 3 presents the test results and discussion. Section 4 summarizes brief conclusions.

2. Principle of Remanence Measurement

In this paper, a fluxgate magnetometer (CTW-6M, resolution: 0.2 nT, range: 100,000 nT) is used to test the remanence of materials commonly used in magnetic shielding devices (concrete materials, composite materials and metal materials), during the test, the material is placed in a magnetic shielding cylinder to cancel the effect of the Earth's magnetic field influences.

The remanence of a material B_r is the absolute value of the test magnetic field B_T minus the background magnetic field B_B:

$$B_r = |B_T - B_B| \tag{1}$$

where B_r is the material remanence, B_T is the test magnetic field, B_B is the background magnetic field (including the environmental magnetic field in the barrel and the sensor bias).

The magnetic shielding cylinder used in the test has a four-layer permalloy structure. The remanence of the magnetic shielding cylinder can be calculated by the magnetic shielding factor S:

$$B_B = \frac{B_0}{S} \tag{2}$$

where B_0 is the Earth's magnetic field.

For the shielding coefficient of a single-layer shielding cylinder, the formula can be calculated as follows [22]:

$$S = \frac{2\mu t R^{\frac{1}{2}}}{L^{\frac{3}{2}}} \tag{3}$$

where S is the shielding coefficient; μ is the relative permeability; R is the radius of the shielding layer; t is the thickness of the shielding layer; L is the length of the shielding layer.

The formula for calculating the shielding factor of the multi-layer magnetic shielding cylinder is:

$$S_z = S_n \prod_{i=1}^{n-1} S_i \left[1 - \left(\frac{D_{i+1}}{D_i}\right)^j\right] \tag{4}$$

where S_z is the n-layer shielding coefficient; S_i is the i-th layer shielding coefficient (the outermost is the first layer).

Since the magnetic shielding cylinder is a long strip structure, there is a gradient in the internal magnetic field. The area with small magnetic field in the magnetic shielding cylinder is determined as the test area through the experimental test. It can be seen from the results that the closer to the shielding layer, the lower the internal remanence, but due to the magnetic concentration of permalloy, it cannot be infinitely close to the cylindrical wall. Therefore, choosing a location 150 mm to 250 mm from the bottom of the magnetic shielding cylinder for testing, as shown in Figure 1.

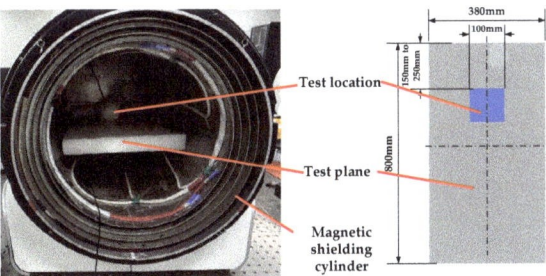

Figure 1. Schematic diagram of the test position in the magnetic shielding cylinder.

When the test area in the magnetic shielding cylinder is determined, the test flow diagram is shown in Figure 2, and the test steps are as follows:

(1) Turn on the fluxgate magnetometer and preheat for 10 min;
(2) Open the door of the magnetic shielding cylinder, place the magnetometer probe on the test platform inside the magnetic shielding cylinder, and move the probe to the center;
(3) Close the door of the magnetic shielding cylinder and test the internal magnetic field BB of the magnetic shielding cylinder;

(4) Open the door of the magnetic shielding cylinder and place the sample to be tested on the head of the magnetometer probe;
(5) Close the door of the magnetic shielding cylinder and test the current situation of the magnetic field Bt inside the magnetic shielding cylinder;
(6) Calculate the material remanence by Formula (1);
(7) Repeat (1)–(6) to test the next sample.

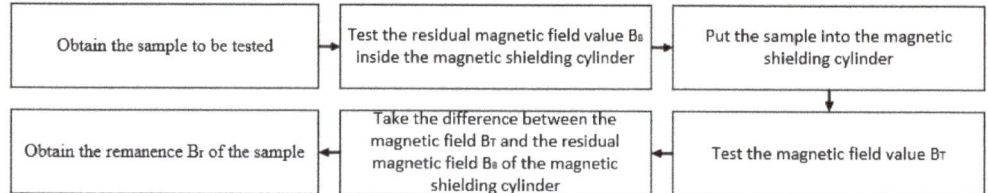

Figure 2. Schematic diagram of test process.

The test device is shown in Figure 3.

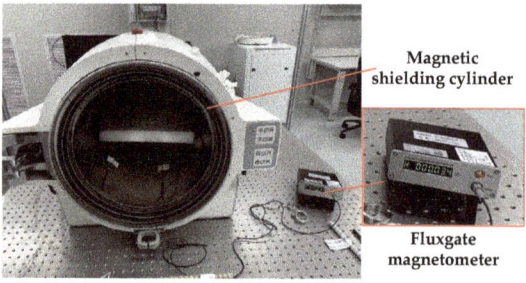

Figure 3. Magnetic shielding cylinder and fluxgate magnetometer.

3. Test Results and Discussion

3.1. Remanence Test of Concrete Component Materials

Concrete is often used as the supporting structure of large magnetic shielding devices. According to building requirements, the remanence test is carried out on some commonly used stones and gravels in concrete. The test samples are shown in Figure 4:

Following the test steps described above, the results are show in Table 1:

Table 1. Concrete material remanence test results.

Test Sample	Test Magnetic Field B_T (nT)	Background Magnetic Field B_B (nT)	Remanence Br (nT)
Sample 1	−27.7	6.4	34.1
Sample 2	24.2	6.4	17.8
Sample 3	−91.6	6.4	98.0
Sample 4	41.8	6.4	0.4
Sample 5	−168.2	6.4	174.6
Sample 6	−1.8	6.4	8.2
Sample 7	−39.2	6.4	45.6
Sample 8	12.7	6.4	6.3
Sample 9	−13.4	6.4	19.8
Sample 10	80.3	6.4	73.9
Sample 11	35.4	6.4	29.0
Sample 12	3.3	6.4	3.1
Sample 13	−1244.2	6.4	1250.6
Sample 14	−1.0	6.4	7.4
Sample 15	−1005.2	6.4	1011.6
Sample 16	5.6	5.3	0.3
Sample 17	19.9	5.3	14.6
Sample 18	−31.2	5.3	36.5
Sample 19	6.5	5.3	1.2
Sample 20	4.3	5.3	1.0

Figure 4. Test samples, sample 1 (sand from Hangzhou), sample 2 (sand from Jiande), sample 3 (sand from Ganjiang), sample 4 (dolomite), sample 5 (tuff), sample 6 (limestone), sample 7 (cement gravel), sample 8 (clay ceramsite from Jiaxing), sample 9 (clay ceramsite from Hubei), sample 10 (mortar block), sample 11 (grey cement mortar block), sample 12 (white cement mortar block), sample 13 (red brick), sample 14 (ceramic sheet), sample 15 (concrete block), sample 16 (white cement), sample 17 (grey cement), sample 18 (fly ash), sample 19 (quartz sand), sample 20 (mineral powder).

3.2. Remanence Testing of Composite Materials

Composite materials are commonly used as reinforcement for large magnetic shielding devices. According to the building requirements, the remanence test is carried out on the composite bars. The test sample is shown in Figure 5:

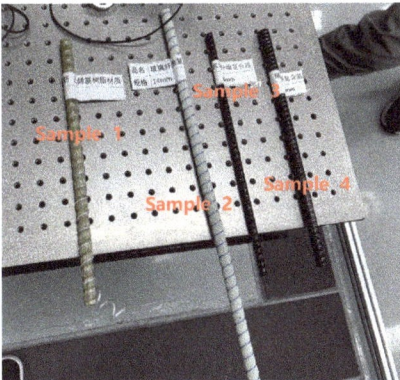

Figure 5. Test samples, sample 1 (vinyl resin), sample 2 (glass fiber), sample 3 (basalt fiber composite bar), sample 4 (carbon fiber composite bar).

Following the test steps described above, the results are show in Table 2:

Table 2. Composite materials remanence test results.

Test Sample	Test Magnetic Field B_T (nT)	Background Magnetic Field B_B (nT)	Remanence Br (nT)
Sample 1	3.5	3.4	0.1
Sample 2	3.8	3.4	0.4
Sample 3	3.8	3.4	0.4
Sample 4	4.1	3.4	0.7

3.3. Remanence Test of Metal Structural Materials

Metal materials are commonly used as internal supporting structures and supporting grids of large magnetic shielding devices. According to the building requirements, the remanence test is carried out on the metal structural materials. The test sample is shown in Figure 6:

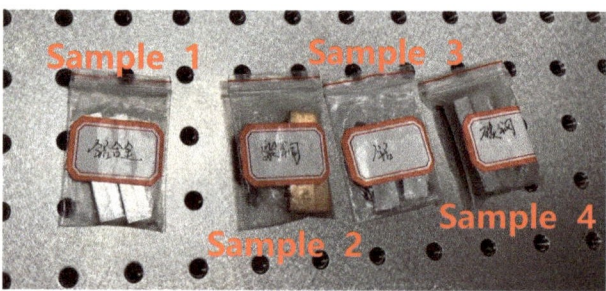

Figure 6. Test samples, sample 1 (aluminum alloy), sample 2 (copper), sample 3 (pure aluminum), sample 4 (carbon steel).

Following the test steps described above, the results are show in Table 3:

Table 3. Metal structural materials remanence test results.

Test Sample	Test Magnetic Field B_T (nT)	Background Magnetic Field B_B (nT)	Remanence Br (nT)
Sample 1	3.8	3.4	0.4
Sample 2	3.9	3.4	0.5
Sample 3	3.7	3.4	0.3
Sample 4	8013.1	3.4	8009.7

3.4. Discussion

Different research objects have different requirements for the internal weak magnetic environment of the magnetic shielding device. For magnetocardiography and magnetoencephalography, the internal magnetic field of the magnetic shielding device is often required to be less than 1 nT. For instruments such as serf gyroscopes, the internal magnetic field of the magnetic shielding device needs to be less than 10 nT. For animal and plant weak magnetic environment experiments, the internal magnetic field of the magnetic shielding device needs to be less than 100 nT. Therefore, the magnetic shielding devices of different objects need to be constructed with appropriate remanence and corresponding materials (such as external support materials, internal structural materials).

According to the above requirements for the use of the internal space of the magnetic shielding device, when selecting an application area, it is common to select a vertical distance (usually 0.2 m to 0.5 m) from the materials described in this paper. According to the test results, if the magnetic field values of 1 nT, 10 nT and 100 nT is to be reached in the magnetic shielding device, the remanence of the supporting materials shall be lower than 5 nT, 50 nT and 140 nT, respectively, in the distance between the above materials.

Therefore, the materials that can be used for ≤1 nT space after comparison are show in Table 4:

Table 4. Materials available inside the space of ≤1 nT.

Test Sample	Remanence Br (nT)
Concrete component material sample 4 (dolomite)	0.4
Concrete component material sample 12 (white cement mortar block)	3.1
Concrete component material sample 16 (white cement)	0.3
Concrete component material sample 19 (quartz sand)	1.2
Concrete component material sample 20 (mineral powder)	1.0
Composite materials sample 1 (vinyl resin)	0
Composite materials sample 2 (glass fiber)	0.4
Composite materials sample 3 (basalt fiber composite bar)	0.4
Composite materials sample 4 (carbon steel)	0.7
Metal structural materials sample 1 (aluminum alloy)	0.4
Metal structural materials sample 2 (copper)	0.5
Metal structural materials sample 3 (pure aluminum)	0.3

The materials that can be used in the space of ≤10 nT are show in Table 5 (Materials other than those listed in Table 4):

Table 5. Materials available inside the space of ≤10 nT.

Test Sample	Remanence Br (nT)
Concrete component material sample 6 (limestone)	8.2
Concrete component material sample 8 (clay ceramsite from Jiaxing)	6.3
Concrete component material sample 14 (ceramic sheet)	7.4
Concrete component material sample 17 (grey cement)	14.6

The materials that can be used in the space of ≤50 nT are show in Table 6 (Materials other than those listed in Tables 4 and 5):

Table 6. Materials available inside the space of ≤100 nT.

Test Sample	Remanence Br (nT)
Concrete component material sample 1 (sand from Hangzhou)	34.1
Concrete component material sample 2 (sand from Jiande)	17.8
Concrete component material sample 7 (cement gravel)	45.6
Concrete component material sample 9 (clay ceramsite from Hubei)	19.8
Concrete component material sample 11 (grey cement mortar block)	29.0
Concrete component material sample 18 (fly ash)	36.5

In summary, the concrete composites with low remanence are concrete composites sample 2 (sand from Jiande), sample 6 (limestone), sample 12 (white cement mortar block), sample 16 (white cement) and sample 20 (mineral powder), the composite material with lower remanence is composite material sample 1 (vinyl resin), and the metal structure material with lower remanence is metal structure material sample 3 (pure aluminum). After obtaining the remanence value of the material, the corresponding magnetic field strength can be calculated through the vacuum permeability μ_0 and substituted into the finite element simulation calculation, which can evaluate the influence of the remanence of the internal materials on the shielding device and improve the accuracy of the simulation calculation of the magnetic shielding device.

4. Conclusions

In order to reduce the adverse effects to the weak magnetic environment caused by the materials of the support and platform (including concrete member materials (stone, sand), composite materials and metal materials) in the extremely weak magnetic field, this paper proposes a method to test the remanence of materials by placing the tested samples in the magnetic shielding cylinder and using a fluxgate magnetometer. This method first tests the internal magnetic field of the magnetic shielding cylinder, and then the magnetic field around the material. The true remanence of a material is the magnetic field around the material minus the internal environmental magnetic field. This method can eliminate the interference of the external magnetic field on the test and obtain more accurate results. Through the test and analysis in this paper, in the 1 nT environment, concrete materials such as dolomite, white cement mortar block and white cement, composite materials such as vinyl resin and glass fiber, and metal materials such as aluminum alloy and copper can be used. In the 10 nT environment, concrete materials such as limestone, clay ceramsite from Jiaxing, and ceramic sheet can be used, and in the 100 nT environment, concrete materials such as cement gravel, grey cement mortar block, and fly ash can be used. Furthermore, the simulation accuracy of magnetic shielding device can be improved by combining test results of material remanence with finite element simulation and provide a basis for the structural support materials and platform materials used for the magnetic shielding device.

Author Contributions: Conceptualization, Y.C. and R.S.; methodology, Y.C.; software, D.K.; validation, Y.C. and Y.L.; formal analysis, Y.C.; investigation, Y.C.; resources, Y.C.; data curation, W.Z.; writing—original draft preparation, Y.C.; writing—review and editing, D.K.; visualization, R.S.; supervision, R.S.; project administration, D.K. All authors have read and agreed to the published version of the manuscript.

Funding: This research received no external funding.

Institutional Review Board Statement: Not applicable.

Informed Consent Statement: Not applicable.

Data Availability Statement: Not applicable.

Conflicts of Interest: The authors declare no conflict of interest.

References

1. Li, J.; Quan, W.; Han, B.; Wang, Z.; Fang, J. Design and Optimization of Multilayer Cylindrical Magnetic Shield for SERF Atomic Magnetometer Application. *IEEE Sens. J.* **2020**, *20*, 1793–1800. [CrossRef]
2. Kvitkovic, J.; Patel, S.; Pamidi, S. Magnetic Shielding Characteristics of Hybrid High-Temperature Superconductor/Ferromagnetic Material Multilayer Shields. *IEEE Trans. Appl. Supercond.* **2017**, *27*, 4700705. [CrossRef]
3. Packer, M.; Hobson, P.J.; Davis, A.; Holmes, N.; Leggett, J.; Glover, P.; Hardwicke, N.L.; Brookes, M.J.; Bowtell, R.; Fromhold, T.M. Magnetic Field Design in a Cylindrical High-Permeability Shield: The Combination of Simple Building Blocks and a Genetic Algorithm. *arXiv* **2021**, arXiv:2107.03170. [CrossRef]
4. Ates, K.; Carlak, H.F.; Ozen, S. Dosimetry Analysis of the Magnetic Field of Underground Power Cables and Magnetic Field Mitigation Using an Electromagnetic Shielding Technique. *Int. J. Occup. Saf. Ergon.* **2022**, *28*, 1672–1682. [CrossRef] [PubMed]
5. Matsuzawa, S.; Kojima, T.; Mizuno, K.; Kagawa, K.; Wakamatsu, A. Electromagnetic Simulation of Low-Frequency Magnetic Shielding of a Welded Steel Plate. *IEEE Trans. Electromagn. Compat.* **2021**, *63*, 1896–1903. [CrossRef]
6. Holmes, N.; Leggett, J.; Boto, E.; Roberts, G.; Hill, R.M.; Tierney, T.M.; Shah, V.; Barnes, G.R.; Brookes, M.J.; Bowtell, R. A Bi-Planar Coil System for Nulling Background Magnetic Fields in Scalp Mounted Magnetoencephalography. *Neuroimage* **2018**, *181*, 760–774. [CrossRef]
7. Yang, J.; Zhang, X.; Han, B.; Wang, J.; Wang, L. Design of Biplanar Coils for Degrading Residual Field in Magnetic Shielding Room. *IEEE Trans. Instrum. Meas.* **2021**, *70*, 6010110. [CrossRef]
8. Yang, K.; Wu, D.; Gao, W.; Ni, T.; Zhang, Q.; Zhang, H.; Huang, D. Calibration of SQUID Magnetometers in Multichannel MCG System Based on Bi-Planar Coil. *IEEE Trans. Instrum. Meas.* **2022**, *71*, 1002209. [CrossRef]
9. Afach, S.; Bison, G.; Bodek, K.; Burri, F.; Chowdhuri, Z.; Daum, M.; Fertl, M.; Franke, B.; Grujic, Z.; Helaine, V.; et al. Dynamic Stabilization of the Magnetic Field Surrounding the Neutron Electric Dipole Moment Spectrometer at the Paul Scherrer Institute. *J. Appl. Phys.* **2014**, *116*, 084510. [CrossRef]
10. Kuchler, F.; Babcock, E.; Burghoff, M.; Chupp, T.; Degenkolb, S.; Fan, I.; Fierlinger, P.; Gong, F.; Kraegeloh, E.; Kilian, W.; et al. A New Search for the Atomic EDM of 129Xe at FRM-II. *Hyperfine Interact* **2016**, *237*, 95. [CrossRef]
11. Tashiro, K.; Wakiwaka, H.; Matsumura, K.; Okano, K. Desktop Magnetic Shielding System for the Calibration of High-Sensitivity Magnetometers. *IEEE Trans. Magn.* **2011**, *47*, 4270–4273. [CrossRef]
12. Fan, W.F.; Quan, W.; Liu, F.; Pang, H.Y.; Xing, L.; Liu, G. Performance of Low-Noise Ferrite Shield in a K-Rb-Ne-21 Co-Magnetometer. *IEEE Sens. J.* **2020**, *20*, 2543–2549. [CrossRef]

13. Borradaile, G.J.; Geneviciene, I. Measuring heterogeneous remanence in paleomagnetism. *Geophys. Res. Lett.* **2007**, *34*, L12302. [CrossRef]
14. Feinberg, J.M.; Solheid, P.A.; Swanson-Hysell, N.L.; Jackson, M.J.; Bowles, J.A. Full vector low-temperature magnetic measurements of geologic materials. *Geochem. Geophys. Geosyst.* **2015**, *16*, 301–314. [CrossRef]
15. Bilardello, D.; Jackson, M.J. A comparative study of magnetic anisotropy measurement techniques in relation to rock-magnetic properties. *Tectonophysics* **2014**, *629*, 39–54. [CrossRef]
16. He, Y.Z. Study on cryogenic remanence measurement technology for chunk permanent magnet. *Acta Phys. Sin.* **2013**, *62*. [CrossRef]
17. Biedermann, A.R.; Jackson, M.; Bilardello, D.; Feinberg, J.M. Anisotropy of Full and Partial Anhysteretic Remanence Across Different Rock Types: 2-Coercivity Dependence of Remanence Anisotropy. *Tectonics* **2020**, *39*, 217502. [CrossRef]
18. Robustelli Test, C.; Zanella, E. Rock Magnetic Signature of Heterogeneities Across an Intraplate Basal Contact: An Example From the Northern Apennines. *Geochem. Geophys. Geosyst.* **2021**, *22*, e2021GC010004. [CrossRef]
19. Lund, S.; Platzman, E. Millennial-Scale Environmental Variability in Late Quaternary Deep-Sea Sediments from the Demerara Rise, NE Coast of South America. *Oceans* **2021**, *2*, 246–265. [CrossRef]
20. Monaico, E.V.; Morari, V.; Kutuzau, M.; Ursaki, V.V.; Nielsch, K.; Tiginyanu, I.M. Magnetic Properties of GaAs/NiFe Coaxial Core-Shell Structures. *Materials* **2022**, *15*, 6262. [CrossRef] [PubMed]
21. Wei, C.; Li, X.; Yang, M.; Ma, Z.; Hou, H. Novel Remanence Determination for Power Transformers Based on Magnetizing Inductance Measurements. *Energies* **2019**, *12*, 4616. [CrossRef]
22. Malkowski, S.; Adhikari, R.; Hona, B.; Mattie, C.; Woods, D.; Yan, H.; Plaster, B. Technique for high axial shielding factor performance of large-scale, thin, open-ended, cylindrical Metglas magnetic shields. *Rev. Sci. Instrum.* **2011**, *82*, 075104. [CrossRef] [PubMed]

Disclaimer/Publisher's Note: The statements, opinions and data contained in all publications are solely those of the individual author(s) and contributor(s) and not of MDPI and/or the editor(s). MDPI and/or the editor(s) disclaim responsibility for any injury to people or property resulting from any ideas, methods, instructions or products referred to in the content.

Article

Transient Magnetic Properties of Non-Grain Oriented Silicon Steel under Multi-Physics Field

Anqi Wang [1], Baozhi Tian [1], Yulin Li [1], Shibo Xu [1], Lubin Zeng [2] and Ruilin Pei [1,*]

[1] School of Electrical Engineering, Shenyang University of Technology, Shenyang 110870, China
[2] Suzhou Inn-Mag New Energy Ltd., Suzhou 215000, China
* Correspondence: peiruilin@inn-mag.com

Abstract: High-speed, high-efficiency and high-power density are the main development trends of high-performance motors in the future. At present, the design accuracy of traditional electric machines is already high enough; however, for the future demand of high performance and utilization in special environments (such as aviation and aerospace fields), more thorough research of materials' performance under multi-physics field (MPF) conditions is still needed. In this paper, a test system that combined temperature, stress and electromagnetic fields along with other fields, at the same time, is built. It can accurately simulate the actual complex working conditions of the motor and explore the dynamic characteristics of non-grain oriented (NGO) silicon steel. The rationality of this method is verified by checking the test result of the prototype, and the calculation accuracy of the motor model is improved.

Keywords: non-grain oriented silicon steel; multi-physics field; high-performance motors; efficiency

Citation: Wang, A.; Tian, B.; Li, Y.; Xu, S.; Zeng, L.; Pei, R. Transient Magnetic Properties of Non-Grain Oriented Silicon Steel under Multi-Physics Field. *Materials* 2022, 15, 8305. https://doi.org/10.3390/ma15238305

Academic Editor: Shenglei Che

Received: 28 October 2022
Accepted: 21 November 2022
Published: 23 November 2022

Publisher's Note: MDPI stays neutral with regard to jurisdictional claims in published maps and institutional affiliations.

Copyright: © 2022 by the authors. Licensee MDPI, Basel, Switzerland. This article is an open access article distributed under the terms and conditions of the Creative Commons Attribution (CC BY) license (https://creativecommons.org/licenses/by/4.0/).

1. Introduction

Silicon steel is one of the most important materials in motor. Under the actual working state of the motor, the core material is in a physical field where temperature, stress, frequency conversion and other external fields are coupled at the same time. For example, when the built-in permanent magnet synchronous motor is running at high speed, the alternating frequency of the magnetic field in iron core is high [1], and the iron loss increases exponentially as the frequency increases [2]. In addition, high frequency will also cause skin effects on the surface of silicon steel sheet, which will further increase the iron loss [3]. The loss of certain grades of silicon steel will increase several times when giving them compressive stress from 0 MPa to 100 MPa [4,5]. Silicon steel material will appear as a dynamic strain aging phenomenon under the action of temperature [6], and its tensile strength will change at the same time. Similarly, the heat dissipation problem and demagnetization problem caused by temperature rise during the operation of permanent magnet synchronous motor indicate that it is necessary to consider the influence of frequency, stress and temperature on the core material.

Traditional motor design and finite element analysis are calculated iteratively through the basic data of materials under power frequency, no external force and constant temperature conditions. However, with the improvement of the requirements for high-performance motors in the future, how to accurately calculate and predict the performance of motors under extreme conditions is particularly important. The magnetic properties and mechanical properties of silicon steel under MPF have gradually become a research hotspot. In Reference [7], the author explored the influence of stress and temperature on the iron loss of silicon steel, and completed the modification of the calculation coefficient of iron loss according to the influence.

Many scholars have studied the effect of a single physical field (SPF) change on silicon steel sheets' magnetic properties. In addition to the excitations, the microstructure of silicon steel will change with the change in stress [8,9], thus causing effects to its magnetic properties. When it is applied with external force, silicon steel will show magneto elastic properties [10] and stress demagnetization [7], which will also promote or deteriorate the magnetization of silicon steel. With the increase in tensile stress, most of the silicon steel will show a trend that its magnetic density firstly increases and then decreases. However, the effect of stress on silicon steel sheet loss is not infinite. When stress reaches a certain value, the loss will remain in a relatively stable numerical range [11].

By applying the temperature field to Epstein coils, Chen's team explored the influence of temperature rise on the magnetic properties of soft magnetic materials, and found that the magnetic density of materials decreased with the increase in temperature [12]. The influence of temperature on the electromagnetic properties of 0.5 mm medium grade NGO silicon steel was studied, and the action mechanism model was proposed [13]. Temperature coefficient K and effective resistivity of silicon steel lamination was subsequently introduced to explore the change trend of iron loss with temperature, and came to a conclusion that the iron loss decreases with temperature rise [14].

In 2016, American scholar G.M.R. Negri [15] conducted experiments involving thermal, metallurgical and electromagnetic measurements on silicon steel, simulated the evolution process of loss using the scalar hysteresis model and predicted the aging of silicon steel based on the Arrhenius formula, providing important significance for motor design. Similarly, Andreas Krings and Oskar Wallmark studied the influence of motor operating temperature on its magnetic performance and proposed a core loss model including temperature influence coefficient [16]. After them, different researchers continuously corrected the temperature coefficient of this iron loss model through experimental measurement, and verified the correctness of the modified model by testing the silicon steel and motor [17,18]. Some scholars have summarized relevant laws by studying the magnetic properties and mechanical properties of NGO silicon steel within a certain temperature range [19–21]. In addition, there are some scholars who also studied the electromagnetic characteristics of oriented silicon steel at different ambient temperatures [22].

In conclusion, more domestic and foreign scholars are paying attention to the characteristics of silicon steel under MPF. However, in the actual motor design process, some designers still rely on the fixed model and basic material data provided by the commercial finite element simulation software and empirical coefficients to design the motor. They still lack the understanding of the nature and change mechanism of the material under the influence of MPF, as well as the final impact on the motor performance.

The first part of this paper summarizes the current research status of the magnetic properties of materials under MPF. In the second part, a high-power density drive motor is designed, and three main factors that affect the characteristics of soft magnetic materials under extreme operating conditions are proposed, which are temperature, stress and frequency. In the third part, a magnetic characteristic test method based on "temperature stress electromagnetic" three-field coupling is proposed, which can proximately simulate the real situation of electrical steel when it runs inside the motor. In the fourth part, 30ADG1500 is used as the experimental object to find out the dynamic characteristics of NGO silicon steel under MPF, and the motor's efficiency is simulated and compared when the influence factors are not considered and when the influence factors are considered. The rationality of the method is verified through the test result of the prototype and the calculation accuracy is improved by this method. Finally, the conclusion is summarized.

2. Motor Design and Multi-Physics Coupling Field Analysis

In order to explore the influence of magnetic characteristics of silicon steel on motor performance, we have designed a new type of high-power density motor based on the actual requirements of engineering, so that electric vehicles can maintain good performance under starting, climbing and braking conditions. Based on the consideration of the motor's performance using flux weakening speed regulation method and the prototype manufacturing capability, we chose the interior permanent magnet rotor, which adopts the design of 48 slots and 8 poles, and the peak speed of the motor can reach 9000 rpm. Detailed parameters of the motor are shown in Table 1.

Table 1. Detailed parameters of the motor.

Parameter	Value
Rated/Peak torque (Nm)	40/150
Rated/Peak power (kW)	20/40
Rated/Peak speed (rpm)	4774/9000
Stack length (mm)	144
Working temperature (°C)	−40~60

In order to further explore what kind of MPF environment the core will be in when the motor is running at high speed, the finite element simulation of the motor was carried out by Motor-CAD and Ansys software. Based on the actual working conditions of electric vehicles, we set the ambient temperature as 60 °C, and predicted the temperature rise and magnetic density of the driving motor at the peak speed, as shown in Figure 1.

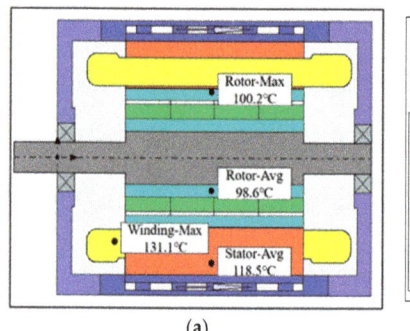 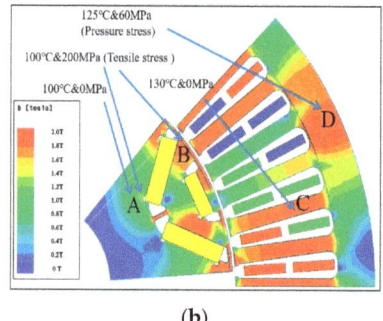 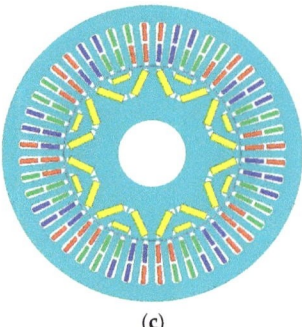

(a) (b) (c)

Figure 1. Temperature distribution and magnetic density distribution @ 9000 rpm. (**a**) Temperature rise simulation (flow rate: 15 L/min); (**b**) finite element simulation; (**c**) finite element model.

When the motor speed reaches 9000 rpm, the magnetic bridge of the rotor will bear about 200 MPa of centrifugal force, which belongs to tensile stress for silicon steel and will affect its loss characteristics. In addition, the fit between the stator and the casing adopts the interference method. According to the interference amount, the calculation shows that the stator will bear about 60 MPa of compressive stress, which will also affect the magnetic characteristics of the silicon steel. Therefore, in the simulation analysis and calculation of the model, we divided the model into four areas according to the actual situation, and applied the characteristic data of silicon steel in different states. The purpose is to improve the accuracy of the calculation.

The color of the magnetic density indicates the degree of saturation of the silicon steel material used in the motor. The red area will heat slightly more than the other areas. As can be seen from Figure 1, the magnetic density of the stator teeth is from 1.8 T to 2 T, the magnetic density of the yoke is mostly between 1 T and 1.2 T, and the average temperature of the iron core is about 120 °C at the peak rotating speed. The magnetic density of most parts of the rotor are from 1 T to 1.4 T, and its maximum temperature is about 100 °C, which is concentrated around the permanent magnet.

In the production process of the motor, the casing and stator usually fixed by interference fit, and the rotor material will be affected by centrifugal force when rotating. Considering those affects, area A of the rotor core is in the physical field of 100 °C and 0 MPa, area B is in the physical field of 100 °C and 200 Mpa of tensile stress (TS), area C of the stator core is in the physical field of about 130 °C and 0 MPa of tensile stress and area D is in the physical field of 125 °C and 60 MPa of compressive stress (PS). Therefore, in order to improve the calculation accuracy, it is necessary to consider the magnetic properties of materials under MPF.

3. Experiments under Varying Ambient Conditions

In the harsh aerospace environment such as extremely low temperature, high temperature or vacuum, the parameters of permanent magnet synchronous motor vary greatly, and silicon steel's characteristic is also prone to nonlinear changes. The coupling relationship among electromagnetic, temperature, fluid, stress and other fields of MPF cannot be ignored. How to accurately test the magnetic properties of silicon steel under MPF and speculate on the motor performance has become a key technical problem.

3.1. Test Platform

In this paper, a test system for magnetic properties of silicon steel based on the MPF environment of "temperature, stress and electromagnetic" is established. The test system is composed of three core modules: electromagnetic characteristic test module, mechanical characteristic test module and adjustable test environment module. The magnetic testing equipment adopted a commercial equipment on the market. In addition, in order to test the characteristics of the sample under more applicable conditions, we also customized a tension machine and a device that can adjust the temperature environment, which uniquely coupled the three parts together. This system can not only test the properties of silicon steel under SPF, but also complete the properties of silicon steel under MPF.

As shown in Figure 2, the test platform can simultaneously change temperature, stress, frequency, magnetic field strength and other major variables. We cut the size and shape of the sample according to the relevant national standards, and wrapped the copper wire around the left and right sides of the sample. During the experiment, the sample is fixed in the equipment with adjustable temperature by the tension machine, and the copper wire is drawn to the magnetic characteristics test equipment. In this way, we can test the magnetic properties of the sample, for example, BH and BP data. The design of the test system not only innovatively coupled multiple devices together to provide different environments for silicon steel testing, but also saved space.

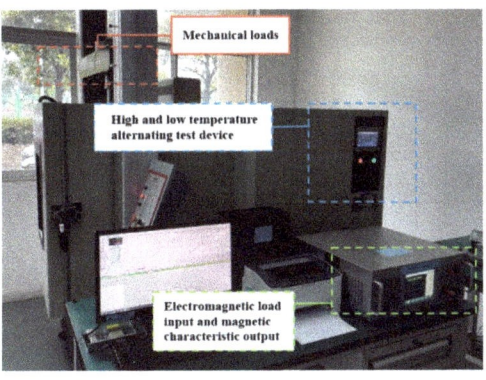 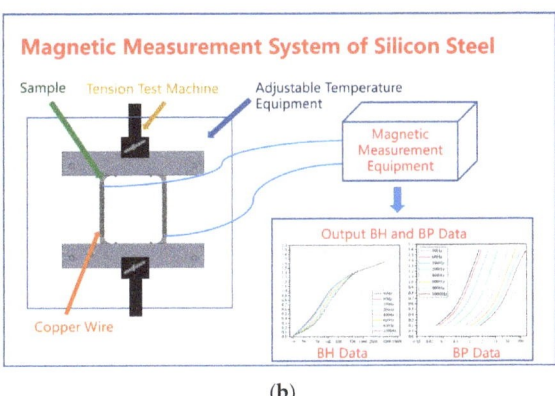

(a) (b)

Figure 2. Test platform. (a) Actual test system; (b) schematic diagram.

When the temperature field is applied, the sample is heated by air heat conduction, and the temperature is fed back by the sensor and adjusted in time. The temperature of the object under test can be varied by contact-free heating or cooling in hot or cold air, so as to avoid interference as much as possible and reduce magnetic leakage of the sample during the test. By applying different voltage, current, frequency and magnetic field intensity to the sample of a commercial magnetic testing equipment of silicon steel, we can obtain the characteristics data of silicon steel samples under different electromagnetic fields. Coupled together, the three devices can be used to load different stress, temperature, electromagnetic and other environments, providing a wider range of support for the magnetic properties of silicon steel testing. Therefore, we can guide the material selection and optimal design of the motor.

3.2. Function of Testing Device

Through the thermal simulation analysis of the motor model based on the Motor-CAD software above, it is not difficult to find that the core will be in the environment of high temperature, high frequency and high stress. Therefore, the magnetic properties of silicon steel under multiple physical fields can more accurately reflect the performance of the motor at high speed. Based on this, the following four variables are set in this paper, and 30ADG1500 NGO silicon steel is used as the research object to analyze its magnetic properties. The test range of the test device in this paper is shown in Table 2 below:

Table 2. Test range of the testing device.

Parameter	Range
Stress	−60–200 MPa
Temperature	−40~150 °C
Frequency	40~1000 Hz
Magnetic field strength	0~10,000 A/m

4. Analysis and Prototype

4.1. Analysis of Magnetic Properties of Materials

4.1.1. Magnetic Properties of Silicon Steel Considering Temperature, Stress and Frequency

It was found that the magnetic properties of silicon steel are affected by both stress and temperature. The stress has an effect on the microstructure of silicon steel, which leads to the difference of hysteresis coefficient of different frequencies in the calculation of iron loss. In addition, temperature affects the calculation coefficient of the eddy current loss at different frequencies. Therefore, for the characteristics of silicon steel under multiple physical fields, we need to consider the influence of stress, temperature and frequency on magnetic characteristics of silicon steel for calculation and analysis. Refer to the following formula for calculating iron loss:

$$P_{Fe} = P_h + P_c = k_h(\sigma) f B_m + k_c(T) f^2 B_m^2 \qquad (1)$$

k_h and k_c are, respectively, the hysteresis loss coefficient and eddy current loss coefficient after data fitting.

In order to clearly express the influence of temperature, stress and frequency on the magnetic properties of silicon steel, this paper selected the experimental results under the magnetic field strength of 1000 A/m to draw curves., as shown in Figure 3. In order to show the magnetic properties of silicon steel with the change in stress and temperature, it is drawn as a function of temperature and two-way stresses.

It can be seen that the magnetic density of NGO silicon steel will decrease with the increase in temperature. Compressive stress will also reduce the magnetic density of materials. However, the magnetic density first increases and then decreases with the increase in tensile stress, and reaches a maximum value near 30 MPa. In addition, the magnetic density is more affected by tensile stress at low temperature; with the increase in temperature, the influence of tensile stress on magnetic density decreases gradually.

When the motor is running at a rated speed, the frequency is about 400 Hz. So, the loss of 30ADG1500 under 1 T and 400 Hz was tested, and the changing trend is shown in Figure 4.

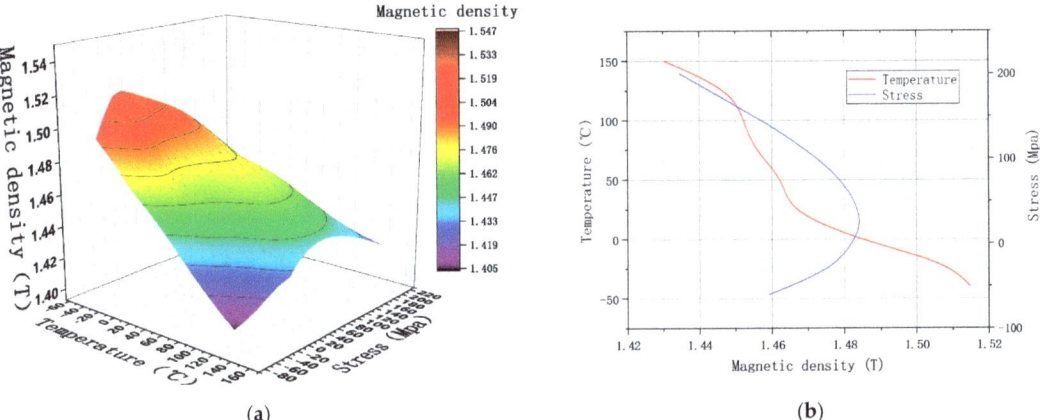

Figure 3. Magnetic density under 1000 A/m. (**a**) The trend of B under MPF; (**b**) B under SPF.

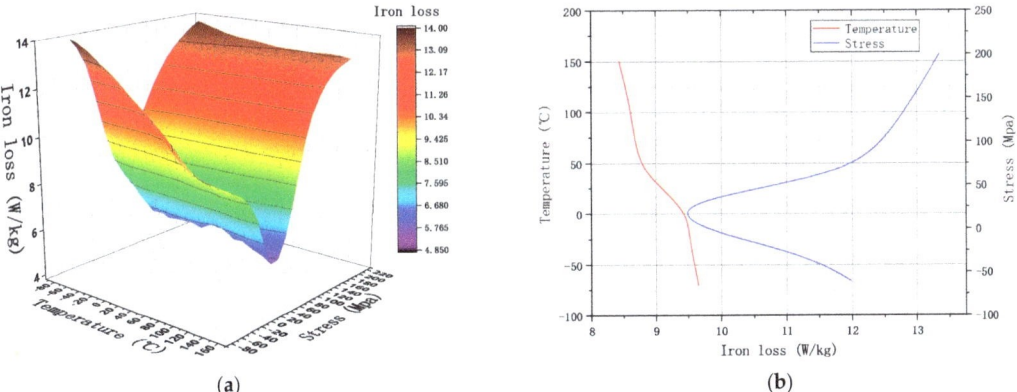

Figure 4. Iron loss under 400 Hz and 1 T. (**a**) The trend of P under MPF; (**b**) P under SPF.

It can be seen that the loss of NGO silicon steel will decrease with the increase in temperature; with the increase in applied stress, it first decreases and then increases, reaching a minimum value around 30 MPa. The trend of loss change is opposite to that of the magnetic density change. At high temperature, the loss is greatly affected by tensile stress, and the influence of tensile stress decreases with the decrease in temperature. However, the compressive stress will make the material loss deteriorate to a large extent, and the deterioration extent will gradually increase with the increase in compressive stress.

As the motor's speed increases, the frequency of the motor increases gradually. When the motor operates at peak value, the operating frequency rises to 800 Hz. In Figure 5, the changing trend of silicon steel's loss with temperature and stress at 1.2 T and 800 Hz was plotted. The changing trend of loss with stress and temperature is generally consistent with Figure 4, but the turning point of tensile stress of loss trend is reduced from 30 MPa to 25 MPa.

From the experimental data above, at different temperatures, their magnetic properties will show different levels of sensitivity to tensile stress. For example, at low temperature, the iron core's magnetic density is more affected by tensile stress, while the loss is less affected; with the increase in temperature, the influence of tensile stress on magnetic density decreases and the influence of tensile stress on loss increases.

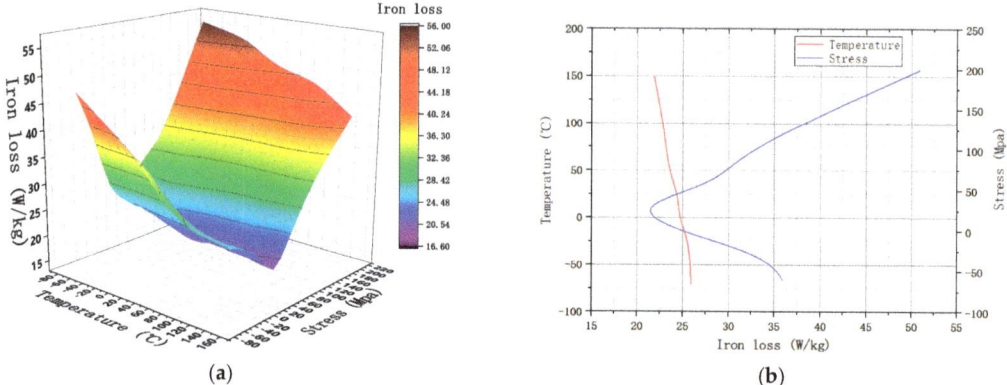

Figure 5. Iron loss under 800 Hz and 1.2 T. (**a**) The trend of P under MPF; (**b**) P under SPF.

4.1.2. Comparison of Magnetic Properties of Materials

Figure 6 shows the comparison of magnetic properties of the NGO silicon steel with and without considering MPF.

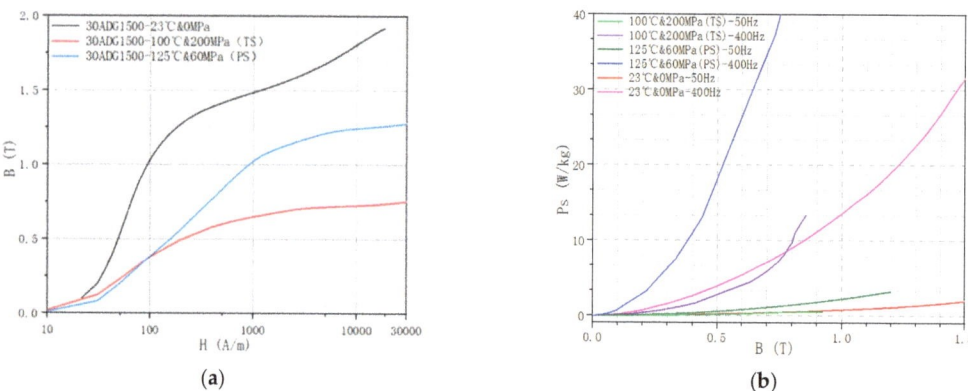

Figure 6. The magnetic characteristic of NGO silicon steel. (**a**) B-H curves; (**b**) B-P curves.

From the above figure, it can be seen that the magnetic properties of silicon steel will change under the action of temperature, stress, frequency and other external fields. In conclusion, in the future design process of high-performance motors, the calculation accuracy of the model will be greatly improved by considering the magnetic characteristics of silicon steel under MPF.

4.2. Analysis of Simulation Results

In this subsection, we replace the A, B, C and D regions of the motor model with the corresponding magnetic characteristics data under high temperature and stress, and compare the motor simulation characteristics of silicon steel under the condition of considering MPF with that of not considering MPF. Based on the data from the test, motor's loss diagram and efficiency comparison diagram are drawn.

Figure 7 shows the analysis of the magnetic characteristic data for 1 T and 400 Hz of the core material tested in the previous article under MPF. At 4775 rpm, motor's iron loss considering single electromagnetic field is 233.46 W, and iron loss considering the coupling of MPF including electromagnetic, stress and temperature is 268.29 W. The iron loss of the motor has increased by 34.83 W, which is about 14.9%. At the speed of 9000 rpm, motor's iron loss considering single electromagnetic field is 466.92 W, and iron loss considering MPF is 536.58 W, the iron loss of the motor has increased by 69.66 W, which is about 14.92%.

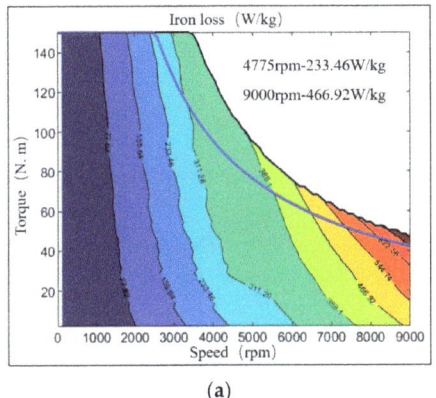
(a)

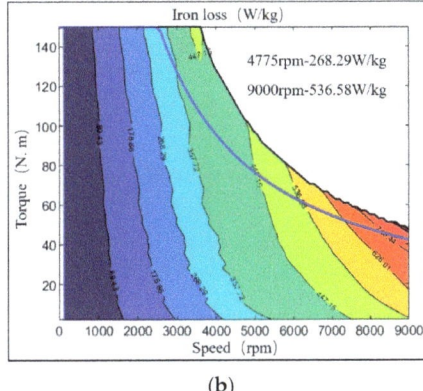
(b)

Figure 7. Comparison of motor's iron loss map. (**a**) Iron loss of SPF; (**b**) iron loss of MPF.

With the increase in rotating speed, the proportion of copper loss decreases and the proportion of iron loss increases, and the influence of the characteristics of iron core material on the motor performance increases gradually. Therefore, for the design of high-speed motors, considering the characteristics of iron core materials under MPF can not only more truly reflect the motor performance, but also greatly improve the calculation accuracy.

The loss of the motor during operation is mainly composed of iron loss, copper loss and mechanical loss. For the high-speed motor, efficiency as a very important index, iron loss plays a more important role. For example, for electric vehicles, the high efficiency of the motor can bring about an increase in the driving range. Therefore, we need to analyze the actual loss of the motor at high speed with the help of the magnetic characteristics data of silicon steel under MPF. In order to verify the reliability of the previous test, we compared and analyzed the efficiency of the motor under SPF and MPF, and drew the efficiency graph pair in the form of contour lines, as shown in Figure 8. The numerical statistics of the efficiency are shown in Table 3.

When the motor runs at the peak speed, the stator and rotor core are in a high-temperature state, and the influence of tensile stress on the loss of silicon steel sheet will increase. It can be seen from the table and contour diagram that for the area where the motor efficiency is greater than 95%, the calculated value of motor efficiency considering the characteristic data of silicon steel in SPF is 26.21%, and the calculated value of motor efficiency considering the characteristic data of silicon steel in MPF is 16.67%. The ratio of the area of the high-efficiency area to the total area decreased by 9.54%. In addition, the maximum efficiency of the motor decreased by 0.31%.

Therefore, in the process of motor design, if only the magnetic characteristics of iron core materials under a SPF are considered, the efficiency of motor simulation will be a bit high, which may possibly lead to the problem of insufficient performance after prototype manufacturing, and have a certain impact on the accuracy of motor design. Therefore, in the design of a high-performance motor, MPF has great practical value for improving the accuracy of motor simulation calculation.

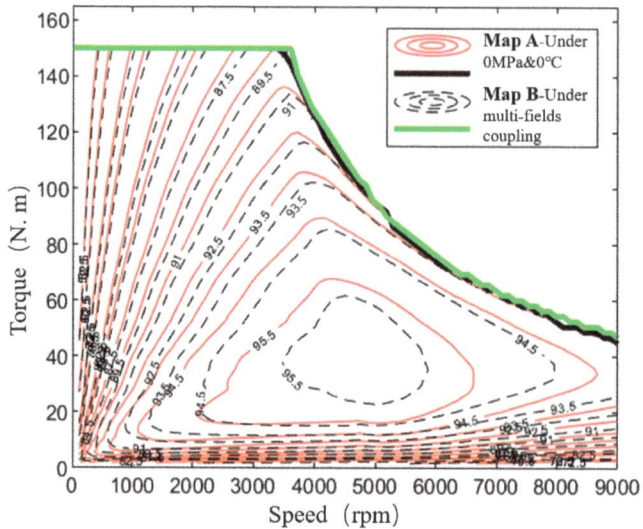

Figure 8. Motor's efficiency map comparison.

Table 3. Comparison of efficiency ratios under two working conditions.

Efficiency	SPF	MPF
Area of efficiency >85%	84.33%	82.55%
Area of efficiency >90%	71.83%	68.38%
Area of efficiency >95%	26.21%	16.67%
Maximum efficiency	96.29%	95.98%

4.3. Prototype Testing

To verify the work of this paper, a 20/40 kW motor prototype was installed on a 0–15,000 rpm test bench for testing, as shown in Figure 9.

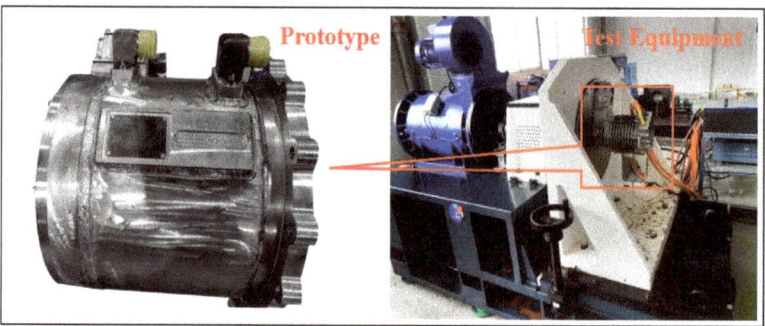

Figure 9. Prototype and test equipment.

4.3.1. Temperature Rise Comparison

The load testing was carried out on the prototype with the flow of cooling water was 15 L/min, and the inlet temperature was 65 °C. The test results were compared with the simulation results as Figure 10.

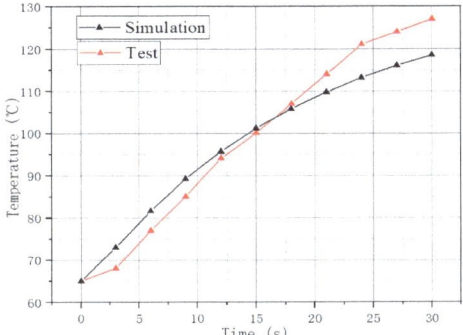

Figure 10. Temperature comparison between simulation and prototype test.

As shown in Figure 10, Under the working condition of 30 s peak operation, the simulated temperature rise is about 55 °C and the measured temperature rise is about 62 °C. Different silicon steels usually have different coatings and will also have different levels of sensitivity to temperature rise. After the research, it was found that the coating of the silicon steel we used would make the temperature rise of the core slightly higher than that of other silicon steels. In addition, other non-standard processes in the production process will also have errors in simulation and testing. Therefore, we should strive to improve the accuracy of processing.

4.3.2. Efficiency Comparison

The efficiency simulation data considering MPF were compared with the measured efficiency data as Figure 11.

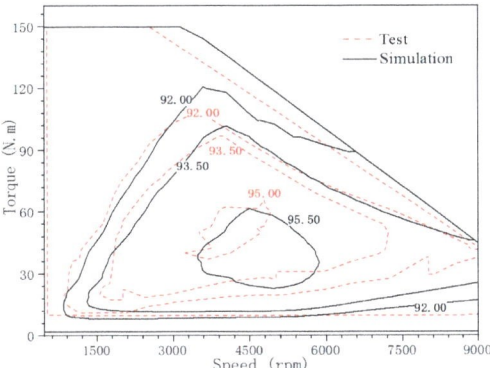

Figure 11. Temperature comparison between simulation and prototype test.

It can be seen from the Figure 11 that the simulation data under the consideration of MPF is more realistic. Although the area of the high-efficiency area of the motor has been reduced, the total error is within 3% and the maximum efficiency point error is within 0.5%. In addition, in the actual production process of the motor, due to other problems such as process and coating, the efficiency of the motor will also decline slightly.

5. Discussion

At present, most motor designs are calculated based on material test data at 50 Hz, 0 MPa and room temperature. However, for future high-performance and special motor requirements, more careful design and precise research under extreme working conditions are still needed. For this reason, the research significance of this paper is to improve the accuracy of model calculations under complex working conditions, and to explore the magnetic characteristics and mechanical properties of silicon steel under MPF from the perspective of material properties testing.

In the experiments carried out in this paper, the main factors such as high temperature, stress and frequency were considered. In the next phase of the test work, we will optimize the function of the test equipment and complete the material characteristics test at extremely low temperature.

6. Conclusions

Based on the analysis of a high-power density motor of an electric vehicle, this paper used 1 T and 400 Hz and 1.2 T and 800 Hz working conditions as examples to explore the characteristics of 30ADG1500 iron core material under the MPF of electromagnetic, stress and temperature fields. A summary of our findings is as follows:

1. The magnetic density decreases with the increase in temperature and compressive stress, and firstly increases and then decreases with the increase in tensile stress. The influence of tensile stress on magnetic density at low temperature is greater than that at high temperature.

2. The iron loss of materials will decrease with the increase in temperature; With the increase in tensile stress, the iron loss firstly decreases and then increases, and the influence increases at high temperature. The compressive stress will only worsen the iron loss situation.

In the process of motor design, with the increase in motor speed, the proportion of iron loss is increasing. The magnetic characteristics of iron core materials play a vital role in motor performance. Based on the experiment, it was found that the iron loss of the motor considering the MPF will increase by around 14.9%, leading to a decrease in the actual efficiency of the motor. The range of efficiency greater than 95% will decrease by 9.54%. Finally, the validity of the conclusion is verified by the prototype test result. Therefore, when designing high-performance motors, it is necessary to consider the characteristics of iron core materials under MPF, such as electromagnetic, stress and temperature fields, to improve the calculation accuracy of the model and increase the reliability of motor design.

Author Contributions: Conceptualization, A.W. and B.T.; methodology, R.P.; software, Y.L.; validation, A.W., B.T. and Y.L.; formal analysis, S.X.; investigation, A.W.; resources, L.Z. and R.P.; data curation, B.T.; writing—original draft preparation, A.W.; writing—review and editing, A.W.; visualization, L.Z. and Y.L.; supervision, L.Z. and R.P.; project administration, R.P. All authors have read and agreed to the published version of the manuscript.

Funding: This research was funded by programme of Scholars of the Xingliao Plan (No. XLYC2002113) and Shenyang University of Technology Interdisciplinary Team Project (No. 100600453).

Institutional Review Board Statement: The study did not involve humans or animals.

Informed Consent Statement: The study did not involve humans.

Data Availability Statement: The data are not publicly available for privacy reasons. The data presented in this study are available from the corresponding author.

Acknowledgments: This work was supported by the programme of Scholars of the Xingliao Plan (No. XLYC2002113) and Shenyang University of Technology Interdisciplinary Team Project (No. 100600453). The authors wish to thank our colleagues Junsheng Shi for his technical support.

Conflicts of Interest: The authors declare no conflict of interest.

References

1. Shen, J.; Li, P.; Hao, H.; Yang, G. Study on electromagnetic losses in high-speed permanent magnet brushless machines-the state of the art. *Zhongguo Dianji Gongcheng Xuebao (Proc. Chin. Soc. Electr. Eng.)* **2013**, *33*, 62–74.
2. Gerada, D.; Mebarki, A.; Brown, N.L.; Gerada, C.; Cavagnino, A.; Boglietti, A. High-speed electrical machines: Technologies, trends, and developments. *IEEE Trans. Ind. Electron.* **2013**, *61*, 2946–2959. [CrossRef]

3. Jiang, S.L.; Zou, J.B.; Xu, Y.X.; Liang, W. Variable coefficient iron loss calculating model considering rotational flux and skin effect. *Proc. CSEE* **2011**, *31*, 104–110.
4. Oda, Y.; Toda, H.; Shiga, N.; Kasai, S.; Hiratani, T. Effect of Si content on iron loss of electrical steel sheet under compressive stress. *IEEE Trans. Magn.* **2014**, *50*, 1–4. [CrossRef]
5. Oda, Y.; Hiratani, T.; Kasai, S.; Okubo, T.; Senda, K.; Chiba, A. Effect of compressive stress on iron loss of gradient Si steel sheet. *Electron. Commun. Jpn.* **2016**, *99*, 74–83. [CrossRef]
6. Qian, K.W.; Li, X.Q.; Xiao, L.G.; Chen, W.Z.; Zhang, H.G.; Peng, K.P. Dynamic strain aging phenomenon in metals and alloys. *J. Fuzhou Univ. (Nat. Sci. Ed.)* **2001**, *26*, 8–23.
7. Liu, G.; Liu, M.; Zhang, Y.; Wang, H.; Gerada, C. High-speed permanent magnet synchronous motor iron loss calculation method considering multi-physics factors. *IEEE Trans. Ind. Electron.* **2019**, *67*, 5360–5368. [CrossRef]
8. Leng, J.C.; Tian, H.X.; Guo, Y.G.; Xu, M.X. Effect of tensile and compressive stresses on magnetic memory signal and its mechanism. *Eng. Sci.* **2018**, *40*, 565–570.
9. Liu, J.; Tian, G.Y.; Gao, B.; Zeng, K.; Qiu, F. Domain wall characterization inside grain and around grain boundary under tensile stress. *J. Magn. Magn. Mater.* **2019**, *471*, 39–48. [CrossRef]
10. Cao, H.; Huang, S.; Shi, W. Influence of core stress on performance of permanent magnet synchronous motor. *J. Magn.* **2019**, *24*, 24–31. [CrossRef]
11. LoBue, M.; Sasso, C.; Basso, V.; Fiorillo, F.; Bertotti, G. Power losses and magnetization process in Fe–Si non-oriented steels under tensile and compressive stress. *J. Magn. Magn. Mater.* **2000**, *215*, 124–126. [CrossRef]
12. Chen, J.; Wang, D.; Cheng, S.; Wang, Y. Measurement and modeling of temperature effects on magnetic property of non-oriented silicon steel lamination. In Proceedings of the 2015 IEEE International Magnetics Conference (INTERMAG), Beijing, China, 11–15 May 2015.
13. Chen, J.; Wang, D.; Cheng, S.; Wang, Y.; Zhu, Y.; Liu, Q. Modeling of temperature effects on magnetic property of nonoriented silicon steel lamination. *IEEE Trans. Magn.* **2015**, *51*, 1–4. [CrossRef]
14. Chen, J.; Wang, D.; Jiang, Y.; Teng, X.; Cheng, S.; Hu, J. Examination of Temperature-Dependent Iron Loss Models Using a Stator Core. *IEEE Trans. Magn.* **2018**, *54*, 1–7. [CrossRef]
15. Negri, G.M.; Sadowski, N.; Batistela, N.J.; Leite, J.V.; Bastos, J.P. Magnetic aging effect losses on electrical steels. *IEEE Trans. Magn.* **2016**, *52*, 1–4. [CrossRef]
16. Krings, A.; Mousavi, S.A.; Wallmark, O.; Soulard, J. Temperature influence of NiFe steel laminations on the characteristics of small slotless permanent magnet machines. *IEEE Trans. Magn.* **2013**, *49*, 4064–4067. [CrossRef]
17. Xue, S.; Feng, J.; Guo, S.; Peng, J.; Chu, W.Q.; Zhu, Z.Q. A new iron loss model for temperature dependencies of hysteresis and eddy current losses in electrical machines. *IEEE Trans. Magn.* **2017**, *54*, 1–10. [CrossRef]
18. Xue, S.; Feng, J.; Guo, S.; Chen, Z.; Peng, J.; Chu, W.Q.; Huang, L.R.; Zhu, Z.Q. Iron loss model under DC bias flux density considering temperature influence. *IEEE Trans. Magn.* **2017**, *53*, 1–4. [CrossRef]
19. Hao, Y.W.; Wang, B. Research on temperature effect of intermediate frequency thin-gauge non-oriented silicon steel. *Electr. Steel* **2020**, *2*, 8–10.
20. Zhang, S.F.; Shen, Z.J.; Li, S.T.; Wu, Y. Effect of different ambient temperatures on magnetic properties of high grade non-oriented silicon steel. *Electr. Mater.* **2018**, 16–18+24. [CrossRef]
21. Xia, X.L.; Wang, L.T.; Pei, Y.H.; Shi, L.F.; Zhang, Z.H. Effect of ambient temperature on mechanical properties of high grade non-oriented silicon steel. *Met. Funct. Mater.* **2017**, *24*, 50–53.
22. Kong, Q.Y.; Cheng, Z.G.; Li, Y.N. Magnetic properties of oriented silicon steel sheet under different ambient temperatures. *High Volt. Eng.* **2014**, *40*, 2743–2749.

Article

Correlation of Magnetomechanical Coupling and Damping in Fe$_{80}$Si$_9$B$_{11}$ Metallic Glass Ribbons

Xu Zhang [1,2,3,*], Yu Sun [1,2,4], Bin Yan [1,2] and Xin Zhuang [1,2,4,*]

1. Aerospace Information Research Institute, Chinese Academy of Sciences, Beijing 100190, China
2. Key Laboratory of Electromagnetic Radiation and Sensing Technology, Chinese Academy of Sciences, Beijing 100190, China
3. School of Physical Science and Technology, Inner Mongolia University, Hohhot 010021, China
4. University of Chinese Academy of Sciences, Beijing 100049, China
* Correspondence: zhangxu2019@imu.edu.cn (X.Z.); zhuangxin@aircas.ac.cn (X.Z.)

Abstract: Understanding the correlation between magnetomechanical coupling factors (k) and damping factors (Q^{-1}) is a key pathway toward enhancing the magnetomechanical power conversion efficiency in laminated magnetoelectric (ME) composites by manipulating the magnetic and mechanical properties of Fe-based amorphous metals through engineering. The k and Q^{-1} factors of FeSiB amorphous ribbons annealed in air at different temperatures are investigated. It is found that k and Q^{-1} factors are affected by both magnetic and elastic properties. The magnetic and elastic properties are characterized in terms of the magnetomechanical power efficiency for low-temperature annealing. The k and Q^{-1} of FeSiB-based epoxied laminates with different stacking numbers show that a −3 dB bandwidth and Young's modulus are expressed in terms of the magnetomechanical power efficiency for high lamination stacking.

Keywords: Fe-based amorphous alloys; magnetomechanical coupling; damping factor

---

Citation: Zhang, X.; Sun, Y.; Yan, B.; Zhuang, X. Correlation of Magnetomechanical Coupling and Damping in Fe$_{80}$Si$_9$B$_{11}$ Metallic Glass Ribbons. *Materials* **2023**, *16*, 4990. https://doi.org/10.3390/ma16144990

Academic Editor: Matteo Cantoni

Received: 11 May 2023
Revised: 1 July 2023
Accepted: 12 July 2023
Published: 14 July 2023

Copyright: © 2023 by the authors. Licensee MDPI, Basel, Switzerland. This article is an open access article distributed under the terms and conditions of the Creative Commons Attribution (CC BY) license (https://creativecommons.org/licenses/by/4.0/).

1. Introduction

In recent years, Fe-based amorphous alloys with significant magnetomechanical effects have been successfully used in laminated magnetoelectric (ME) composites, which usually consist of piezoelectric and magnetostrictive layers [1–5]. The ME energy coupling of laminated piezoelectric/magnetostrictive composites can be described as magnetic energy converted to mechanical energy and then to electric energy through the strain/stress coupling between these two types of materials [2]. When a magnetic field is applied to ME composites, their ferromagnetic layers shrink or stretch due to the magnetostrictive effect; then, the resulting strain/stress is transferred to the piezoelectric material, leading to a voltage change [2]. Devices based on laminated ME structures have shown promising potential for use in magnetic sensors, acoustically driven antennas and power conversion devices due to their high energy conversion efficiency [1–4,6–11].

As a key component of ME structures, Fe-based amorphous alloys should possess a relatively high conversion efficiency between magnetic power and mechanical power. To develop such materials, the requirement is to balance two or more parameters, as relevant parameters in materials usually impact one another to a significant degree [1]. However, keeping a parameter unchanged while improving other correlated parameters is normally quite a challenging task, even more so to balance two main parameters in some energy-converting materials. Specifically, the magnetomechanical power conversion efficiency (η), defined by the ratio between the converted power and the input power and that can be expressed as $\eta = \frac{1}{1+1/(k^2 Q)}$ [4,12], in magnetostrictive materials is based on two mutually exclusive, but related, parameters—the magnetomechanical coupling coefficient (k) and damping factor (Q^{-1}) [2,4]. The k factor quantifies the ratio of the magnetic energy that is converted into mechanical energy or vice versa over a period in magnetostrictive devices [13–15]. The Q factor, a reciprocal of the damping factor, is another important parameter that quantifies the ratio between the power storage and the power loss [12,14]. The two factors can be used to characterize the magnetomechanical properties in Fe-based amorphous alloys [4]. The k and Q factors accompany each other in most magnetomechanical transduction materials. To increase η, k^2Q needs to be maximized, although efforts to increase k often lead to a reduction in Q and vice versa. For example, the traditional annealing procedure usually causes a similar trend in changes to the k^2 and Q^{-1} factors; this implies a small η value [1,2]. To enhance the efficiency a step forward, and to minimize the inevitable correlation between the two factors, further clarity is needed on the correlation in variations between k and Q^{-1}.

Over recent decades, the influence of annealing on magnetic properties of Fe-based amorphous ribbons has been vastly studied [4,16–18]. It has been found that the annealing procedure can adjust the saturation magnetic flux density (B_s), coercivity (H_c), magnetic permeability (μ_r) and the k, Q and η factors, but variations to these parameters are rarely independent [1,4,8,11–13,19–35]. Thus, an opportunity has arisen for researchers to explore ways to cooperatively elevate two or more parameters while avoiding negatively impacting others [1–3].

In this study, the influence of heat treatment on the magnetomechanical properties for FeSiB amorphous ribbons is investigated. The correlation between the k and Q (or Q^{-1}) factors of such ribbons annealed at temperatures far below the crystallization temperature is discussed. The magnetic and magnetomechanical properties of single-foil ribbons and epoxy–ribbon laminates are also measured and compared in this work.

2. Experimental Method

The samples used in this work to perform the experiments were Fe-based amorphous ribbons with a nominal composition of $Fe_{80}Si_9B_{11}$ (at%). The length, width and thickness of a single ribbon were 40 mm, 5 mm and ~25 µm, respectively. The ribbons were annealed at different temperatures for 20 min in air using a muffle furnace. The annealed ribbons were inserted into a rectangular winding coil with an approximate size of 40 × 10 × 10 mm^3 and were tightly wound into six layers using 32-AWG magnetic wires in copper. When amorphous ribbons are driven by a magnetic excitation near a mechanical resonant frequency, their impedance (and inductive component) is extremely sensitive to the applied DC field [5,7,36,37]. The inductance and the impedance of the coils with inserted ribbons were measured with a high-precision impedance analyzer (HP 4294A). The resonant frequencies (f_r at maximum impedance) and antiresonant frequencies (f_a at minimum impedance) on the impedance spectrum were also measured with the impedance analyzer under a DC magnetic bias field (H_{dc}) along the longitudinal direction of the ribbons [6,38]. It is worth mentioning that the antiresonance occurred at a frequency above the actual mechanical resonance frequency when the impedance contribution from the ribbon minimized the total impedance [12]. The values of k were determined using resonant and antiresonant frequencies, while Q could be directly measured using the two poles (f_1 at maximum inductance, f_2 at minimum inductance) on the inductance spectrum [4,12,38]. From the frequencies at f_r, f_a, f_1 and f_2, the values of k and Q were calculated using the following dependences: $k = \sqrt{(\pi^2/8)(1 - f_r^2/f_a^2)}$ and $Q = f_r/(f_2 - f_1)$, where $\Delta f = f_2 - f_1$ also corresponds to the −3 dB bandwidth near the resonance [12]. A schematic diagram of the experiment is shown in Figure 1.

The magnetic permeability μ_r of the Fe-based amorphous ribbon was an order of 10^4–10^5 [39]. When cut into the rectangular shape, the apparent magnetic permeability (μ_{app}) along the long axis of the ribbon was several hundred due to the demagnetization field [40]. Furthermore, because the inductance of the winding coil was almost constant for a fixed frequency, this caused the measured inductance (L) values of the winding coil with an inserted Fe-based ribbon to increase with the increase in μ_{app}, i.e., $L \propto \mu_{app}$. The values of Young's modulus (E) for the ribbons with lengths matching the coils' should have been approximately proportional to the square of the antiresonant frequency f_a or the square of the resonant frequency f_r [12,14,34], i.e., $E \propto f_a^2 (or\ f_r^2)$. The magnetic hysteresis loops of the ribbons annealed at different temperatures were measured using the vibrational sample measurement system (VSM, MicroSense EZ-7). The samples with dimensions of approximately $5 \times 5 \times 0.025$ mm^3 were prepared together in a furnace with the 40 mm long ribbons for the VSM measurements.

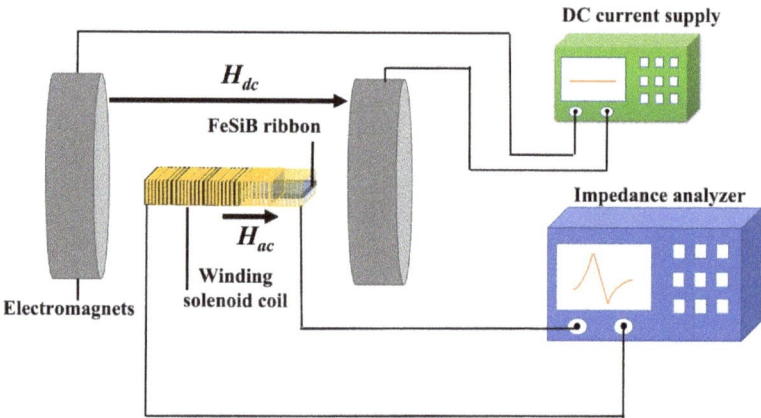

Figure 1. Schematic diagram of the experiment.

3. Theoretical Analysis and Measurement

3.1. Eddy Current Loss

The equivalent input loss factor (ζ) is defined as the inverse of the product of k^2 and Q. It quantifies the ratio between the power conversion and the power loss in the magnetic-to-mechanical energy conversion. Following previous investigations [4,12,32,34,41], ζ could be written as

$$\zeta = \frac{1}{k^2 Q} = c_e \mu_0 \chi f_r + \frac{1}{k^2 Q_0}, \quad (1)$$

where c_e is the loss coefficient for the eddy current in an individual ribbon, χ is the magnetic susceptibility of the ribbons and Q_0 is the quality factor under magnetostriction-free conditions. From the right-hand side of Equation (1), the first term c_e represents the energy loss for the dynamic magnetization procedure, which was affected by the magnetic properties of the ribbon; the second part represents loss not related to the eddy current. In single ribbons, the eddy current loss dominated after the heat treatment under an annealing temperature far below the crystallization temperature. Following Herzer et al. [42], the formula for c_e was expressed as

$$c_e = \frac{(\pi t)^2}{6 \rho_{el}} \left(1 + \frac{w^2}{(w \cos\beta + t)^2} \frac{m^2}{1-m^2} \right), \quad (2)$$

where t is the ribbon thickness, ρ_{el} is the electric resistivity and β is the angle between the average magnetic anisotropy and the ribbon direction. $m = J_H / J_s$ is the average longitudinal magnetization (J_H) normalized to the saturation magnetization (J_s). w is the magnetic domain width; the formula was given as [42]

$$w^2 = \frac{8 L_{ex} t}{N_{zz} \sin^2\beta + N_{yy} \cos^2\beta}, \quad (3)$$

where $L_{ex} = \sqrt{A/K}$ is the magnetic exchange length. A is defined as the exchange stiffness and K is the anisotropy constant. N_{zz} and N_{yy} are the demagnetization factors along the longitudinal and width directions of the ribbon, respectively. With the help of Equations (2) and (3), the contribution of the eddy current in Equation (1) could be rewritten as

$$\xi_{eddy} = \frac{(\pi t)^2}{6\rho_{el}}\mu_0\chi f_r + \frac{(\pi t)^2}{6\rho_{el}}\frac{m^2}{1-m^2}\frac{w^2}{(w\cos\beta+t)^2}\mu_0\chi f_r. \quad (4)$$

Taking into account that the average anisotropy angle β was close to zero for ribbons annealed at low temperatures, Equation (4) was rewritten as

$$\xi_{eddy} = \frac{(\pi t)^2}{6\rho_{el}}\mu_0\chi f_r + \frac{\pi^2}{6\rho_{el}}\frac{m^2}{1-m^2}\left(\frac{1}{t}+\frac{1}{w}\right)^{-2}\mu_0\chi f_r. \quad (5)$$

Since the magnetic domain width was usually much larger than the ribbon thickness in low-temperature annealed samples, the assumption $w \gg t$ could be considered in the calculation. Thus, Equation (5) could be rewritten as

$$\xi_{eddy} = \frac{1}{k^2 Q} = \frac{(\pi t)^2}{6\rho_{el}}\left(1+\frac{m^2}{1-m^2}\right)\mu_0\chi f_r. \quad (6)$$

The equivalent loss factor induced by the eddy current loss for the ribbons annealed at low temperature was dominated by the variation in χ and f_r for the ribbons.

3.2. Magnetic and Magnetomechanical Properties

It was assumed that the hysteresis loops of the magnetic materials could provide important information on the magnetic properties of the materials. Figure 2a shows the DC magnetic hysteresis loops of the samples annealed in air at different annealing temperatures (T_{AN}) for 20 min. The loops showed small values of remnant magnetization and H_c for all the samples, suggesting that the samples retained good soft magnetic properties after the heat treatment in air. As shown in the inset in Figure 2a, the values of H_c mostly stayed close to 0.1 Oe for T_{AN} = 370 °C to 450 °C. However, H_c began to grow sharply when T_{AN} exceeded 450 °C. In this case, the change in the chemical concentration causing the ordered clusters and the subsequent topological and chemical long-range orderings on the surfaces of the ribbons resulted in variations in the values of H_c [28,30,31,33,41,43]. When T_{AN} was above 500 °C, H_c showed a high increase, which suggested further deterioration in the soft magnetic properties. This was due to (1) an increase in the fraction of the surface crystallization at high T_{AN} and (2) the film of boron oxides that formed due to the excessive presence of boron atoms that were separated from the α-Fe crystallites, since the α-Fe crystallites had a much lower solubility effect on B than that of amorphous Fe [4,22,27,28,30,33]. Consequently, the magnetic domains in the amorphous remainders were thinned and turned toward the out-of-plane direction through compression stress induced by the surface crystallization and surface oxidation films [4,22,27,28,30,33], leading to an increasing in H_c.

In our previous work [4], we reported on the T_{AN} dependency of k and Q for single FeSiB ribbons; the value of k reached its maximum at approximately a T_{AN} of 430 °C, while Q showed a minimal value at approximately a T_{AN} of 410 °C. The k^2 and Q^{-1} evolutions after annealing at various temperatures T_{AN} are shown in Figure 2b. In the low T_{AN} region, below 400 °C, with increasing T_{AN}, both k^2 and Q^{-1} increased at almost the same pace, indicating that k^2 and Q^{-1} were correlated in the low annealing temperature region. For higher T_{AN}, from 400 °C to 500 °C, the variation in k^2 and Q^{-1} diverged, suggesting that the correlation level of the two parameters dropped significantly. Furthermore, as a result of the collective changes in k^2 and Q^{-1} for low T_{AN}, below 400 °C, the equivalent input loss ξ should have stayed invariable for low T_{AN}, below 400 °C, as reported by our earlier investigations [4].

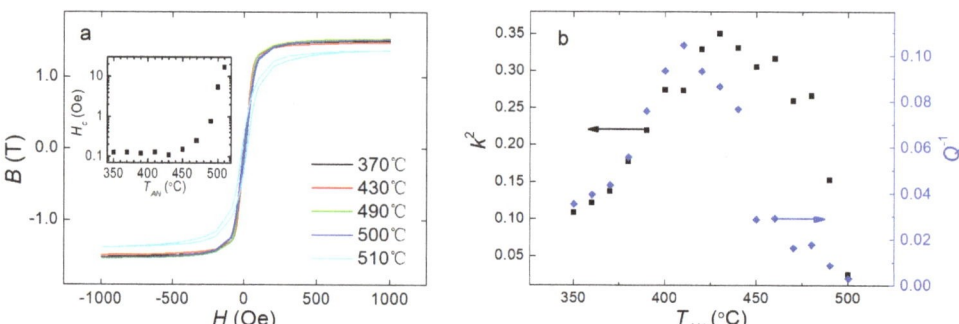

Figure 2. (**a**) The hysteresis loops of FeSiB amorphous ribbons obtained after annealed at different temperatures, from 370 °C to 510 °C, for 20 min in air; the inset figure is the annealing temperature (T_{AN}) dependence of coercivity H_c. (**b**) Damping factors (Q^{-1}) and the square of magnetomechanical coupling (k) as a function of T_{AN} for 20 min in air. k was acquired at its maximal values with external DC magnetic bias field H_{dc}. The values of Q^{-1} were measured under the same H_{dc} when k was maximized.

3.3. Softening of Magnetic and Elastic Properties

Figure 3 shows the measured values of the inductance (L) at 10 kHz after isothermal annealing at various temperatures T_{AN}. It could be observed that L increased with T_{AN} in the region from 350 °C to 410 °C and then decreased for T_{AN} from 420 °C to 490 °C. As mentioned above in the Section 2, the variation in L values suggested a collective change in the apparent magnetic permeability (μ_{app}) of the ribbon [44], i.e., $L \propto \mu_{app}$.

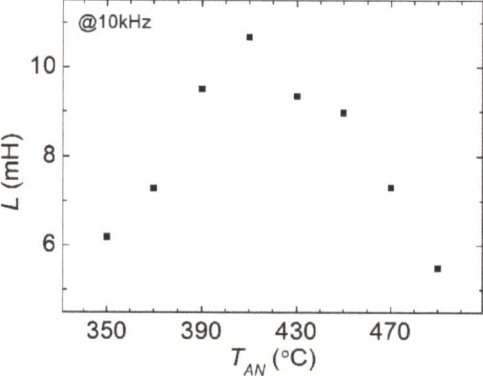

Figure 3. Variations in inductance (L) of the winding coil with an inserted FeSiB ribbon annealed for 20 min at different annealing temperatures T_{AN}, from 350 °C to 490 °C.

Figure 4 shows the T_{AN} dependence of resonant f_r and antiresonant f_a. It could be stated that in the low T_{AN} region, both f_r and f_a decreased with the increase in T_{AN}, reaching a minimal value at 410 °C. Then, they rose as T_{AN} continued to increase. The values of Young's modulus (E) in the ribbons with lengths matching those of the coils should have been approximately proportional to the square of *the* antiresonant frequency f_a or the square of the resonant frequency f_r [12,14,34], i.e., $E \propto f_a^2 (or\ f_r^2)$. By comparing the behavior of L and f_a (f_r) versus T_{AN}, it could be seen that the two parameters exhibited a respective extreme value as the function of T_{AN}. However, the curves of L and f_a (f_r) showed a lag of 20 °C versus T_{AN} to reach their maximum/minimum values.

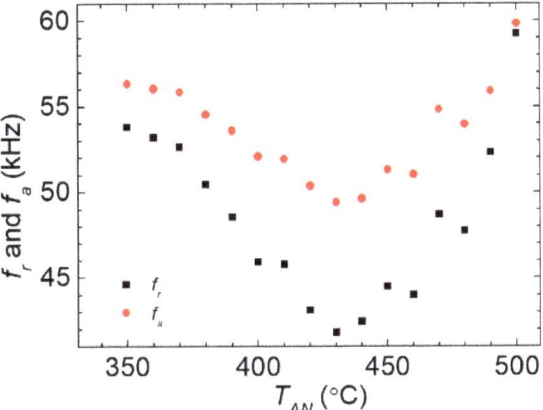

Figure 4. The variation in the resonant frequencies (f_r) and antiresonant frequencies (f_a) of the winding coil with an inserted FeSiB ribbon annealed for 20 min at different annealing temperatures T_{AN}, from 350 °C to 500 °C. The f_r and f_a on the motion impedance curve were measured with the impedance analyzer under a DC magnetic bias field (H_{dc}) along the longitudinal direction of the ribbons.

By comparing the T_{AN} dependence of L with the T_{AN} dependence of the Q^{-1} factors, it could be observed that the variations in L and Q^{-1} with the increase in T_{AN} were almost the same below 400 °C; f_a (f_r) had a similar profile to the k factor. According to Equation (6), the trends of k and Q were recognized as the competition between the magnetic and mechanical properties, χ and f_a (f_r). Since the L curve and Q^{-1} curve shared the same trend and the same extreme value point of T_{AN} below 410 °C while the f_a (f_r) curve and the k curve showed an inverse trend and the same extreme value point of T_{AN} below 430 °C, it was probable that L or μ_{app} were dominant in the values of the Q^{-1} factors, while f_a (f_r) or E had a dominative effect on the k factor for the low T_{AN} below 410 °C (for L and Q) or 430 °C (for f_a and k). Moreover, because k and E were related to each other, the annealing experience through α relaxation had a significant influence on the softening of elasticity below 430 °C [4]. For higher T_{AN}, because of the emergence of the surface crystallization and the long-range chemical orderings with B oxidation during the annealing procedure, the competition of magnetism and elasticity became much more complex, leading to the divergence between the k^2 curve and Q curves, as well as between the f_a (f_r) curve and the L curve versus T_{AN} [4,28,30,31,33,35,41].

Below 410 °C-T_{AN}, the β relaxation in the Fe-deficient zones was dominant and led to the chemical short-range ordering (CSRO) in Fe-rich zones with doped B atoms through diffusion [4]. In addition, the topological short-range ordering (TSRO) also occurred at this low T_{AN} region as another result of β relaxation, leading to the release of internal stress, both of which weakened the pinning effect of the magnetic domain through defects in the sample [4]. The expanding speed of the magnetic domain increased, therefore, i.e., L (μ_{app}) increased. For T_{AN}, from 450 °C to 500 °C, the sharp decrease in L (μ_{app}) may have occurred due to the emergence of surface crystallization together with more severe surface oxidation (such as Fe-O, B-O, silica) that led to the appearance of the grain boundary, as well as the magnetic anisotropy deviating from the long axis direction, both of which inhibited the movement or rotation of the magnetic moment of the magnetic domain, resulting in the smaller variation rate of the magnetic moment. The increase in H_c above 450 °C also coincided with the decrease in L (μ_{app}), suggesting the deterioration of the soft magnetic properties in the high T_{AN} region.

The decrease in E in the low T_{AN} region suggested that the deformation quantity of the ribbon decreased with the increasing T_{AN}. According to previous studies [4,35], as a result of β relaxation for low T_{AN}, besides the increase in the Fe–Fe bond caused by the CSRO, the TSRO could also be triggered through β relaxation close to the cluster–matrix boundaries.

The TSRO initiated a decrease in the ductility of the sample; thus, the sample became "more flexible" when stretched. Consequently, E (f_a and f_r) decreased after annealing at a relatively low temperature, and reached its minimum at 430 °C, as shown in Figure 4. The E (f_a and f_r) vs. T_{AN} curve switched to an increasing trend when T_{AN} exceeded 430 °C. This could be attributed to the apparency of α relaxation, which was more intense and usually occurred at a high T_{AN}. This α relaxation further enhanced the diffusion of the atoms, affecting the atomic orderings in a larger scale and usually reshaping the clusters. Fe or metalloid atoms experiencing long-term relaxation in the B-rich area led to the generation of some clusters with high elasticity, which increased the values of E. In addition, because the annealing procedure in this article was performed in air, the oxidation was stronger and penetrated deeper into the ribbons at a high T_{AN}; this also had a strong influence on the rigidity of the ribbon [4,20].

From Figures 3 and 4, it could be obtained that the variation trends in magnetic susceptibility χ (χ was also approximately proportional to L in Figure 3) and resonant (antiresonant) frequency f_r (f_a) were quite similar, but opposite for low T_{AN} from 350 °C to 400 °C; the product of χ and f_r (f_a) remained close to the constant. Moreover, the measured η, relating to k^2Q, was nearly a constant value in this region of T_{AN}, following our previous investigations [4,8]. Therefore, according to Equation (1), it could be inferred that the loss factor c_e that represented the eddy current loss should have been mostly unchanged for low T_{AN} from 350 °C to 400 °C for the single FeSiB ribbons. This was consistent with our analysis in the theoretical section as well.

When T_{AN} approached the crystallization temperature (T_x) of the FeSiB glassy metals, the size of the magnetic domain decreased due to the change in surface stress [4,31,33]. According to Herzer et al. [4,12,32,34,41,43], the loss factor c_e is related to the width of the magnetic domain w, following Equation (2). Therefore, the collectively changing behavior of k^2 and Q^{-1} suggested that the improvement in the soft magnetic properties may have derived from the increase in the number of activated magnetic units rather than variations in the width of the magnetic domains in the low T_{AN} region. Thus, the values of ξ for the ribbons annealed at low T_{AN} remaining constant had more to do with the increase in the quantity of magnetic units rather than the change in the domain size.

3.4. Time–Temperature Equivalence for Annealing

Figure 5a,b display the variations in the k, k^2Q and Q factors for different annealing times t_{AN} at T_{AN} ranging from 470 °C to 490 °C, respectively. A significant decrease in the k factor with the increase in the Q factor occurred with the increase in t_{AN} for a T_{AN} of 470 °C and 490 °C. The curve for k and Q showed a cross profile at t_{AN} = 40 min and 15 min, respectively. After the ribbons were annealed for 40 min at T_{AN} = 470 °C and 15 min at T_{AN} = 490 °C, k decreased from approximately 60% to 40%, while the values for Q increased close to 200. The equivalent input loss factors ξ showed a maximum value when the t_{AN} was 20 min for both samples, which suggested the optimal annealing time (20 min) for annealing at 470 °C and 490 °C. Based on the data mentioned above, it was implied that the decrease in k was caused by the surface oxidation and surface crystallization that induced the increase in coercivity through the out-of-plane magnetic anisotropy. The increase in Q factors might have been due to the long-range orderings that took place on the surface regions of the ribbons.

The Q factor represents the quality of mechanical performance in our FeSiB ribbons; a high value in the Q factor indicates a high ratio between the storage power and power loss. Figure 5a,b show that the overall trend of the k, k^2Q and Q curves at 470 °C T_{AN} was similar to that at 490 °C T_{AN}, separately. Moreover, the increase in T_{AN} from 470 °C to 490 °C narrowed the T_{AN} window before the magnetomechanical performance visibly deteriorated. In terms of the η factors, it was similar to the increase in T_{AN} for a constant t_{AN} or to increase in t_{AN} at a fixed T_{AN}, which suggested that T_{AN} and t_{AN} had an equal effect on the η factor in the heat treatment procedure for the FeSiB ribbons to some extent. However, the equivalence of t_{AN} and T_{AN} in terms of the magnetomechanical power conversion efficiency seemed to exist only at relatively higher T_{AN}; we did not observe this equivalent with T_{AN} below 400 °C.

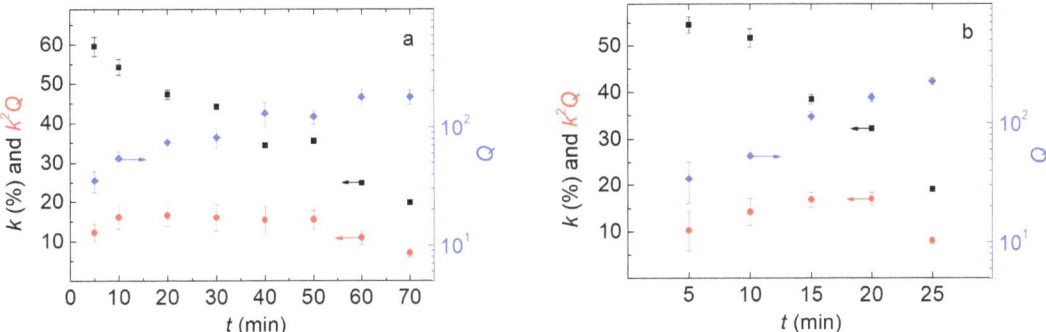

Figure 5. The magnetomechanical coupling coefficient (k) for FeSiB ribbons as a function of the annealing time (t_{AN}) at annealing temperature of (**a**) 470 °C and (**b**) 490 °C. The blue and red curves represent the values of the quality factors (Q) and efficiency factors (k^2Q) associated with the maximum values of coupling coefficient k, respectively.

3.5. Magnetomechanical Properties in Epoxy–Ribbon Composites

A schematic diagram of a laminated composite consisting of magnetostrictive amorphous FeSiB ribbons bonded with epoxy resin is given in Figure 6a. Several FeSiB foils were fabricated using hot-pressing techniques and an A–B part epoxy. The ratio between the epoxy (A part) and the curing agent (B part) should be 3:1, but in this case the curing part is a little bit less than the should-be value. Figure 6b is a photo of the laminated FeSiB composites; we applied a DC magnetic field along the laminated composite, as shown in the upper part of Figure 6b. The ratio between the quantity of the magnetostrictive layers and the quantity of epoxy resin varied during the investigation of the changes in k, Q and k^2Q in the FeSiB laminated composites with different foil numbers. The results are given in Figure 7a. It was found that as the number of FeSiB layers increased, the k factors increased slightly and then decreased rapidly, while the Q factors showed an overall increasing trend to a maximum close to 400, corresponding to a foil number from 18 to 21 in the FeSiB laminates. The efficiency factor k^2Q first increased in the presence of more FeSiB ribbon layers and then reached a relatively stable value at approximately 25, corresponding to an FeSiB foil number from 12 to 21. Figure 7b shows the trend of the −3 dB bandwidth (Δf) and E for the FeSiB laminated composites with different foil numbers.

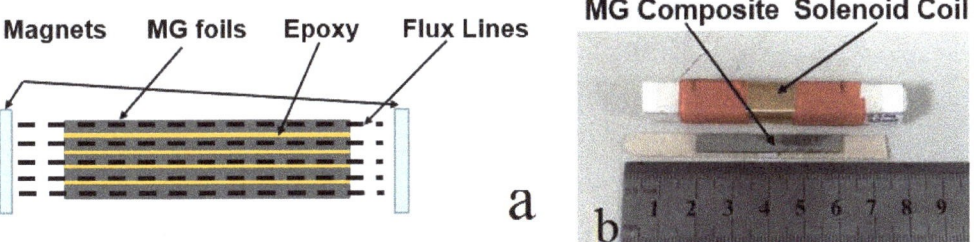

Figure 6. (**a**) Schematic diagram of a laminated composite consisting of magnetostrictive amorphous FeSiB ribbons bonded with epoxy resin. (**b**) Photo of laminated metallic glass (FeSiB) ribbon composites and the winding coil used to generate the AC magnetic field.

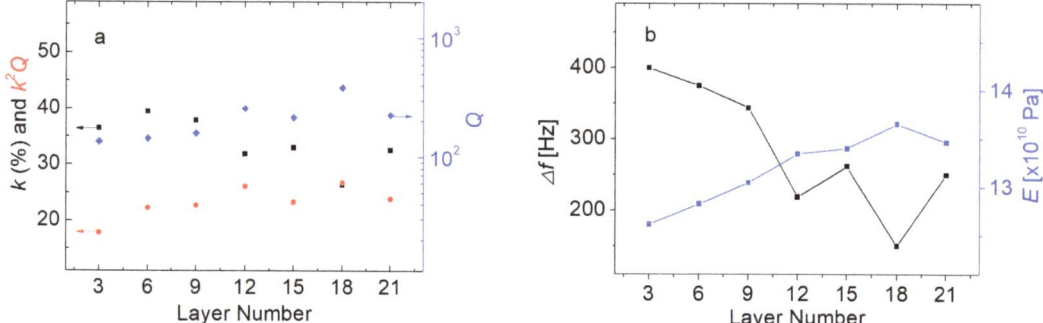

Figure 7. (a) The black curve is the magnetomechanical coupling factor (k) for laminated metallic glass composites consisting of magnetostrictive amorphous FeSiB ribbons as a function of the FeSiB layer number. The blue and red curves represent the values of the quality factors (Q) and efficiency coefficients (k^2Q) associated with the maximum values of coupling coefficient k, respectively. (b) The trend of the −3 dB bandwidth (Δf) and Young's modulus (E) for laminated FeSiB ribbon composites as a function of the FeSiB layer number, where $\Delta f = f_2 - f_1$ (f_1 at maximum inductance, f_2 at minimum inductance).

In Figure 7a, it was observed that when the foil number was small, there was no obvious similarity or correlation between the k and Q factors as the FeSiB layer number varied. The k and Q in the laminated composites with nine layers, six layers and three layers of FeSiB ribbons were examples. However, for the laminated composites with more foil number of 12 layers, 15 layers, 18 layers and 21 layers, the data suggest that there was a mutually exclusive relationship between the k and Q factors. This caused the k^2Q factors to reach a relatively stable state, while the k^2Q factors did not continue to increase with the increase in FeSiB layers. For the record, we needed to particularly emphasize that the ratio between the epoxy and curing agents was *not* 3:1. The amount of curing agent was reduced, giving a ratio of more than 3:1 between the epoxy and the curing agent, so the epoxy resin was not fully solidified for the laminated samples. Thus, the mechanical loss due to the interfriction increased compared to the samples with fully cured epoxy. Unlike the eddy current loss in the single-foil ribbons, the dominant loss in the FeSiB laminates was ascribed to the mechanical loss that triggered the temperature rise when the laminated composites were driven under high-power conditions.

According to previous research [4,31], as T_{AN} approached T_x, the size of the magnetic domain decreased due to the change in surface stress. As mentioned above, the efficiency of a single-foil ribbon experienced a low T_{AN} that remained constant, and this was more due to the increase in the number of magnetic units rather than the change in the magnetic domain size. In contrast, the increase in the k^2Q factors at T_{AN} above 450 °C was due to the reduction in magnetic domain size. Following Equation (1), because the k^2Q factors of the laminated FeSiB composites were approximately constant for a high layer number of laminates, and because the variation trend of E (associated with f_r and f_a) and the bandwidth (associated with χ) of the laminated FeSiB composites remained consistent with each other for all the foil numbers, the c_e of the laminated composites should have also remained unchanged. As the foil number of the FeSiB laminates decreased, the volume proportion of magnetostrictive materials (FeSiB) in the laminated composite gradually increased. Using a similar analysis principle to Figure 7a, the reason for the constant values in the k^2Q factors for the ribbons with foil numbers ranging from 12 to 21 was probably due to the increase in the number of ribbons rather than the change in the relative fraction of the ribbons. The reason for the change in the k^2Q factors for the laminates with foil numbers ranging from 3 to 12 was due to the variation in the relative volume fraction of FeSiB ribbons rather than a variation in the ribbon number. This indicated that the change in the k^2Q factors between the laminates with ribbon numbers ranging from 3 to 12 derived from the relative size of the magnetic units, which probably corresponded to the relative volume fraction of the FeSiB ribbons in the laminates, while the constant k^2Q factors between the 12-layer and 21-layer laminates might have arisen from the change in the number of magnetic units rather than the change in the relative volume of magnetic units [45].

4. Conclusions

In summary, we investigated the influence of magnetic and elastic properties on magnetomechanical coupling factors (k) and damping factors (Q^{-1}) by measuring the resonant and antiresonant frequencies from the motion impedance and inductance spectrum of a winding coil with an annealed FeSiB ribbon. We could draw the following conclusions:

1. Through annealing under different temperatures, it was found that the dynamic magnetic and elastic properties in the $Fe_{80}Si_9B_{11}$ ribbons varied correlatively with the annealing temperature. However, the resonant frequency reached its minimum value at 430 °C, with a lag of 20 °C, when the magnetic parameter reached its minimum at 410 °C, coinciding with the behaviors of k and Q^{-1} when the annealing temperature changed.

2. It was found that the annealing temperature and annealing time equally impacted the heat treatment procedure for the FeSiB ribbons in maximizing the magnetomechanical power efficiency. However, the equivalence was correct only when the annealing temperature was high enough for the long-range orderings to occur on the surface region of the ribbons.

3. For the FeSiB laminated composites, when the foil number ranged from 12 to 21, the trends of k and Q were synchronously related but opposite. This suggested that the behavior of the k and Q factors in the laminates with foil numbers ranging from 12 to 21 was at some point in analogy with the behaviors of the k and Q^{-1} factors in the single-foil FeSiB ribbons with various annealing temperatures ranging from 350 °C to 400 °C.

Author Contributions: Conceptualization, X.Z. (Xin Zhuang); methodology, X.Z. (Xin Zhuang) and X.Z. (Xu Zhang); software, X.Z. (Xin Zhuang) and X.Z. (Xu Zhang); validation, X.Z. (Xin Zhuang) and X.Z. (Xu Zhang); formal analysis, X.Z. (Xin Zhuang) and X.Z. (Xu Zhang); investigation, X.Z. (Xin Zhuang) and X.Z. (Xu Zhang); resources, X.Z. (Xin Zhuang) and X.Z. (Xu Zhang); data curation, X.Z. (Xin Zhuang) and X.Z. (Xu Zhang); writing—original draft preparation, X.Z. (Xu Zhang); writing—review and editing, X.Z. (Xin Zhuang), X.Z. (Xu Zhang), Y.S. and B.Y.; visualization, X.Z. (Xin Zhuang) and X.Z. (Xu Zhang); supervision, X.Z. (Xin Zhuang); project administration, X.Z. (Xin Zhuang) and X.Z. (Xu Zhang); funding acquisition, X.Z. (Xin Zhuang) and X.Z. (Xu Zhang). All authors have read and agreed to the published version of the manuscript.

Funding: This research was funded by the National Key R & D Program of China, no. 2021YFA0716500, the Natural Science Fund of Inner Mongolia Autonomous Region, no. 2021BS05010, and the talent program with the Chinese Academy of Sciences.

Institutional Review Board Statement: Not applicable.

Informed Consent Statement: Not applicable.

Data Availability Statement: Not applicable.

Conflicts of Interest: The authors declare no conflict of interest.

References

1. Channagoudra, G.; Dayal, V. Magnetoelectric Coupling in Ferromagnetic/Ferroelectric Heterostructures: A Survey and Perspective. *J. Alloys Compd.* **2022**, *928*, 167181. [CrossRef]
2. Leung, C.M.; Li, J.; Viehland, D.; Zhuang, X. A Review on Applications of Magnetoelectric Composites: From Heterostructural Uncooled Magnetic Sensors, Energy Harvesters to Highly Efficient Power Converters. *J. Phys. D Appl. Phys.* **2018**, *51*, 263002. [CrossRef]
3. Narita, F.; Fox, M. A Review on Piezoelectric, Magnetostrictive, and Magnetoelectric Materials and Device Technologies for Energy Harvesting Applications. *Adv. Eng. Mater.* **2018**, *20*, 1700743. [CrossRef]
4. Zhuang, X.; Xu, X.; Zhang, X.; Sun, Y.; Yan, B.; Liu, L.; Lu, Y.; Zhu, W.; Fang, G. Tailoring the Magnetomechanical Power Efficiency of Metallic Glasses for Magneto-Electric Devices. *J. Appl. Phys.* **2022**, *132*, 104502. [CrossRef]

5. Gutiérrez, J.; Barandiarán, J.M.; Nielsen, O.V. Magnetoelastic Properties of Some Fe-Rich Fe-Co-Si-B Metallic Glasses. *Phys. Status Solid* **1989**, *111*, 279–283. [CrossRef]
6. Leung, C.M.; Zhuang, X.; Xu, J.; Li, J.; Zhang, J.; Srinivasan, G.; Viehland, D. Enhanced Tunability of Magneto-Impedance and Magneto-Capacitance in Annealed Metglas/PZT Magnetoelectric Composites. *AIP Adv.* **2017**, *8*, 055803. [CrossRef]
7. Stoyanov, P.G.; Grimes, C.A. A Remote Query Magnetostrictive Viscosity Sensor. *Sens. Actuators A Phys.* **2000**, *80*, 8–14. [CrossRef]
8. Sun, Y.; Zhang, X.; Wu, S.; Zhuang, X.; Yan, B.; Zhu, W.; Dolabdjian, C.; Fang, G. Magnetomechanical Properties of Fe-Si-B and Fe-Co-Si-B Metallic Glasses by Various Annealing Temperatures for Actuation Applications. *Sensors* **2023**, *23*, 299. [CrossRef]
9. Qiao, K.; Hu, F.; Zhang, H.; Yu, Z.; Liu, X.; Liang, Y.; Long, Y.; Wang, J.; Sun, J.; Zhao, T.; et al. Unipolar Electric-Field-Controlled Nonvolatile Multistate Magnetic Memory in FeRh/(001)PMN-PT Heterostructures over a Broad Temperature Span. *Sci. China Phys. Mech. Astron.* **2021**, *65*, 217511. [CrossRef]
10. Pal, M.; Srinivas, A.; Asthana, S. Enhanced Magneto-Electric Properties and Magnetodielectric Effect in Lead-Free (1-x)0.94Na$_{0.5}$Bi$_{0.5}$TiO$_3$-0.06BaTiO$_3$–x CoFe$_2$O$_4$ Particulate Composites. *J. Alloys Compd.* **2022**, *900*, 163487. [CrossRef]
11. Li, T.X.; Li, R.; Lin, Y.; Bu, F.; Li, J.; Li, K.; Hu, Z.; Ju, L. Study of Enhanced Magnetoelectric Coupling Behavior in Asymmetrical Bilayered Multiferroic Heterostructure with Two Resonance Modes. *J. Alloys Compd.* **2022**, *895*, 162674. [CrossRef]
12. Kaczkowski, Z. Magnetomechanical Properties of Rapidly Quenched Materials. *Mater. Sci. Eng. A* **1997**, *226–228*, 614–625. [CrossRef]
13. Kabacoff, L.T. Thermal, Magnetic, and Magnetomechanical Properties of Metglas 2605 S2 and S3. *J. Appl. Phys.* **1982**, *53*, 8098–8100. [CrossRef]
14. Savage, H.; Clark, A.; Powers, J. Magnetomechanical Coupling and ΔE Effect in Highly Magnetostrictive Rare Earth—Fe$_2$compounds. *IEEE Trans. Magn.* **1975**, *11*, 1355–1357. [CrossRef]
15. Kaczkowski, Z.; Vlasák, G.; Švec, P.; Duhaj, P. Magnetostriction and Magnetomechanical Coupling of Heat-Treated Fe54Ni20Nb3Cu1Si13B9 Metallic Glass. *Mater. Sci. Eng. A* **2004**, *375–377*, 1062–1064. [CrossRef]
16. Chen, H.S.; Leamy, H.J.; Barmatz, M. The Elastic and Anelastic Behavior of a Metallic Glass. *J. Non Cryst. Solids* **1971**, *5*, 444–448. [CrossRef]
17. Morito, N.; Egami, T. Correlation of the Shear Modulus and Internal Friction in the Reversible Structural Relaxation of a Glassy Metal. *J. Non Cryst. Solids* **1984**, *61–62*, 973–978. [CrossRef]
18. Soshiroda, T.; Koiwa, M.; Masumoto, T. The Internal Friction and Elastic Modulus of Amorphous Pd–Si and Fe–P–C Alloys. *J. Non Cryst. Solids* **1976**, *22*, 173–187. [CrossRef]
19. Xu, D.D.; Zhou, B.L.; Wang, Q.Q.; Zhou, J.; Yang, W.M.; Yuan, C.C.; Xue, L.; Fan, X.D.; Ma, L.Q.; Shen, B.L. Effects of Cr Addition on Thermal Stability, Soft Magnetic Properties and Corrosion Resistance of FeSiB Amorphous Alloys. *Corros. Sci.* **2018**, *138*, 20–27. [CrossRef]
20. Lin, J.; Li, X.; Zhou, S.; Zhang, Q.; Li, Z.; Wang, M.; Shi, G.; Wang, L.; Zhang, G. Effects of Heat Treatment in Air on Soft Magnetic Properties of FeCoSiBPC Amorphous Core. *J. Non Cryst. Solids* **2022**, *597*, 121932. [CrossRef]
21. Murugaiyan, P.; Mitra, A.; Jena, P.S.M.; Mahato, B.; Ghosh, M.; Roy, R.K.; Panda, A.K. Grain Refinement in Fe-Rich FeSiB(P)NbCu Nanocomposite Alloys through P Compositional Modulation. *Mater. Lett.* **2021**, *295*, 129852. [CrossRef]
22. Fang, Z.; Nagato, K.; Liu, S.; Sugita, N.; Nakao, M. Investigation into Surface Integrity and Magnetic Property of FeSiB Metallic Glass in Two-Dimensional Cutting. *J. Manuf. Process.* **2021**, *64*, 1098–1104. [CrossRef]
23. Wang, C.; Guo, Z.; Wang, J.; Sun, H.; Chen, D.; Chen, W.; Liu, X. Industry-Oriented Fe-Based Amorphous Soft Magnetic Composites with SiO$_2$-Coated Layer by One-Pot High-Efficient Synthesis Method. *J. Magn. Magn. Mater.* **2020**, *509*, 166924. [CrossRef]
24. Cheng, Y.Q.; Ma, E. Atomic-Level Structure and Structure-Property Relationship in Metallic Glasses. *Prog. Mater. Sci.* **2011**, *56*, 379–473. [CrossRef]
25. Czyż, O.; Kusiński, J.; Radziszewska, A.; Liao, Z.; Zschech, E.; Kąc, M.; Ostrowski, R. Study of Structure and Properties of Fe-Based Amorphous Ribbons after Pulsed Laser Interference Heating. *J. Mater. Eng. Perform.* **2020**, *29*, 6277–6285. [CrossRef]
26. Zhang, J.; Shan, G.; Li, J.; Wang, Y.; Shek, C.H. Structures and Physical Properties of Two Magnetic Fe-Based Metallic Glasses. *J. Alloys Compd.* **2018**, *747*, 636–639. [CrossRef]
27. Li, Y.; Chen, W.-Z.; Dong, B.-S.; Zhou, S.-X. Effects of Phosphorus and Carbon Content on the Surface Tension of FeSiBPC Glass-Forming Alloy Melts. *J. Non Cryst. Solids* **2018**, *496*, 13–17. [CrossRef]
28. Lopatina, E.; Soldatov, I.; Budinsky, V.; Marsilius, M.; Schultz, L.; Herzer, G.; Schäfer, R. Surface Crystallization and Magnetic Properties of Fe84.3Cu0.7Si4B8P3 Soft Magnetic Ribbons. *Acta Mater.* **2015**, *96*, 10–17. [CrossRef]
29. Ok, H.N.; Morrish, A.H. Origin of the Perpendicular Anisotropy in Amorphous Fe82B12Si6 Ribbons. *Phys. Rev. B* **1981**, *23*, 2257–2261. [CrossRef]
30. Gemperle, R.; Kraus, L.; Kroupa, F.; Schneider, J. Influence of Surface Oxidization on Induced Anisotropy of Amorphous (FeNi) PB Wires. *Phys. Status Solidi (a)* **1980**, *60*, 265–272. [CrossRef]
31. Ok, H.N.; Morrish, A.H. Amorphous-to-Crystalline Transformation of Fe82B12Si6. *Phys. Rev. B* **1980**, *22*, 3471–3480. [CrossRef]
32. Keupers, A.; De Schepper, L.; Knuyt, G.; Stals, L.M. Chemical and Topological Short Range Order in Amorphous Fe40Ni38Mo4B18 (Metglas 2826MB). *J. Non Cryst. Solids* **1985**, *72*, 267–278. [CrossRef]
33. Morito, N. Surface Crystallization Induced by Selective Oxidation of Boron in Fe-B-Si Amorphous Alloy. *Key Eng. Mater.* **1990**, *40–41*, 63–68. [CrossRef]

34. Herzer, G. Magnetomechanical Damping in Amorphous Ribbons with Uniaxial Anisotropy. *Mater. Sci. Eng. A* **1997**, *226–228*, 631–635. [CrossRef]
35. Tong, X.; Zhang, Y.; Wang, Y.; Liang, X.; Zhang, K.; Zhang, F.; Cai, Y.; Ke, H.; Wang, G.; Shen, J.; et al. Structural Origin of Magnetic Softening in a Fe-Based Amorphous Alloy upon Annealing. *J. Mater. Sci. Technol.* **2022**, *96*, 233–240. [CrossRef]
36. Herrero-Gómez, C.; Marín, P.; Hernando, A. Bias Free Magnetomechanical Coupling on Magnetic Microwires for Sensing Applications. *Appl. Phys. Lett.* **2013**, *103*, 142414. [CrossRef]
37. Hernando, A.; Madurga, V.; Barandiarán, J.M.; Liniers, M. Anomalous Eddy Currents in Magnetostrictive Amorphous Ferromagnets: A Large Contribution from Magnetoelastic Effects. *J. Magn. Magn. Mater.* **1982**, *28*, 109–116. [CrossRef]
38. Kaczkowski, Z.; Vlasák, G.; Švec, P.; Duhaj, P.; Ruuskanen, P.; Barandiarán, J.M.; Gutiérrez, J.; Minguez, P. Influence of Heat-Treatment on Magnetic, Magnetostrictive and Piezomagnetic Properties and Structure of $Fe_{64}Ni_{10}Nb_3Cu_1Si_{13}B_9$ Metallic Glass. *Mater. Sci. Eng. A* **2004**, *375–377*, 1065–1068. [CrossRef]
39. Herzer, G. Modern Soft Magnets: Amorphous and Nanocrystalline Materials. *Acta Mater.* **2013**, *61*, 718–734. [CrossRef]
40. Séran, H.C.; Fergeau, P. An Optimized Low-Frequency Three-Axis Search Coil Magnetometer for Space Research. *Rev. Sci. Instrum.* **2005**, *76*, 044502. [CrossRef]
41. Egami, T. Structural Relaxation in Amorphous $Fe_{40}Ni_{40}P_{14}B_6$ Studied by Energy Dispersive X-ray Diffraction. *J. Mater. Sci.* **1978**, *13*, 2587–2599. [CrossRef]
42. Herzer, G. Effect of Domain Size on the Magneto-Elastic Damping in Amorphous Ferromagnetic Metals. *Int. J. Mater. Res.* **2022**, *93*, 978–982. [CrossRef]
43. Egami, T. Structural Relaxation in Amorphous Alloys—Compositional Short Range Ordering. *Mater. Res. Bull.* **1978**, *13*, 557–562. [CrossRef]
44. Setiadi, R.N.; Schilling, M. Sideband Sensitivity of Fluxgate Sensors Theory and Experiment. *Sens. Actuators A Phys.* **2019**, *285*, 573–580. [CrossRef]
45. Hasegawa, R.; Ramanan, V.R.V.; Fish, G.E. Effects of Crystalline Precipitates on the Soft Magnetic Properties of Metallic Glasses. *J. Appl. Phys.* **1982**, *53*, 2276–2278. [CrossRef]

Disclaimer/Publisher's Note: The statements, opinions and data contained in all publications are solely those of the individual author(s) and contributor(s) and not of MDPI and/or the editor(s). MDPI and/or the editor(s) disclaim responsibility for any injury to people or property resulting from any ideas, methods, instructions or products referred to in the content.

Article

Investigation and Application of Magnetic Properties of Ultra-Thin Grain-Oriented Silicon Steel Sheets under Multi-Physical Field Coupling

Zhiye Li [1], Yuechao Ma [1], Anrui Hu [1], Lubin Zeng [2], Shibo Xu [1] and Ruilin Pei [1,*]

1 School of Electrical Engineering, Shenyang University of Technology, Shenyang 110807, China
2 Suzhou Inn-Mag New Energy Ltd., Suzhou 215000, China
* Correspondence: peiruilin@sut.edu.cn

Abstract: Nowadays, energy shortages and environmental pollution have received a lot of attention, which makes the electrification of transportation systems an inevitable trend. As the core part of an electrical driving system, the electrical machine faces the extreme challenge of keeping high power density and high efficiency output under complex workin g conditions. The development and research of new soft magnetic materials has an important impact to solve the current bottleneck problems of electrical machines. In this paper, the variation trend of magnetic properties of ultra-thin grain-oriented silicon steel electrical steel (GOES) under thermal-mechanical-electric-magnetic fields is studied, and the possibility of its application in motors is explored. The magnetic properties of grain-oriented silicon steel samples under different conditions were measured by the Epstein frame method and self-built multi-physical field device. It is verified that the magnetic properties of grain-oriented silicon steel selected within 30° magnetization deviation angle are better than non-grain-oriented silicon steel. The magnetic properties of the same ultra-thin grain-oriented silicon steel as ordinary non-oriented silicon steel deteriorate with the increase in frequency. Different from conventional non-grain-oriented silicon steel, its magnetic properties will deteriorate with the increase in temperature. Under the stress of 30 Mpa, the magnetic properties of the grain-oriented silicon steel are the best; under the coupling of multiple physical fields, the change trend of magnetic properties of grain-oriented silicon steel is similar to that of single physical field, but the specific quantitative values are different. Furthermore, the application of grain-oriented silicon steel in interior permanent magnet synchronous motor (IPM) for electric vehicles is explored. Through a precise oriented silicon steel motor model, it is proved that the magnetic flux density of stator teeth increases by 2.2%, the electromagnetic torque of motor increases by 2.18%, and the peak efficiency increases by 1% after using grain-oriented silicon steel. In this paper, through the investigation of the characteristics of grain-oriented silicon steel, it is preliminarily verified that grain-oriented silicon steel has a great application prospect in the drive motor (IPM) of electric vehicles, and it is an effective means to break the bottleneck of current motor design.

Keywords: grain-oriented silicon steel; magnetization angle; multi-physical field coupling; drive motor; torque density; flux density; core loss

Citation: Li, Z.; Ma, Y.; Hu, A.; Zeng, L.; Xu, S.; Pei, R. Investigation and Application of Magnetic Properties of Ultra-Thin Grain-Oriented Silicon Steel Sheets under Multi-Physical Field Coupling. *Materials* **2022**, *15*, 8522. https://doi.org/10.3390/ma15238522

Academic Editor: Shenglei Che

Received: 28 October 2022
Accepted: 23 November 2022
Published: 29 November 2022

Publisher's Note: MDPI stays neutral with regard to jurisdictional claims in published maps and institutional affiliations.

Copyright: © 2022 by the authors. Licensee MDPI, Basel, Switzerland. This article is an open access article distributed under the terms and conditions of the Creative Commons Attribution (CC BY) license (https://creativecommons.org/licenses/by/4.0/).

1. Introduction

At present, after more than two hundred years of development, electrical machine theory has reached a very mature stage. It is an effective method to use new soft magnetic materials to break through the bottleneck of electrical machine performance [1]. Grain-oriented silicon steel is widely used in transformers and large electrical machines. In recent years, with the strict requirements for the performance of the drive electrical machine, the application of grain-oriented silicon steel in the drive electrical machine has gradually developed [2–5]. In the axial-flux electrical machine, an electrical machine hub with axial flux is designed in the literature [6], which adopts yokeless and segmented armature (YASA) structure. In order to improve the magnetic flux density of stator tooth, the stator adopts grain-oriented silicon steel, which makes the motor have higher torque density, higher efficiency and larger torque at low speed, and the effect is remarkable. In order to alleviate the problem of high iron loss during overload operation, the axial laminated flux-switching permanent magnet electrical machine (ALFSPMM) is designed with grain-oriented silicon steel, and its excellent magnetic property in the rolling direction is used to improve its torque characteristics. However, this type of motor experiences difficulties in the assembly process [7,8]. For yokeless axial flux motors, the anisotropy of grain-oriented silicon steel can be used to improve the electrical machine performance, and the output torque is increased by 3%, meanwhile, the iron consumption is reduced by 10% [9]. In other reluctance electrical machines, the high permeability of grain-oriented silicon steel along the rolling direction is also used to match the torque characteristics of the reluctance electrical machine, and both peak torque and constant torque are improved by the stator tooth splicing method [3].

Radial flux electrical machines also favor using grain-oriented silicon steel, which is applied to stator teeth in a unique splicing method. In literature [10], the stator adopts splicing layered design, which divided each layer into six sections. Adjacent layers have a different rolling direction, and they were laminated interlaced. The magnetic field distribution and the electrical machine short circuit test show that its overall loss is reduced, but the electrical machine's performance still can be improved. In reference [11,12], orientated silicon steel is applied to the stator core and compared with the prototype by adopting five different electrical machine topologies. According to the simulation results, the distortion rate of the electrical machine back electromotive force decreases by 0.6% at least, the output torque increases by more than 2.3%, and the total stator losses decreases by 54.2% after using the grain-oriented silicon steel. The weight reduction is realized under the same output torque. In the fractional slot concentrated winding electrical machine, the segmented grain-oriented silicon steel stator structure is used to provide a more accurate modeling method to analyze the essential reasons for the torque improvement. By comparing the two-dimensional finite element simulation with the prototype, it is found that the electromagnetic torque under the same peak current is increased by 4%, and the torque fluctuation is significantly reduced from 1.6% to 0.6%. From the change in magnetic circuit parameters, the parasitic air gap generated by splicing has a great influence on the magnetic circuit of the motor. The current research is still in the simulation stage [13–15]. The hybrid steel splicing method was applied to traction electrical machine in literature [16,17]. Based on the comparison of the performance of grain-oriented silicon steel and non-grain-oriented silicon steel in the electrical machine, it was analyzed that under the same conditions, the maximum output torque of stator core with grain-oriented silicon steel splicing could reach 195 Nm (4.3% increase), and the rated efficiency could be increased by 2.7%. The peak efficiency can be increased by 1.5%. Similarly, in literature [18], the good performance of grain-oriented silicon steel splicing electrical machine was verified based on the grain characteristics of grain-oriented silicon steel with clear grain boundaries observed in the metallographic experiment and the comparative analysis of thermal simulation. With the gradual deepening of research, literature [19,20] began to study the influence of yoke splicing shapes on electrical machine performance, and pointed out that circular splicing was an optimal scheme, which reduced electrical machine iron loss by 36.5%. The unique Epstein measurement structure was used to separate and calculate the loss in the stator core, and the effects of different stitching shapes, stitching angles and the depth of the yoke on the performance of the electrical machine were further studied. The results show that using splicing teeth can significantly improve the torque under heavy load condition, but the modeling method still needs to be improved. In terms of finite element modeling, the accurate acquisition of material magnetic property parameters helps to improve the modeling accuracy, which, combined with the mathematical model, can effectively predict the magnetic properties of the interior permanent magnet motor, but failed when applied to the grain-oriented silicon steel motor [21].

Therefore, it is very important to accurately model and simulate the actual operating conditions of the electrical machine by using grain-oriented silicon steel in stator yoke splicing. In this paper, in order to analyze the possibility of application of grain-oriented silicon steel in IPM more accurately, the magnetic performance measurement under the coupling of thermal-mechanical-electric-magnetic multi-physical field and the performance comparison of two kinds of soft magnetic materials at different angles are carried out. In addition, a variety of properties of grain-oriented silicon steel motors are analyzed by using more refined multi-angle modeling.

2. Materials and Methods

Compared with conventional non-grain-oriented electrical steels, grain-oriented electrical steels exhibit excellent magnetic properties with low iron loss and high saturation flux density along the rolling direction (RD), but the magnetic properties deteriorate gradually as the magnetization direction deviates. In this paper, the properties of 0.08 mm grain-oriented silicon steel (GO) H-80 provided by Huaci Technologies Co., Ltd. (Shenzhen, China) and 0.3 mm non-grain-oriented silicon steel (NGO) provided by Angang Steel Company Limited (Anshan, China) are investigated in the background of electric vehicle drive motor application. Because of the complex service conditions, a multi-physical field non-standard ring sample method was designed to measure the magnetic properties of grain-oriented silicon steel under thermal-mechanical-electric-magnetic multi-physical fields. The magnetic properties of grain-oriented silicon steel at different angles were tested by Epstein's square circle method. The response of grain-oriented silicon steel to external physical fields was explored by comparison and analysis, and its variation law was revealed and applied to the motor.

2.1. Non-Standard Ring Sample Method

As shown in Figure 1, we present a schematic diagram of the whole system for the measurement of magnetic properties under single field action and thermal-mechanical-electric-magnetic multi-field coupling by non-standard ring sample method. Among them, Figure 1b shows the test sample of grain-oriented silicon steel. The primary and secondary windings both require 200 turns. The equipment in Figure 1a can apply different temperatures through the insulation thermostat device, and the sample can be fixed on the fixture to achieve stress loading. By connecting the primary and secondary winding to the variable frequency electromagnetic excitation source, the sample can obtain a specific frequency electromagnetic field. The device can realize single physical field loading and multi-physics coupling loading. Through this device, the material response state under the actual operating conditions of the motor under the multi-physical field coupling is simulated, and the magnetic performance change in the material under the multi-physical field coupling condition is explored, which provides a strong guiding basis for the optimal design of the subsequent motor. The thermal insulation thermostat can realize the temperature field change of $-75\sim200\ °C$, and the fixture can apply $0\sim200$ Mpa stress to the sample. Combined with the excitation current at different frequencies, the magnetic characteristics measurement under the coupling of thermal-mechanical-electric-magnetic four fields can be realized.

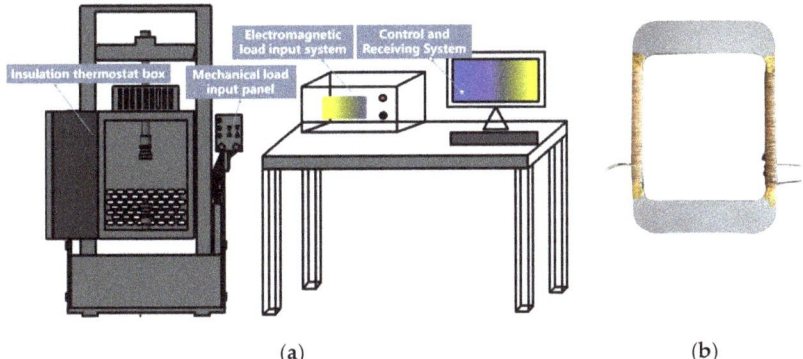

(a)   (b)

Figure 1. The overall schematic diagram of the multi-physics coupling test system. (**a**) Multi-physics coupling test system; (**b**) grain-oriented silicon steel stress test specimen.

2.2. Epstein Frame Method

In order to obtain the accurate magnetic properties of grain-oriented silicon steel with different magnetization angles, we use the Epstein square circle method specified by the International Electrotechnical Commission (IEC). The test conditions were 20 °C, 0 Mpa laboratory environment, and the magnetization characteristics of grain-oriented silicon steel at different magnetization angles were measured by the Epstein square circle method. As shown in Figure 2a, the Epstein frame measuring instrument consists of a primary winding, a secondary winding and a specimen as a core, which as a whole form a no-load transformer. The skeleton supporting the four coil windings consists of a hard insulating material. To verify the magnetization characteristics of grain-oriented silicon steel with different magnetization angles, the specimen is deviated from the rolling direction by a certain angle, and the shearing method is shown in Figure 2b, with α representing the different angles made with the shearing direction and the rolling direction. Taking the rolling direction, i.e., the magnetization direction or the deviation angle of 0° as an example, a group of specimens were measured through a set of coils in parallel direction, whose rolling direction was parallel to the magnetic circuit direction, and another group of specimens were perpendicular to the rolling direction. The final plotted magnetization and loss curves of the grain-oriented silicon steel at different angles is shown below.

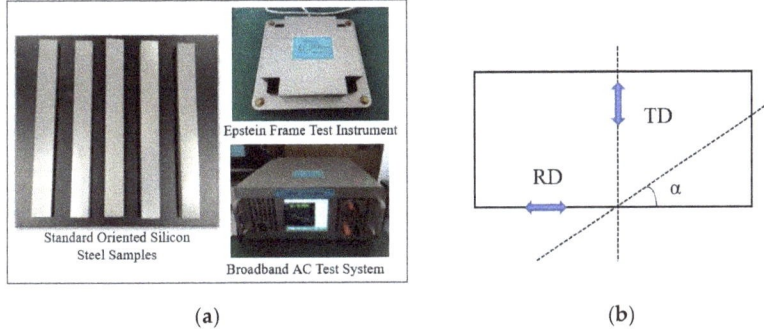

(a)   (b)

Figure 2. Test equipment and samples. (**a**) Overall schematic diagram of Epstein magnetic measurement system; (**b**) shearing of grain-oriented silicon steel sheet.

2.3. Segmented Modeling of Grain-Oriented Silicon Steel Electrical Machine

The current mainstream commercial finite element simulation software uses the magnetic permeability tensor method to model grain-oriented silicon steel. In the 2D simulation, an elliptical model is constructed as the magnetic field anisotropy curve of grain-oriented silicon steel by obtaining the rolling direction and transverse magnetization characteristic curves and using interpolation fitting. This method defaults to the transverse direction as the worst magnetization direction of the magnetic field of the grain-oriented silicon steel, however, it is known from the basic material science principle of the grain-oriented silicon steel that this method has a large error. In the 3D simulation calculation, the magnetic permeability of the normal direction is obtained on the basis of 2D to form the magnetic permeability matrix, and then mathematical interpolation is performed to calculate the mathematical model as follows [22]:

$$B = \begin{bmatrix} \mu_x & 0 & 0 \\ 0 & \mu_y & 0 \\ 0 & 0 & \mu_z \end{bmatrix} \times H \tag{1}$$

where H is the magnetic field intensity, B is the magnetic flux density, μ_x, μ_y, μ_z are the permeability of the rolling direction, transverse direction and normal direction, respectively.

In view of the inaccuracy of the current mathematical model of grain-oriented silicon steel, which will lead to the calculation error of the core and the poor guidance to the actual production, this paper uses the experimental data of the magnetic properties of grain-oriented silicon steel under multiple magnetization angles in the laboratory, and uses a splicing modeling method to improve the modeling accuracy of grain-oriented silicon steel segmented motor [13,23,24]. Because the stator uses a modular grain-oriented silicon steel punching sheet, the magnetic properties of grain-oriented silicon steel at different magnetization angles are very different. In order to reflect the material properties of different regions of the stator core more accurately, a segmented modeling method is used as shown in Figure 3. Among them, T0 is parallel to the rolling direction, and its magnetization angle is 0°. In this paper, the motor model is 48 slots and eight poles, and six slots form a module. The punching and installation of the stator core are shown in Figure 4. There is a difference of 7.5° between adjacent teeth. The magnetization angle of T_1 is 7.5°, and the magnetization angles of T_2 and T_3 are 15° and 22.5°. Referring to the modeling literature of grain-oriented silicon steel motor [13], the yoke magnetization angle is calculated by the average value of the inflow angle and outflow angle of the magnetic field line. The difference between Y_1 and T_0 is 90°, and the difference between Y_1 and T_1 is 82.5°. Therefore, the magnetization angle of Y_1 is 86.25°, and the magnetization angles of Y_2 and Y_3 are 78.75° and 71.25°, respectively.

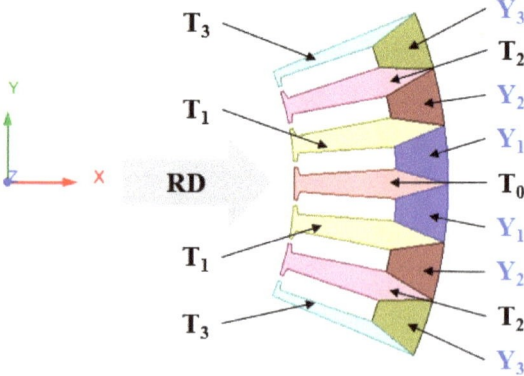

Figure 3. Schematic diagram of segmented modeling of grain-oriented silicon steel stator core.

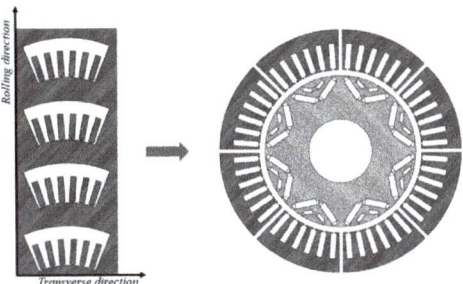

Figure 4. Example of GO laminations cutting to be installed in a 48 s/8 p IPM modular machine.

3. Results

3.1. Test Results of Non-Standard Ring Sample Method

3.1.1. Magnetic Characteristics of Grain-Oriented Silicon Steel under Temperature Field

Drive motors are located in transportation carriers such as electric vehicles, where the installation space is small and the ambient temperature is high, in addition to the electrical machine itself is a heat source. So, the effect of temperature on the magnetic properties of the core must be accurately predicted. The flux density and loss variation in grain-oriented silicon steel under different temperature is shown in Figure 5. The saturation magnetic flux density decreases significantly with the change in temperature in the test range, and the loss increases with a rising temperature. As can be seen in Figure 5a, the initial magnetic permeability is less affected by temperature, and the degradation of permeability accelerates with the enhancement of the magnetic field; in Figure 5b, it can be seen that, contrary to the pattern of ordinary non-grain-oriented silicon steel affected by temperature, the loss of ultra-thin grain-oriented silicon steel increases with the increase in temperature.

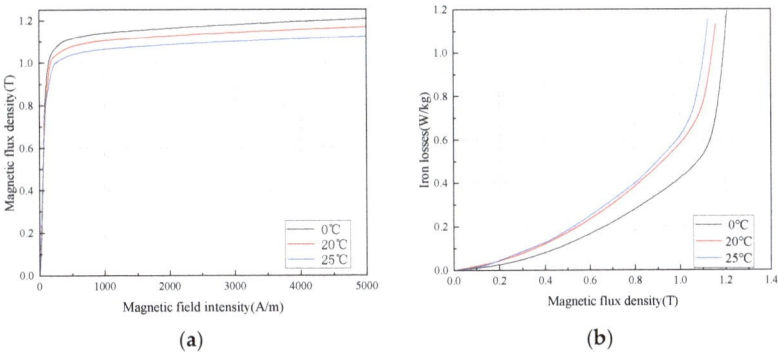

Figure 5. Magnetic properties of grain-oriented silicon steel at different temperature fields (**a**) magnetization curves of H-80 @ 50 Hz and 0 Mpa (**b**) iron loss curves of H-80 @50 Hz and 0 Mpa.

The magnetization process is that the magnetic moment of each magnetic domain formed by spontaneous magnetization in ferromagnetic materials is transferred to the direction of the external magnetic field or close to the direction of the external magnetic field by adding an external magnetic field, which shows the process of magnetism. Therefore, this process is also the change process of magnetic domain structure under the action of external magnetic field. The change in magnetic domain structure is carried out by two magnetization modes: domain wall movement and spontaneous magnetization vector rotation (domain rotation) in the magnetic domain. It will be subjected to resistance during domain wall movement or magnetic domain rotation. The source of this resistance includes thermal strain, so as the temperature increases, the resistance of the magnetic domain becomes larger, resulting in a decrease in the saturation magnetic flux density. The classical core loss model includes hysteresis loss and eddy current loss. The hysteresis loss is related to the grain size and is less affected by temperature. The eddy current loss is related to the resistivity. As the temperature increases, the resistivity becomes larger, and the eddy current loss becomes smaller theoretically. This is the reason the core loss of ordinary non-grain-oriented silicon steel decreases with increasing temperature. From Figure 5b, it can be found that the classical loss model is no longer applicable, so the core loss model needs to be corrected to:

$$P_{core-loss} = P_h + P_e + P_a \quad (2)$$

Among them, $P_{core\text{-}loss}$ is core loss, P_h is hysteresis loss, P_e is eddy current loss, P_a is anomalous loss. Experiments show that for grain-oriented silicon steel, the increase in temperature increases the sum of abnormal loss and eddy current loss, so the core loss increases overall.

3.1.2. Magnetic Characteristics of Grain-Oriented Silicon Steel under Stress Field

When the electrical machine is operating, the stator is fed with a three-phase alternating current to produce a rotating magnetic field, and the rotor is embedded in a permanent magnet to produce an excitation field. The forces are mutual and the stator core is subjected to the same magnitude of tangential force when the rotor is rotating. In addition, the magnetic field is a space vector, so there are also radial electromagnetic and axial forces, and there are often stresses between the stator and the casing. It is known from the book [25] that stress affects the magnetization behavior of the magnetic domains, so in Figure 6, we tested the effect of stress on the magnetic flux density and core loss. It should be noted here that only tensile stresses were tested in this experiment, compressive stresses and more accurate testing methods will continue to be studied by the research team in the future.

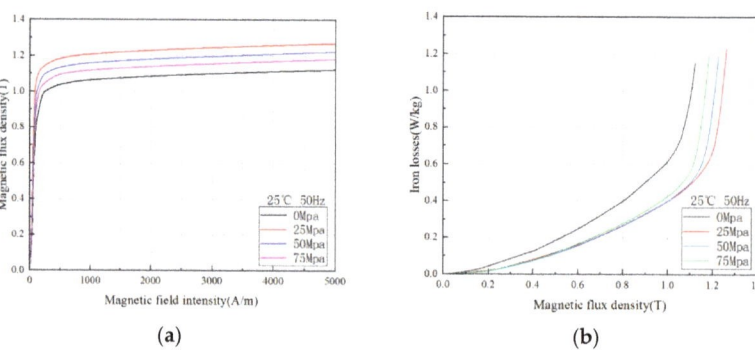

Figure 6. Magnetic properties of grain-oriented silicon steel under different stress fields (**a**) magnetization curves of H-80 material @25 °C and 50 Hz under different stresses; (**b**) iron loss curves of H-80 material @25 °C and 50 Hz under different stresses.

In Figure 6a, it can be seen that compared with the 0 Mpa stress-free condition, with the increase in tensile stress, the saturated magnetic flux density of grain-oriented silicon steel increases first and then decreases, and it can also be found in Figure 6b that the core loss decreases first and then increases with the increase in tensile stress. This trend can be more intuitively reflected in three-dimensional surface diagram. This change rule is based on the principle of materials science. When ferromagnetic materials are magnetized, elastic stress is generated due to magnetostriction. When it is magnetized to elongate and is limited to not elongate, pressure is generated inside; otherwise, it produces tension. If the material is subjected to external stress or internal stress at the same time, there is an energy coupled by magneto strictive λ_s and stress σ in the material. This energy is called magnetoelastic energy E_σ, which is related to the size and direction of λ_s and stress σ. For a certain magnet of λ_s, under the action of uniaxial stress σ, when the angle between stress and magnetization is θ, the magnetoelastic energy:

$$E_\sigma = \frac{3}{2}\lambda_s \sigma \sin\theta \quad (3)$$

For materials with $\lambda_s > 0$, such as Fe-Si alloy such as silicon steel, E_σ is the lowest when θ = 0° under tension (σ > 0). At this time, the direction of magnetization turns to the direction of tension, which promotes magnetization. Under pressure (σ < 0), E_σ is the lowest when θ = 90°. At this time, the direction of magnetization turns perpendicular to the pressure, hindering magnetization. Even if the tensile stress promotes magnetization, the magnetic domain structure of the material is fixed and cannot be stretched indefinitely, so there is an optimal tensile stress point.

This rule is exciting, because in the application process, we can carry out core stress analysis, select the best stress working point of the material, and fully develop the advantages of grain-oriented silicon steel.

3.1.3. Magnetic Properties of Grain-Oriented Silicon Steels at High Frequencies

With the higher speed of the drive motor, the frequency of the electrical machine is getting higher and higher, so the characteristics of the grain-oriented silicon steel under variable frequency conditions are necessary to be investigated. The experimental results are shown in Figure 7, where it can be observed that the saturation flux density of the grain-oriented silicon steel increases with increasing frequency and the losses show an increasing trend, but the growth is slower at lower frequencies and increases faster at higher frequencies, which is also in accordance with Equation (4) [26]. It can be seen that the core loss has a certain exponential relationship with the magnetic density and frequency, so the higher the frequency is, the faster the loss increases. It is worth noting that Formula (4) considers the core loss superposition of two magnetization angles of grain-oriented silicon steel. In fact, the magnetic properties of grain-oriented silicon steel in different directions are different, so a more accurate iron loss model will be a future research direction.

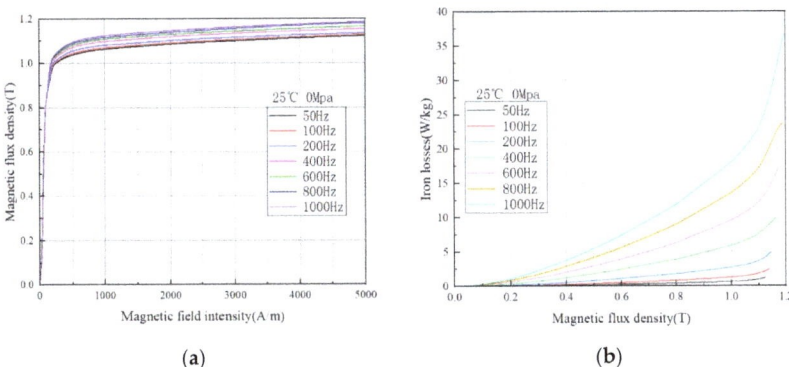

Figure 7. Magnetic properties of grain-oriented silicon steel at different frequencies (**a**) magnetization curves of H-80 material @25 °C and 0 Mpa (**b**) iron loss curves of H-80 material @25 °C and 0 Mpa.

$$P_{core-loss} = P_h + P_e + P_a = \sum_{i=1}^{N} \begin{cases} (K_{hRD} \cdot B_{RD}^{1.6} + K_{hTD} \cdot B_{TD}^{1.6}) \cdot f^1 \\ (K_{eRD} \cdot B_{RD}^{2} + K_{eTD} \cdot B_{TD}^{2}) \cdot f^2 \\ P_a \end{cases} \quad (4)$$

Among them, N is the number of core segments, K_{hRD}, K_{hTD}, K_{eRD} and K_{eTD} are the hysteresis loss coefficient in the rolling direction, the transverse hysteresis loss coefficient, the eddy current loss coefficient in the rolling direction and the transverse eddy current loss coefficient, respectively. They are related to the material properties and can be obtained by fitting the material test data.

3.1.4. Magnetic Characteristics of Grain-Oriented Silicon Steel under Multi-Physical Field Coupling

The operating environment of the electrical machine is in a multi-field coupled space such as thermal-force-electrical-magnetic, etc. In order to achieve the accuracy of the electrical machine simulation model, the realistic characterization of the silicon steel material in multi-physics fields is especially important [27]. Under the multi-physics test platform built in the laboratory, the team measured the loss variation pattern and the flux density of 1000 A/m at 50 Hz for 400 Hz and 1 T ultra-thin grain-oriented silicon steel under multi-field coupling. From Figure 8a, it can be seen that with the increase in stress at a certain temperature, the core loss of grain-oriented silicon steel decreases and then increases, and at about 30 Mpa, the loss is the smallest; with the increase in temperature at a certain stress, the core loss of grain-oriented silicon steel increases slightly, and the higher the temperature, the less obvious the change trend. In Figure 8b, the flux density increases and then decreases with the increase in stress at a certain temperature, and the flux density is maximum at about 35 Mpa; the flux density shows a decreasing trend with the increase in temperature at a certain stress. Comparing the magnetic properties of grain-oriented silicon steel under single physical and multi-physical field coupling conditions, it can be found that the change rules of the two are similar, but the specific quantity changes are still different, which is the influence of multi-physical field coupling. We will continue to study the coupling mechanism in depth.

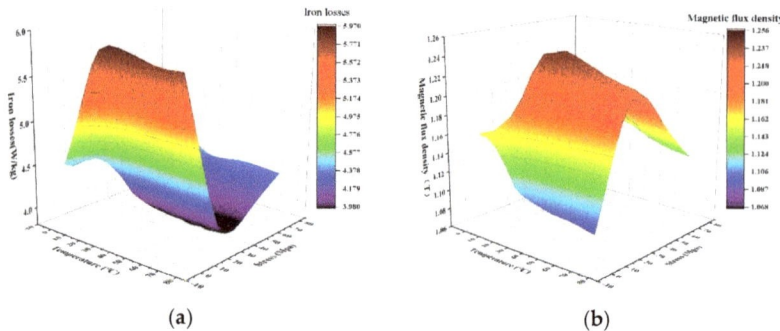

Figure 8. Magnetic properties of grain-oriented silicon steel under multi-physical field coupling (**a**) loss variation in H-80 under tensile stress @1 T and 400 Hz; (**b**) magnetic density variation in H-80 under tensile stress@1000 A/m and 50 Hz.

3.2. Experimental Results of Epstein Frame Method

3.2.1. Magnetic Characteristics of Grain-Oriented Silicon Steel at Different Angles

The Epstein frame is one of the most accurate methods to test the magnetic properties of silicon steel. In order to understand the variation in the magnetic properties of the experimental grain-oriented silicon steel from the rolling direction, we tested the magnetic properties of the samples in the range of 0–90°. Figure 9 shows some experimental results. It should be noted that the magnetic properties of a grain-oriented silicon steel within 0–360° can be determined only by measuring 0–90°, because 90–180° and 0–90° are symmetrical, and 180° is a change period. Figure 9a shows the change in saturation magnetic induction intensity with increasing deviation angle. It can be seen that the performance of grain-oriented silicon steel decreases seriously with the increase in deviation angle, and the deterioration is more serious after 30°. In addition, it can be clearly seen from the local amplification diagram that the permeability decreases with the increase in the deviation angle, which is the reason for the change in its macroscopic characterization (magnetization curve and iron loss curve); Figure 9b shows the loss change, which can be found to be continuously deteriorating, and the deterioration is also serious after 30°.

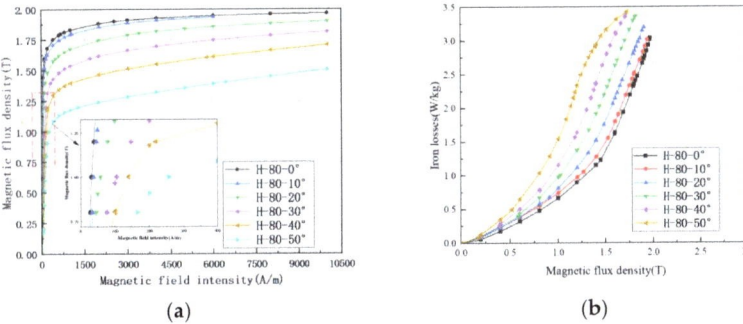

Figure 9. Magnetic characteristics of grain-oriented silicon steel at different angles (**a**) magnetization curves of H-80 material at different angles @50 Hz; (**b**) iron loss curves of H-80 material at different angles @50 Hz.

The ultra-thin grain-oriented silicon steel studied is Si-Fe alloy, and the crystal is a cubic structure. According to the principle of metal materials, [100] that is, the edge length direction is the easy magnetization direction; [111] and the body diagonal direction is the difficult magnetization direction. This magnetic difference in different crystal axis directions is called magneto crystalline anisotropy. The reason for magneto crystalline anisotropy is that the regular arrangement of atoms or ions in the crystal causes an inhomogeneous electrostatic field with spatial periodic changes, which changes the orbital angular momentum of electrons in the atom, but its average change may be zero. So, the atomic magnetic moment shown in the crystal is mainly its total electron spin magnetic moment. On one hand, the electron is affected by the non-uniform static electromagnetic field of the spatial periodic change, and at the same time, there is an exchange effect between the electron orbits of the adjacent atoms. The orbital motion of the electron is coupled with its spin. Thus, the magnetic moment has different energy levels in different directions of the crystal.

The magneto crystalline anisotropy energy E_k is usually expressed as a power series of the direction cosine of the saturation magnetization vector M_s relative to the main crystal axis. For a cubic crystal, due to its cubic symmetry, E_k is:

$$E_k = K_0 + K_1 \left(\alpha_1^2 \alpha_2^2 + \alpha_2^2 \alpha_3^2 + \alpha_3^2 \alpha_1^2 \right) + K_1 \alpha_1^2 \alpha_2^2 \alpha_3^2 + \cdots \quad (5)$$

where α_1, α_2 and α_3 are direction cosines of Ms relative to three [100] axes; K_0, K_1 and K_2 are the crystal anisotropy constants, which vary with the material and temperature. The sign and value of K_0, K_1 and K_2 determine the easy magnetization direction and difficult magnetization degree of the material. Since K_0 is independent of the magnetization direction, and in most cases, K_2 is too small to be ignored, E_k is mainly represented by K_1. Grain-oriented silicon steel (this material) $K_1 > 0$, [100] the direction of the lowest energy, so this direction is the easy magnetization direction. According to Equation (5), the change trend of E_k with the magnetization angle can be achieved, and E_k is the lowest in the 0° direction parallel to the rolling direction in the magnetization direction. At this time, the saturation flux density is the largest, the core loss is the lowest, and the permeability is also the highest. The calculation of E_k can explain the experimental results in Figure 9. With the deviation of magnetization angle, the magnetic properties of grain-oriented silicon steel gradually deteriorate. In the direction of about 55°, the E_k is the largest, the corresponding permeability is the lowest, and the magnetic flux density is the lowest at the magnetic field strength of 4000 A/m. The multi-angle test results of grain-oriented silicon steel are consistent with the calculation results of Formula (5).

In the previous paragraphs, the effect of different frequencies on grain-oriented silicon steel was discussed in the non-standard ring method, and it is impossible to compare the effect of different frequencies at different angles. Therefore, the effect of frequency on the magnetic properties of ultra-thin grain-oriented silicon steel at multiple angles was further measured using the Epstein frame method, and the results are shown in Figure 10. The amount of data is large, and the results are only listed for the cases of 50 Hz and 400 Hz according to the application equipment of this grain-oriented silicon steel. It is not difficult to find that the variation pattern in Figure 10a under different deviation angles is the same as that in Figure 9a, as the magnetization angle increases, the saturation magnetic flux density decreases. The magnetic flux density is basically the same at both frequencies. In Figure 10b it is obvious that the losses are affected by the frequency and show an order of magnitude increase, much larger than the effect of the different deviation angles.

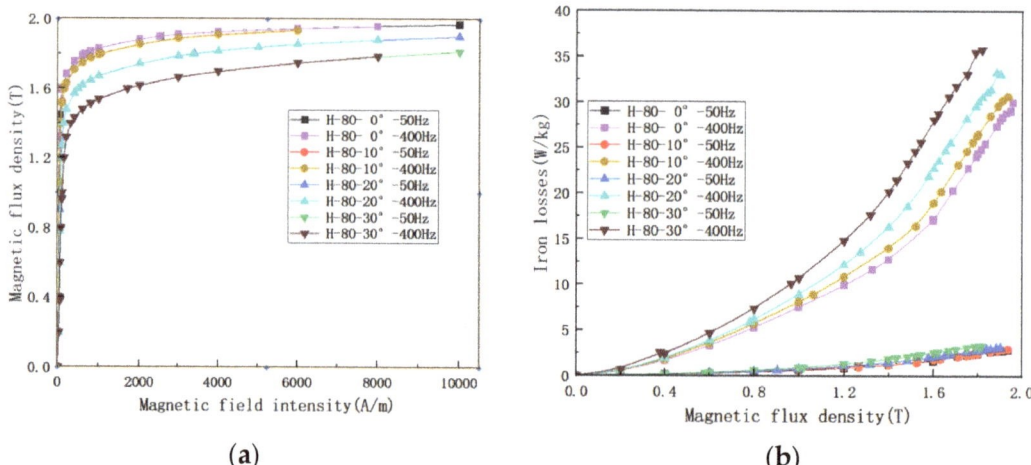

Figure 10. Magnetic characteristics of grain-oriented silicon steel at different angles (**a**) magnetization curves of H-80 material at different angles @400 Hz; (**b**) iron loss curves of H-80 material at different angles @400 Hz.

3.2.2. Magnetic Flux Density Variation Diagram of Two Materials at Different Angles

Finally, by performing the experiments, the variation in magnetic flux density of grain-oriented silicon steel and non-grain-oriented silicon steel with deviation angle α within 180 ° was compared under the condition of 50 Hz and 4000 A/m. As shown in Figure 11, the magnetic flux density of the ultra-thin grain-oriented silicon steel selected in the experiment is higher than that of the non- grain-oriented silicon steel within 30°. Therefore, using the grain-oriented silicon steel to replace the non- grain-oriented silicon steel within 30° will bring obvious advantages to the motor core. It should be emphasized here that the advantages of different grain-oriented silicon steel angle (grain-oriented silicon steel magnetic properties better than the same level of non-grain-oriented silicon steel magnetic properties of the angle) is different, according to the laboratory 's previous range of 20~30°.

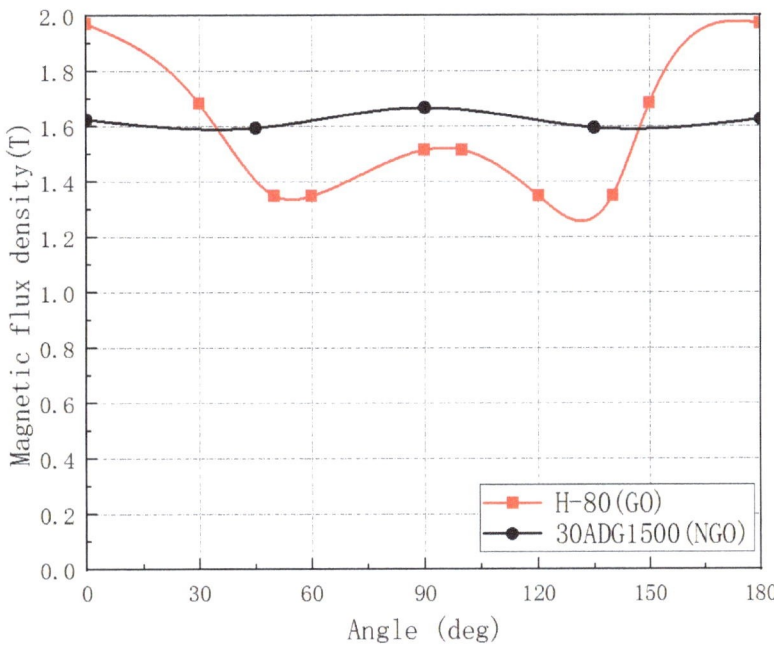

Figure 11. Comparison of saturation flux density of oriented and non-grain-oriented silicon steel with different magnetization angles.

3.3. Drive Electrical Machine Based on Grain-Oriented Silicon Steel

The reference model for the design of the grain-oriented silicon steel electrical machine is a 48-slot, eight-pole drive electrical machine for electric vehicles, which has the basic performance parameters required, as shown in Table 1.

Table 1. Electrical machine performance parameters.

Performance	Parameters
The rated line voltage	350 VDC
Rated current	260 A
Peak power	≥145 kW
Speed	4500 rpm
Peak torque	340 Nm
The efficiency of	96%

The conventional non-grain-oriented silicon steel electrical machine cannot meet the critical performance requirements of this drive electrical machine, so this paper attempts to use ultra-thin grain-oriented silicon steel in the stator core. Faced with the specificity of the superior performance of grain-oriented silicon steel in specific directions, the angular range of grain-oriented silicon steel superior to non-grain-oriented silicon steel was determined based on material tests, as shown in Figure 4, and the electrical machine stator was determined to be divided into six sections (all six sections are identical, using grain-oriented silicon steel punching sheets), and further finite element precision modeling of the designed electrical machine in the form of iron core splicing was performed. The detailed settings of finite element numerical calculation are shown in Table 2.

Table 2. Finite Element Numerical Calculation Detailed Settings.

	NGO-IPM	GO-IPM
Total number of elements	3668	3578
Solver Time Step	0.00011 s	0.00011 s
Solver Boundaries	Vector Potential (Value: 0) Independent Dependent (Bdep = −Bind)	Vector Potential (Value: 0) Independent Dependent (Bdep = −Bind)

The reference electrical machine model is shown in Figure 12a. The electrical machine requires high power density and high torque density, so the non-grain-oriented silicon steel stator core is close to saturation to achieve the performance requirements under the limited constraints. According to the working principle of the electrical machine, the stator flux density is the largest at the teeth, and the magnetic flux density at the yoke is relatively low, so how to improve the saturation flux density at the teeth is the key point. According to the magnetic field distribution characteristics of the electrical machine, the investigated alternative model electrical machine stator core uses a 60° fan-shaped oriented silicon steel sheet, with the center tooth parallel to the rolling direction, i.e., the center tooth is 0° magnetization characteristic of the grain-oriented silicon steel, and the outermost layer of the whole fan-shaped sheet does not exceed 30°, which is just within the dominant range of the grain-oriented silicon steel. Applying the topological simulation model as in Figure 12b, it can be found that the magnetic flux density is increased at the same tooth position compared with Figure 12a, and it is increased by 2.20% at the center tooth.

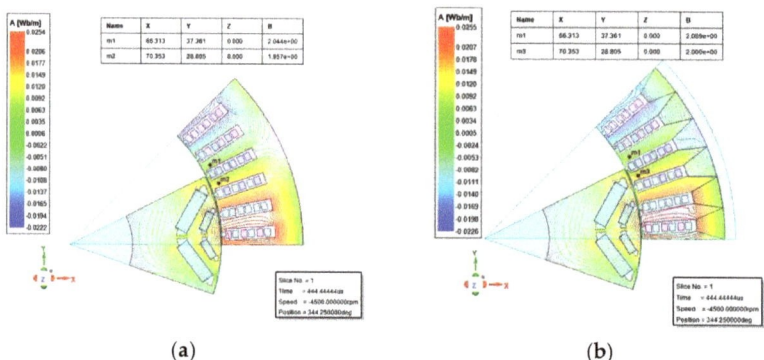

(a)   (b)

Figure 12. Electrical machine finite element simulation model. (**a**) Traditional electrical machine modeling reference model (NGO-IPM) (**b**) multi-angle spliced grain-oriented silicon steel electrical machine model (GO-IPM).

Because the two motor models only have different stator materials and other structural parameters are exactly the same, the end effect is ignored, so only the 2D finite element numerical calculation is carried out. In the follow-up work, we are ready to manufacture the prototype, which will take into account more parameters affecting the performance of the motor, such as the parasitic air gap problem at the splicing of adjacent grain-oriented silicon steel stator teeth and the mathematical model for more accurate prediction of actual motor performance.

Figure 13 shows the torque comparison of the two motors. Compared with the non- grain-oriented silicon steel stator, the result torque of the grain-oriented silicon steel segmented stator can reach 343.38 Nm, while the non-grain-oriented silicon steel stator structure can only reach 336.05 Nm even if it approaches saturation and reaches the limit, which cannot meet the peak torque of 340 Nm. However, it cannot be ignored that the torque ripple is more than 3% in the stator structure of grain-oriented silicon steel. One of the reasons is that the splicing material modeling is adopted, which will not occur in engineering production.

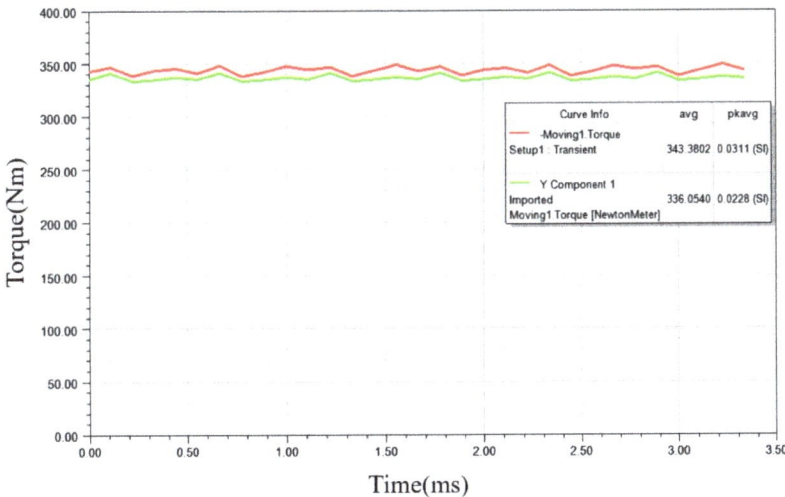

Figure 13. Torque comparison of non-grain-oriented silicon steel electrical machine (green line) and grain-oriented silicon steel electrical machine (red line).

The speed of an electric vehicle drive motor varies during its operation, so, in order to reflect the superiority of the segmented grain-oriented silicon steel stator more clearly and to fully reflect the electrical machine performance, a map chart comparison of torque-speed efficiency was presented. Under the condition of utilizing grain-oriented silicon steel, the range of efficiency greater than 97% is significantly enhanced, and the maximum efficiency is up to 98%, which shows the particular advantage of using grain-oriented silicon steel in drive motors.

As a matter of fact, the drive motor is in a temperature-stress-electric-magnetic multi-field coupling situation during operation; however, in the electrical machine research stage, the data of material modeling often only considers the electromagnetic field, thus causing wrong evaluation of the product. By building a multi-physics field experimental platform independently, our experimental team measured the performance of grain-oriented silicon steel under multi-field coupling and compared it with the standard Epstein frame method measurement results, the data comparison is shown in Figure 14. The multi-physics field coupling experimental conditions are 75 °C, 75 Mpa and 400 Hz, the conventional conditions are room temperature, no stress and 400 Hz, the purpose is to verify the real operation of the electrical machine conditions and the errors formed by inaccurate material modeling during the R&D phase.

In fact, the drive motor is in a temperature-stress-electric-magnetic multi-field coupling during operation, however, in the motor development stage, the material modeling data often only consider the electromagnetic field, resulting in the wrong assessment of the product. By independently building a multi-physical field experimental platform, the experimental team measured the performance of grain-oriented silicon steel under multi-field coupling, and compared it with the measurement results of the standard Epstein square ring method; see Figure 15. The metaphysical field coupling experimental conditions are 75 °C, 75 Mpa and 400 Hz, conventional conditions are room temperature, no stress and 400 Hz, the purpose is to verify the error caused by inaccurate material modeling under the real operating conditions of the motor and the research and development stage.

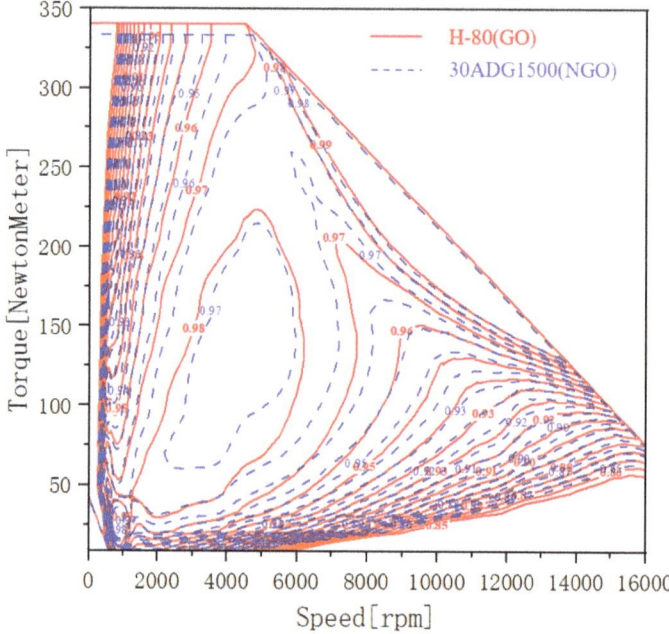

Figure 14. Comparison of efficiency map between traditional non-grain-oriented silicon steel motor model and grain-oriented silicon steel motor model.

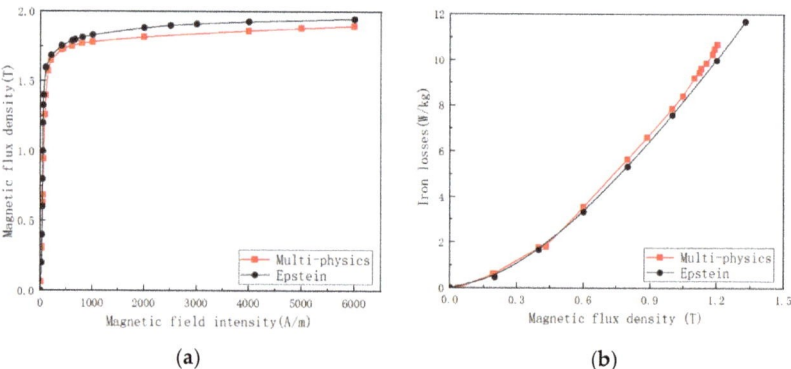

Figure 15. Comparison of magnetic properties of grain-oriented silicon steel under multi-physical field coupling and conventional measurements. (**a**) magnetization curves of H-80 material @75 °C, 75 Mpa and 400 Hz. (**b**) iron loss curves of H-80 material at different angles @@75 °C, 75 Mpa and 400 Hz.

4. Conclusions

Compared with non-grain-oriented silicon steel, grain-oriented silicon steel has obvious advantages and drawbacks. This paper comprehensively explores the magnetic property variations law of an ultra-thin grain-oriented silicon steel under different physical fields and multi-field coupling conditions, and verifies its application possibilities in electric vehicle drive motors. Under the temperature field, with the increase in temperature, the magnetic induction strength decreases and the loss increases (the loss of ordinary non-grain-oriented silicon steel decreases); under the stress field, with the increase in tensile stress, the magnetic induction strength increases and then decreases, and the loss decreases and then increases, but both are better than the non-stressed state. Under the electromagnetic field of different frequencies, the magnetic induction decreases as the frequency increases, and the loss gradually deteriorates, and the higher the frequency, the more serious the deterioration; under the conditions of temperature-stress-electrical-magnetic multi-field coupling, when the stress is certain, the temperature has less effect on the loss, the magnetic induction strength decreases slightly with the increase in temperature, and the temperature is certain with the increase in stress, the performance of grain-oriented silicon steel first improves and then deteriorates, and in the 30–40 MPa range it is performs the best. The magnetic properties of the grain-oriented silicon steel are seriously affected by the deviation angle, which is similar to the literature, and it is worth mentioning that the magnetic properties of this ultra-thin grain-oriented silicon steel are better than those of the non-grain-oriented silicon steel in the range of 30°.

In terms of application, our laboratory intends to develop a spliced grain-oriented silicon steel electric electrical machine product because ordinary non-grain-oriented silicon steel cannot meet the high-performance requirements of the drive motor. In order to improve the accuracy of finite element modeling, this paper adopts a design model based on multi-angle data stitching. The conventional model is unable to meet the performance index when the flux density reaches the limit. After adopting the grain-oriented silicon steel, the flux density of the teeth is significantly improved, and then the torque is also improved by 2%, breaking the performance bottleneck, and its efficiency can reach 98%, expanding the high efficiency area nearly twice. The initial verification of the orientation of silicon steel in the drive motor has a greater application perspective, is an effective method to break the current electrical machine design bottleneck.

Author Contributions: Conceptualization, Z.L. and Y.M.; methodology, Z.L.; software, Z.L. and Y.M.; validation, Z.L., A.H., Y.M., L.Z. and R.P.; formal analysis, Z.L.; investigation, A.H., Y.M. and L.Z.; resources, Z.L. and A.H.; data curation, A.H. and R.P.; writing—original draft preparation, Z.L. and A.H.; writing—review and editing, Z.L., Y.M. and S.X.; supervision, L.Z. and R.P.; project administration, Z.L., Y.M. and A.H. All authors have read and agreed to the published version of the manuscript.

Funding: This research was funded by programme of Scholars of the Xingliao Plan (No.XLYC2002113) and Shenyang University of Technology Interdisciplinary Team Project (No. 100600453).

Institutional Review Board Statement: Not applicable.

Informed Consent Statement: Not applicable.

Data Availability Statement: Not applicable.

Conflicts of Interest: The authors declare no conflict of interest.

References

1. Krings, A.; Boglietti, A.; Cavagnino, A.; Sprague, S. Soft magnetic material status and trends in electric machines. *IEEE Trans. Ind. Electron.* **2016**, *64*, 2405–2414. [CrossRef]
2. Taghavi, S.; Pillay, P. A novel grain-oriented lamination rotor core assembly for a synchronous reluctance traction motor with a reduced torque ripple algorithm. *IEEE Trans. Ind. Appl.* **2016**, *52*, 3729–3738. [CrossRef]
3. Wang, S.; Ma, J.; Liu, C.; Wang, Y.; Lei, G.; Guo, Y.; Zhu, J. Design and performance analysis of a novel PM assisted synchronous reluctance machine. *Int. J. Appl. Electromagn. Mech.* **2021**, *67*, 131–140. [CrossRef]
4. Taghavi, S.; Pillay, P. A sizing methodology of the synchronous reluctance motor for traction applications. *IEEE J. Emerg. Sel. Top. Power Electron.* **2014**, *2*, 329–340. [CrossRef]
5. Taghavi, S. Design of Synchronous Reluctance Machines for Automotive Applications. Ph.D. Thesis, Concordia University, Montréal, QC, Canada, 2015.
6. Talebi, D.; Gardner, M.C.; Sankarraman, S.V.; Daniar, A.; Toliyat, H.A. Electromagnetic design characterization of a dual rotor axial flux motor for electric aircraft. In Proceedings of the 2021 IEEE International Electric Machines & Drives Conference (IEMDC), Hartford, CT, USA, 17–20 May 2021.
7. Wang, T.; Liu, C.; Xu, W.; Lei, G.; Jafari, M.; Guo, Y.; Zhu, J. Fabrication and experimental analysis of an axially laminated flux-switching permanent-magnet electrical machine. *IEEE Trans. Ind. Electron.* **2016**, *64*, 1081–1091. [CrossRef]

8. Ma, J.; Li, J.; Fang, H.; Li, Z.; Liang, Z.; Fu, Z.; Qu, R.; Xiao, R. Optimal design of an axial-flux switched reluctance electrical machine with grain-oriented electrical steel. *IEEE Trans. Ind. Appl.* **2017**, *53*, 5327–5337. [CrossRef]
9. Geng, W.; Wang, Y.; Wang, J.; Hou, J.; Guo, J.; Zhang, Z. Electromagnetic Analysis and Efficiency Improvement of Axial-Flux Permanent Magnet Electrical machine with Yokeless Stator by Using Grain-oriented Silicon Steel. *IEEE Trans. Magn.* **2021**, *58*, 1–5. [CrossRef]
10. Mallard, V.; Demian, C.; Brudny, J.F.; Parent, G. The use of segmented-shifted grain-oriented sheets in magnetic circuits of small AC electrical machines. *Open Phys.* **2019**, *17*, 617–622. [CrossRef]
11. Ozdincer, B.; Aydin, M. Design of Innovative Radial Flux Permanent Magnet Electrical Machine Alternatives with Non-Oriented and Grain-Oriented Electrical Steel for Servo Applications. *IEEE Trans. Magn.* **2021**, *58*, 1–4. [CrossRef]
12. Son, J.C.; Lim, D.K. Novel Stator Core of the Permanent Magnet Assisted Synchronous Reluctance Electrical machine for Electric Bicycle Traction Electrical Machine Using Grain-Oriented Electrical Steel. In Proceedings of the 2021 24th International Conference on Electrical Electrical Machines and Systems (ICEMS), Gyeongju, Korea, 31 October–3 November 2021; pp. 1215–1218.
13. Maraví-Nieto, J.; Azar, Z.; Thomas, A.S.; Zhu, Z.Q. Utilisation of grain oriented electrical steel in permanent magnet fractional slot-Modular electrical machines. *J. Eng.* **2019**, *2019*, 3682–3686. [CrossRef]
14. Munteanu, A.; Nastas, I.; Simion, A.; Livadaru, L.; Virlan, B.; Nacu, I. A New Topology of Fractional-Slot Concentrated Wound Permanent Magnet Synchronous Electrical machine with Grain-Oriented Electric Steel for Stator Laminations. In Proceedings of the 2021 International Conference on Electromechanical and Energy Systems (SIELMEN), Iasi, Romania, 6 October 2021; pp. 349–352.
15. Hayslett, S.; Strangas, E. Analytical design of sculpted rotor interior permanent magnet machines. *Energies* **2021**, *14*, 5109. [CrossRef]
16. Gao, L.; Zeng, L.; Yang, J.; Pei, R. Application of grain-oriented electrical steel used in super-high speed electric electrical machines. *AIP Adv.* **2020**, *10*, 015127. [CrossRef]
17. Pei, R.; Zeng, L.; Li, S.; Coombs, T. Studies on grain-oriented silicon steel used in traction electrical machines. In Proceedings of the 2017 20th International Conference on Electrical Electrical Machines and Systems (ICEMS), Sydney, Australia, 11–14 August 2017; pp. 1–5.
18. Pei, R.; Xu, J.; Gao, L.; Zheng, H.; Zhou, X.; Yuan, Y.; Zeng, L. Studies of magnetic and thermodynamic characteristics iron core under high frequencies for traction electrical machines. *AIP Adv.* **2020**, *10*, 035107. [CrossRef]
19. Hu, A.; Zhang, X.; Wang, L.; Zhang, H.; Xu, S.; Pei, R. Research on New Design and Analysis of Stator Tooth Splicing Structure Using Anisotropic Electrical Steel in an Electric Electrical machine. In Proceedings of the 2022 IEEE 5th International Electrical and Energy Conference (CIEEC), Nanjing, China, 27–29 May 2022; pp. 4341–4346.
20. Rebhaoui, A.; Randi, S.A.; Demian, C.; Lecointe, J.P. Analyse of Flux Density and Iron Loss Distributions in Segmented Magnetic Circuits Made with Mixed Electrical Steel Grades. *IEEE Trans. Magn.* **2021**, *58*, 2000711. [CrossRef]
21. Elsherbiny, H.; Szamel, L.; Ahmed, M.K.; Elwany, M.A. High Accuracy Modeling of Permanent Magnet Synchronous Motors Using Finite Element Analysis. *Mathematics* **2022**, *10*, 3880. [CrossRef]
22. Nakata, T.; Fujiwara, K.; Takahashi, N.; Nakano, M.; Okamoto, N. An improved numerical analysis of flux distributions in anisotropic materials. *IEEE Trans. Magn.* **1994**, *30*, 3395–3398. [CrossRef]
23. Liu, J.; Basak, A.; Moses, A.J.; Shirkoohi, G.H. A method of anisotropic steel modelling using finite element method with confirmation by experimental results. *IEEE Trans. Magn.* **1994**, *30*, 3391–3394. [CrossRef]
24. Aggarwal, A.; Meier, M.; Strangas, E.; Agapiou, J. Analysis of Modular Stator PMSM Manufactured Using Oriented Steel. *Energies* **2021**, *14*, 6583. [CrossRef]
25. He, Z. Current situation and prospect of electrical steel (continued). *China Metall.* **2001**, 15–17.
26. Krings, A.; Soulard, J. Overview and comparison of iron loss models for electrical machines. *J. Electr. Eng.* **2010**, *10*, 8.
27. Wang, A.; Tian, B.; Li, Y.; Xu, S.; Zeng, L.; Pei, R. Transient Magnetic Properties of Non-Grain Oriented Silicon Steel under Multi-Physics Field. *Materials* **2022**, *15*, 8305. [CrossRef]

Article

Study of High-Silicon Steel as Interior Rotor for High-Speed Motor Considering the Influence of Multi-Physical Field Coupling and Slotting Process

Deji Ma [1], Baozhi Tian [1], Xuejie Zheng [2], Yulin Li [1], Shibo Xu [1] and Ruilin Pei [1,*]

[1] Department of Electric Engineering, Shenyang University of Technology, Shenyang 110870, China
[2] Suzhou Inn-Mag New Energy Ltd., Suzhou 215000, China
* Correspondence: peiruilin@sut.edu.cn

Abstract: Currently, high-speed motors usually adopt rotor structures with surface-mounted permanent magnets, but their sheaths will deteriorate performance significantly. The motor with interior rotor structure has the advantages of high power density and efficiency. At the same time, high silicon steel has low loss and high mechanical strength, which is extremely suitable for high-speed motor rotor core material. Therefore, in this paper, the feasibility of using high silicon steel as the material of an interior rotor high-speed motor is investigated. Firstly, the magnetic properties of high silicon steel under multi-physical fields were tested and analyzed in comparison with conventional silicon steel. Meanwhile, an interior rotor structure of high-speed motor using high silicon steel as the rotor core is proposed, and its electromagnetic, mechanical, and thermal properties are simulated and evaluated. Then, the experimental comparative analysis was carried out in terms of the slotting process of the core, and the machining of the high silicon steel rotor core was successfully completed. Finally, the feasibility of the research idea was verified by the above theoretical analysis and experimental characterization.

Keywords: motor; interior rotor; high-silicon steel; multi-physical field; slotting

Citation: Ma, D.; Tian, B.; Zheng, X.; Li, Y.; Xu, S.; Pei, R. Study of High-Silicon Steel as Interior Rotor for High-Speed Motor Considering the Influence of Multi-Physical Field Coupling and Slotting Process. *Materials* **2022**, *15*, 8502. https://doi.org/10.3390/ma15238502

Academic Editor: Shenglei Che

Received: 28 October 2022
Accepted: 25 November 2022
Published: 29 November 2022

Publisher's Note: MDPI stays neutral with regard to jurisdictional claims in published maps and institutional affiliations.

Copyright: © 2022 by the authors. Licensee MDPI, Basel, Switzerland. This article is an open access article distributed under the terms and conditions of the Creative Commons Attribution (CC BY) license (https://creativecommons.org/licenses/by/4.0/).

1. Introduction

In recent years, the need for more efficient electrical applications is increasing due to the growing environmental concerns and the consequent gradual transition to a decarbonized society [1]. High-speed permanent magnet motors are used in a variety of applications due to their high efficiency and power density, such as aircraft generators, flywheel energy storage systems, high-speed spindles, turbomolecular pumps, air compressors, blowers, turbochargers, and microturbines. However, a large centrifugal force is generated when the rotor runs at high speed, which can cause irreversible mechanical damage to the motor rotor and thus affect the reliable operation of the motor [2]. In order to avoid the influence of centrifugal forces, a sleeve structure is usually applied on the outside of the motor rotor, currently the mainstream sleeve materials are metal and carbon fiber. However, with metal sleeves, eddy currents in the sleeves produce large eddy current losses, which can have a deteriorating effect on the efficiency of the motor. High eddy currents may also lead to overheating of the rotor to the point of irreversible demagnetization of the permanent magnets. If carbon fiber winding is used to protect the rotor from centrifugal forces, it results in a relatively large motor air gap length which affects the performance of the motor. Meanwhile, the heat dissipation capacity of the rotor is reduced, again leading to irreversible demagnetization of the permanent magnets [3–6]. These problems can be avoided by the rotor's interior permanent magnet design with no sleeve structure [7].

During the operation of the motor, the centrifugal force affects the rotor structure, which requires the material of the interior rotor to have good mechanical strength while considering the magnetic properties [8]. Kasai analyzed the mechanical strength and magnetic properties of a high-silicon steel (10JNEX900) and found that both properties of this type of steel are superior to those of conventional steels, which makes it compatible with the requirements imposed by interior permanent magnet motor rotors [9]. This study used high silicon steel as the rotor material of a high-speed motor and demonstrated the feasibility of its application [10]. Figure 1 shows the yield strength and iron loss of several materials. Although 1K101 amorphous material has high yield strength and low loss properties, its performance is severely affected by processing and cannot be mass-produced [7]. Materials, such as Vacoflux48 and Vacoflux17, have low yield strengths and are not suitable for rotor core materials of high-speed motors. Compared to high strength steel 20SW1200H and high silicon steel two materials in mechanical properties and loss performance, traditional silicon steel also does not have any advantage. Although the high-strength steel 20SW1200H and high silicon steel 10JNEX900 performance are similar, their mechanical properties and loss performance still have gaps.

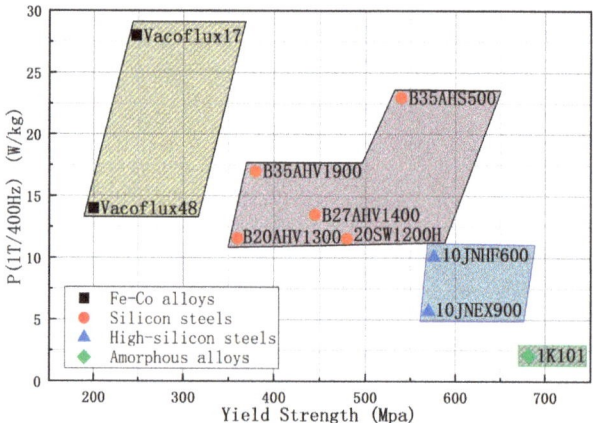

Figure 1. Comparison of yield strength and core loss for different silicon steels.

Guangwei Liu's team studied that during motor operation, the silicon steel is subjected to centrifugal force and a certain temperature rise, which causes a certain change in the magnetic properties of the rotor core [11]. Andreas Krings and Oskar Wallmark et al. studied the loss of iron cores as affected by temperature [12] and Junquan Chen et al. developed a model for silicon steel loss considering the effect of temperature [13]. Later, a large number of experiments were conducted, and the temperature coefficients of the loss model were corrected according to the results to verify the correctness of the model [14,15]. Above studies show that the core of the motor is under the operating conditions of coupled multi-physical fields, such as temperature, stress, and electromagnetism, but there is no literature focusing on the magnetic performance of high silicon steel under multi-physical coupled fields.

In addition to the above, the machining process of the motor core also has a non-negligible impact on the performance of the motor [16]. Meanwhile, the literature [17] also points out that the selection of the motor rotor material should determine not only the most suitable material to withstand the stress, but also the best processing technique for the rotor material, as well as the main factors affecting the fatigue resistance of the selected material. Yingzhen Liu's study showed that the core processing process can have some deteriorating effects on the performance of conventional silicon steel cores [18]. Design with multiple slots inside the rotor should consider the impact of the cutting process on the core material more than the surface-mounted design with a sleeve. The stamping process has the advantages of low cost and high efficiency and is the primary choice for mass production of electric motors. Laser cutting is mainly used for the process of some special purpose motors [16]. The internal microstructure of the cutting part was observed by Aroba Saleem et al. [19] and its effect was studied in [20,21]. In addition, wire electrical discharge machining is also suitable for the processing of high silicon steel cores due to its precise machining accuracy and low deterioration effect on the core. Studying the effect of core processing on high silicon steel cores to find out the suitable slotting for interior rotors will also affect the performance of motors to some extent.

In Section 2 of this paper, the mechanical properties of high silicon steel (10JNEX900) and high strength steel (20SW1200H) of the same thickness are tested. A set of "electric-magnetic-thermal-stress" multi-physical field coupling magnetic property testing devices was built to investigate the different advantages of two materials, as well as to investigate and analyze their magnetic property change law under the multi-physical coupling field. In Section 3, a 90-kW high-speed interior permanent magnet motor is designed, the effect of the two materials on the motor performance under the rotor core of the high-speed motor is calculated by simulation, and the performance differences of electromagnetic, stress, and temperature are compared and analyzed. Section 4 compares the performance of the cores processed by the two different cutting processes and explores the extent of their influence on the core performance. Moreover, we analyze the mechanism related to this influence, then conclude the most suitable cutting process for high silicon steel. Finally, the feasibility of high-silicon steel for high-speed motor rotor is verified by analyzing high-silicon steel rotor in three different dimensions.

2. Material Electromagnetic Performance Evaluation

High-Si steels are Fe-Si alloys with a Si mass fraction of 6.5%, where the Si atoms are uniformly diffused inside the material by chemical vapor deposition. The resistivity of the silicon steel material is greatly enhanced by the addition of more Si elements. High-silicon steel sheet has the advantages of low iron consumption, low hysteresis, and high magnetic permeability and is suitable for use as an iron core in electric motors.

High-silicon steel is prepared by chemical vapor deposition (CVD) method, and the process is shown in Figure 2. The $SiCl_4$ gas is used to react with the silicon steel strip at high temperature to form a layer of Fe_3Si, and then the furnace temperature is increased to diffuse Si into the interior of the strip, and the CVD deposition is followed by high temperature diffusion annealing under reducing atmosphere protection to achieve a composition of 6.5% Si [22].

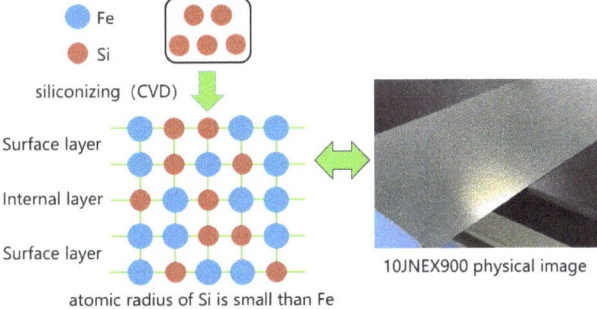

Figure 2. 10JNEX900 CVD process and physical schematic diagram.

2.1. Experiment Method

In this experiment, the magnetic properties of two kinds of silicon steel sheets, 6.5% Si and 20SW1200H, were tested under the conditions of temperature-tension stress-electromagnetic field coupling, using a three-part coupling of silicon steel sheet electromagnetic characteristics test module, mechanical property test module and test environment adjustment module. The principle of the multi-physics field coupled test system is shown in Figure 3.

Figure 3. Multi-physics field coupled magnetic energy measurement system.

The experimental sample in this study is a non-standard circular sample, the length of the magnetic circuit of the sample is set to 625 mm, the primary winding and secondary winding are both 200 turns, and the cross-sectional area of the test sample is calculated based on the theoretical thickness and the lamination factor. The test specimen is fixed to the instrument by the auxiliary tooling. The top and bottom of the sample are laminated with oriented silicon steel which has high flux density and low iron loss, and the loss effect on the two wide sides of the sample is thus negligible. The physical diagram of test specimen and device is shown in Figure 4.

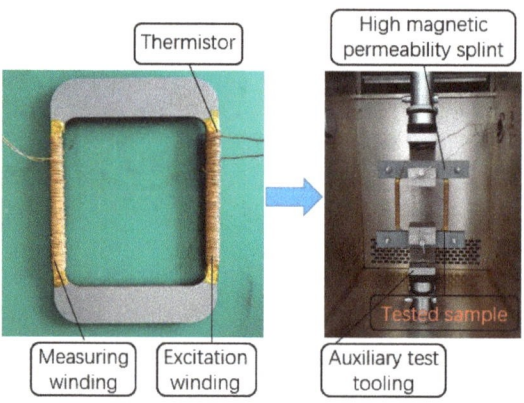

Figure 4. Physical diagram of test specimen and process.

2.2. Effect of Temperature on Magnetic Properties

Figure 5 demonstrates the effect of temperature on the magnetic properties of the two silicon steel materials under no stress. The saturation flux density of 10JNEX900 at 50 Hz shows a clear tendency to decrease with increasing temperature. The magnetic permeability of the material is maximum at 0 °C. Its permeability also decreases with increasing temperature. The magnetic flux density of 10JNEX900 at 50 Hz and 1000 A/m decreased by 5.07% when the temperature increased from 0 °C to 100 °C. On the contrary, the saturation flux density of 20SW1200H is less affected by temperature, and its permeability also changes slightly with the increase in temperature. When the magnetic field strength is larger than 250 A/m, the saturation flux density shows a decreasing trend with the increase in temperature. The flux density of 20SW1200H at 50 Hz and 1000 A/m decreases by 1.8% when the temperature increases from 0 °C to 100 °C. The reason for this phenomenon is mainly due to the continuous increase in the speed of molecular motion as the temperature increases. This phenomenon prevents the movement of magnetic domains and domain walls during the magnetization process, leading to a decrease in the magnetic permeability of the material, which in turn reduces the saturation magnetic flux density of the material [23].

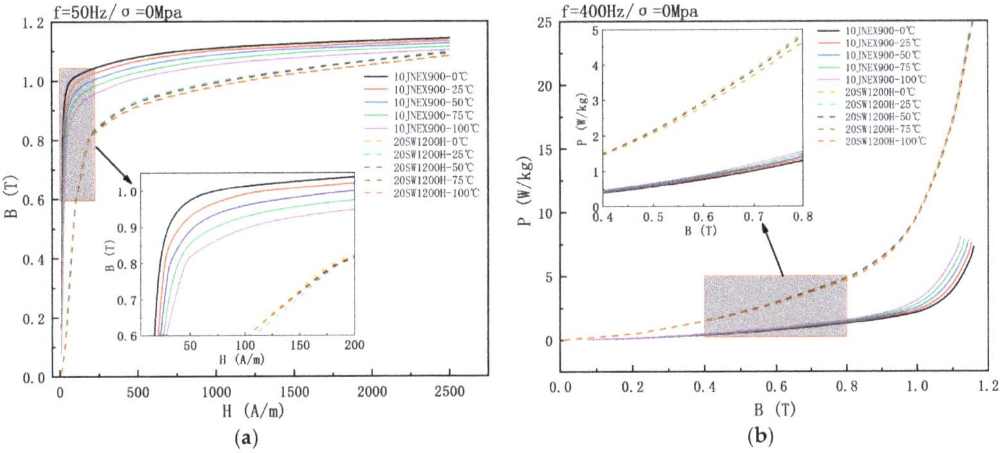

Figure 5. Effect of temperature on magnetic properties of two silicon steels at 0 Mpa. (**a**) B-H curve. (**b**) B-P curve.

In terms of iron loss, the loss of 10JNEX900 at 400 Hz shows a significant increasing trend with the increase of temperature, while the loss of 20SW1200H shows a slight decreasing trend with the increase of temperature. When the temperature increases from 0 °C to 100 °C, the loss of 10JNEX900 at 400 Hz and 1 T increases by 42.7%, and the loss of 20SW1200H at 400 Hz and 1 T decreases by 2.01%. The loss of 20SW1200H decreases with temperature mainly because the resistivity of the material increases with the temperature, so the eddy current loss decreases too. The opposite result of 10JNEX900 is due to its special preparation process (CVD). When the temperature increases, the internal stress inside the material will lead to an increase in hysteresis loss, and the increase in hysteresis loss is greater than the decrease in eddy current loss, so the total loss of the material increases [22].

2.3. Effect of Stress on Magnetic Properties

Figure 6 demonstrates the effect of tensile stress on the magnetic properties of the two silicon steel materials at 25 °C. The saturation flux density of 10JNEX900 at 50 Hz shows a decreasing trend with the increase of tensile stress, but the curves at 1000 A/m attachment 25 Mpa and 50 Mpa are crossed. The permeability of 10JNEX900 is almost unaffected by the tensile stress, while the saturation flux density and permeability of 20SW1200H are both greatly affected by increasing tensile stress, and the saturation flux density decreases with higher tensile stress. The magnetic flux density of 10JNEX900 at 50 Hz and 1000 A/m decreased by 3.6% when the stress was increased from 0 Mpa to 75 Mpa.

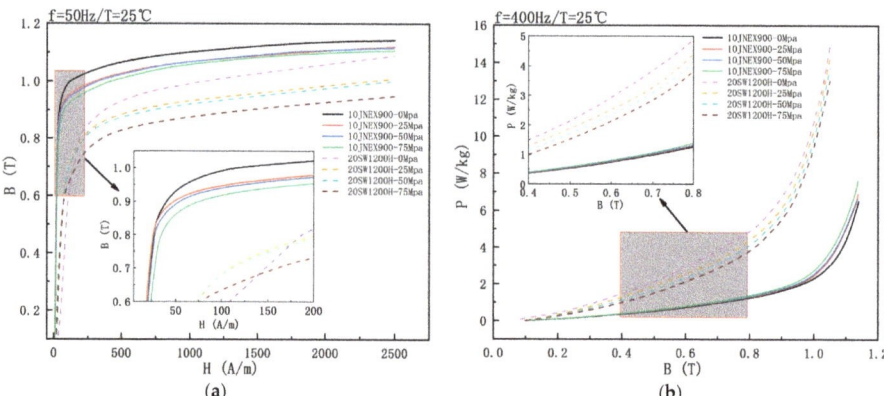

Figure 6. Effect of tensile stress on magnetic properties of two silicon steels at 25 °C. (**a**) B-H curve. (**b**) B-P curve.

The loss of 10JNEX900 at 400 Hz shows a significant increasing trend with the increase of tensile stress, while the loss of 20SW1200H shows a significant decreasing trend with the increase of temperature. When the stress increases from 0 Mpa to 75 Mpa, the loss of 10JNEX900 at 400 Hz and 1 T increases by 17.58%, and the loss of 20SW1200H at 400 Hz and 1 T decreases by 16.52%. The main reason for this phenomenon in 10JNEX900 is the presence of stresses that hinder the movement of the magnetic domains and the walls of the domains increasing the energy loss and reducing the saturation flux density. In contrast, the high yield strength capability of high-strength silicon steels is obtained partly by the thermal expansion and infiltration process with lower dislocations, which induce stronger residual internal stresses and thus prevent the formation of 180° magnetic domains under tensile stress [8]. Therefore, the loss of high-strength silicon steel 20SW1200H is reduced.

2.4. Effect of Temperature-Stress Coupling on Magnetic Properties

Figure 7 shows the variation of saturation flux density and loss of 10JNEX900 under the condition of multi-physics field coupling. The magnetic properties of 10JNEX900 vary with temperature and tensile stress as analyzed above. The lowest point of its magnetic flux density occurs at 100 °C and 75 Mpa, and the lowest point of loss occurs at 0 °C and 0 Mpa. The flux density and loss of 10JNEX900 increased by 8.42% and 98.21% from the lowest point to the highest point, respectively. Therefore, it can be concluded that the deterioration of material loss is more obvious in the case of coupled temperature and tensile stress.

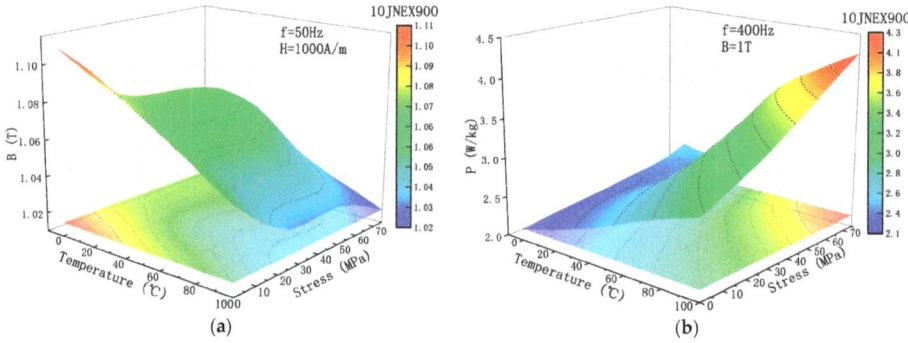

Figure 7. Effect of multi-physical field coupling on the magnetic properties of 10JNEX900. (**a**) Trend of B. (**b**) Trend of P.

3. Motor Finite Element Analysis

In order to evaluate the performance of high-silicon steel 10JNEX900 as an interior rotor for high-speed motors, this section analyzes the performance of rotors made from 10JNEX900 and 20SW1200H in terms of electromagnetic, mechanical, and thermal properties utilizing simulation calculations.

3.1. Main Structural Parameters of Motor

Figure 8 illustrates the structure of the permanent magnet synchronous motor proposed in this paper. The motor's pole-slot ratio is 8 poles and 48 slots, and the rotor's permanent magnet structure is designed as an interior U-shape, with a maximum motor speed of 19,000 rpm. The main performance parameters of the motor are shown in Table 1.

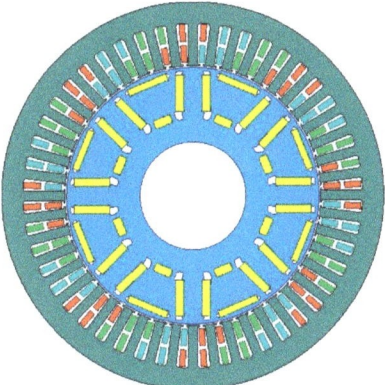

Figure 8. Model diagram of permanent magnet motor.

Table 1. Some performance parameters of permanent magnet motor.

Parameter	Value	Parameter	Value
Rated power/kW	90	Peak power/kW	210
Rated speed/rpm	9549	Peak speed/rpm	19,000
DC bus voltage/V	540	Peak current/A	500

3.2. Motor Electromagnetic Performance Analysis

Figure 9 shows the flux density distribution for the two material motors. In this section, the electromagnetic performance of two motors is analyzed using the finite element analysis software Maxwell. The results show that the flux density at the rotor spacer bridge of the 10JNEX900 motor is slightly lower than that of the 20SW1200H motor. Figure 10 shows the loss distribution for the two material motors under loaded condition. Although the rotor loss of the PM synchronous motor is lower, it can still be seen that the loss at the outer edge of the rotor of the 10JNEX900 motor is slightly lower than that of the 20SW1200H motor. All of the above reasons are because the 10JNEX900 has a larger saturation flux density and permeability and has lower losses compared to the 20SW1200H, but with less impact. This will give the motor a higher operating efficiency.

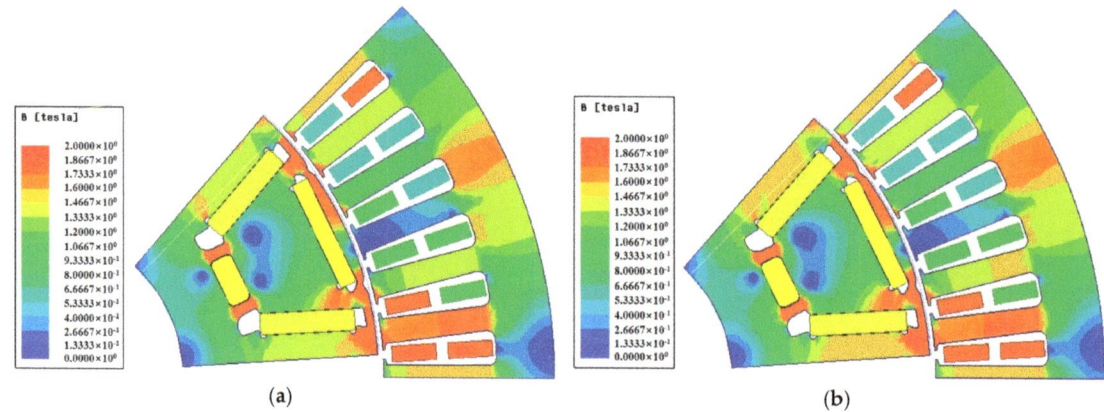

Figure 9. Flux density distribution of two materials motor under loaded condition. (a) 10JNEX900. (b) 20SW1200H.

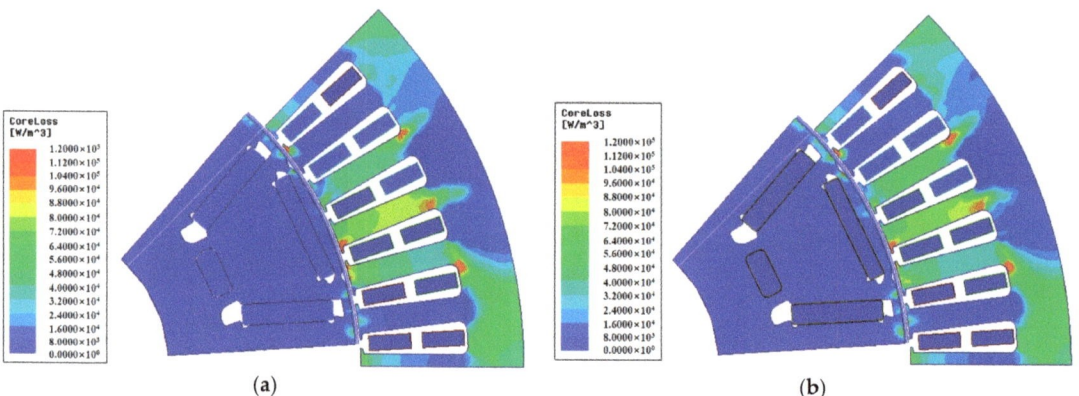

Figure 10. Iron loss distribution of two material motor under loaded condition. (a) 10JNEX900. (b) 20SW1200H.

Figure 11 shows the back EMF waveforms of the two motors during no-load operation. The difference between the two no-load counter potential cases is very small. This indicates that for the same motor, the change in rotor material has less effect on the motor's no-load back EMF.

Figure 12 shows the load air-gap flux density waveforms for both motors. The overall difference between the two is small, but the air-gap flux density of the 10JNEX900 motor is greater than that of the 20SW1200H motor when the electrical angle is between 100° and 250°. This is mainly due to the above-mentioned reasons. The 10JNEX900 has better magnetic properties compared to 20SW1200H. As a result, the 10JNEX900 motor also has better overload performance.

The MAP comparison of motor efficiency for the two rotor core materials is shown in Figure 13. There is a difference in the maximum operating efficiency and the percentage of high efficiency area between the two motors. However, when the rotor core is 10JNEX900, the maximum operating efficiency of the motor operation reaches 98.1%, which is 0.1% higher compared to 98.06% of 20SW1200H. The area of high-efficiency zones with efficiency >90% also increased from 92.1% to 92.43%. Therefore, compared to the conventional high-strength silicon steel 20SW1200H, 10JNEX900 has a greater advantage in terms of motor operating efficiency in the case of finite element analysis.

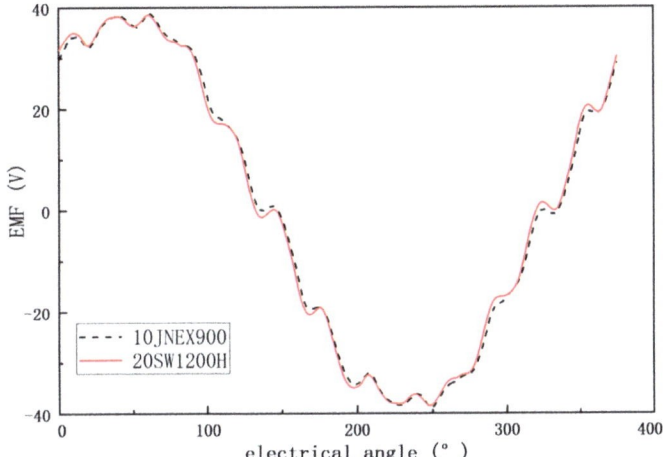

Figure 11. Comparison of no-load back EMF of 10JNEX900 and 20SW1200H motors.

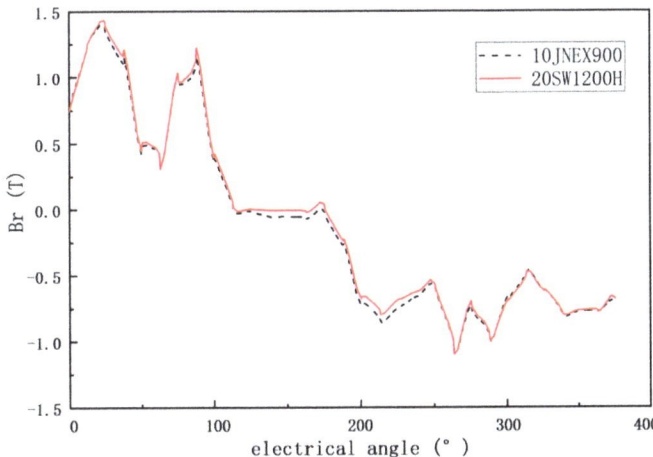

Figure 12. 10JNEX900 and 20SW1200H motor load air-gap flux density comparison.

3.3. Rotor Mechanical Stress Analysis

In order to verify whether the rotor can safely withstand the centrifugal force caused by the maximum speed of the motor, the centrifugal force of the motor with two different rotor materials at maximum speed was analyzed in this paper using ANSYS software, as shown in Figure 14. It is obvious that the maximum stress points on the rotor are at the magnetic isolation bridge. The maximum stress value of 10JNEX900 motor is 451.43 Mpa, while the yield strength of 10JNEX900 is 570 Mpa. With a safety margin of 1.2 times considered, the 10JNEX900 fully meets the requirements in terms of mechanical performance. The maximum stress value of 20SW1200H motor is 451.75 Mpa, while the yield strength of the material is 480 Mpa. Therefore, for the same structure of the rotor, 20SW1200H can meet the requirements of the motor in terms of mechanical performance, but there is no outflow of sufficient safety margin. Since the mass density of the two materials is different, the distribution of stresses in the 10JNEX900 rotor is lower for the same rotor structure.

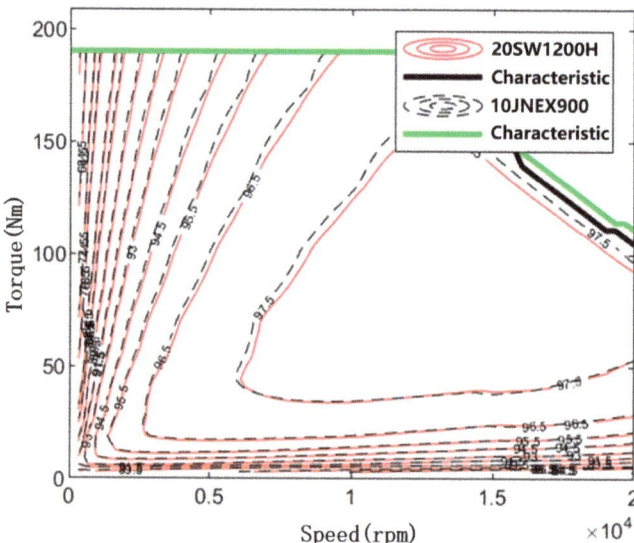

Figure 13. 10JNEX900 and 20SW1200H motor efficiency MAP comparison.

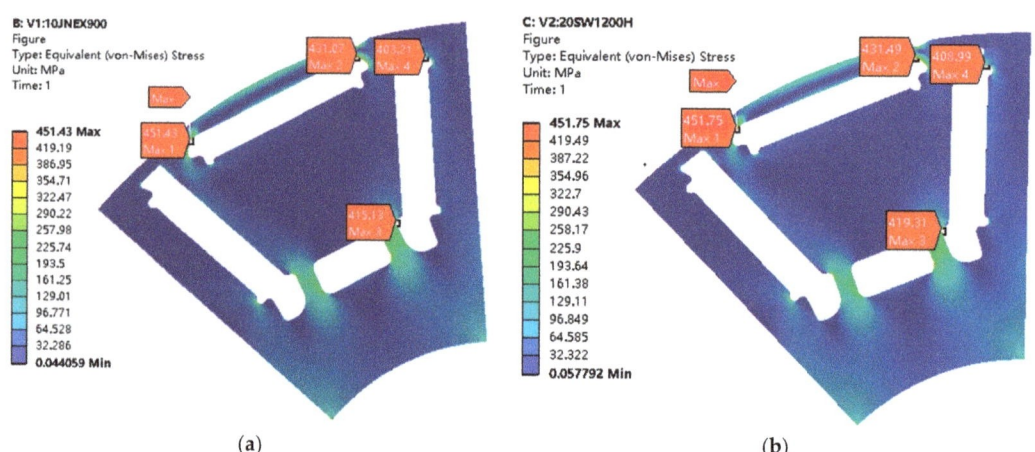

Figure 14. Mechanical stress distribution of rotor cores of two materials at 19,000 r/min. (**a**) 10JNEX900. (**b**) 20SW1200H.

3.4. Rotor Temperature Rise Performance Analysis

Because this motor runs at high power and speed condition, the temperature rise of the motor rotor will also be higher. In the above loss and efficiency analysis, the 10JNEX900 motor shows greater advantages. The cooling method of the motor is water-cooled. In order to analyze the temperature rise of these two motors at the maximum speed, this paper uses ANSYS software to perform accurate simulations on the 3D model of the motor.

Regarding the details of the simulation analysis, the ambient temperature and the inlet water temperature were set to 30 °C, and the inlet water velocity at the inlet end of the motor housing was 1 m/s. Figure 15 shows the rotor temperature distribution of the motor for both materials at the peak motor speed. It can be clearly seen that the overall temperature distribution of the 20SW1200H motor rotor is higher than that of the 10JNEX900, and the highest temperature point is 3.33 °C higher. This is due to the fact that the 10JNEX900 core produces lower iron consumption when the motor is running at peak speed, so the rotor temperature will be lower. The motor rotor has the highest temperature at the outer diameter edge and the lowest temperature at the inner diameter edge. This is due to the fact that, in addition to the heat generated by the motor rotor itself, the heat generated by the windings is transferred through the air gap to the outer surface of the rotor and further to the inside of the motor rotor. The inner diameter of the motor is at the direct connection of the rotor, and the heat of the rotor will be transferred through the rotor.

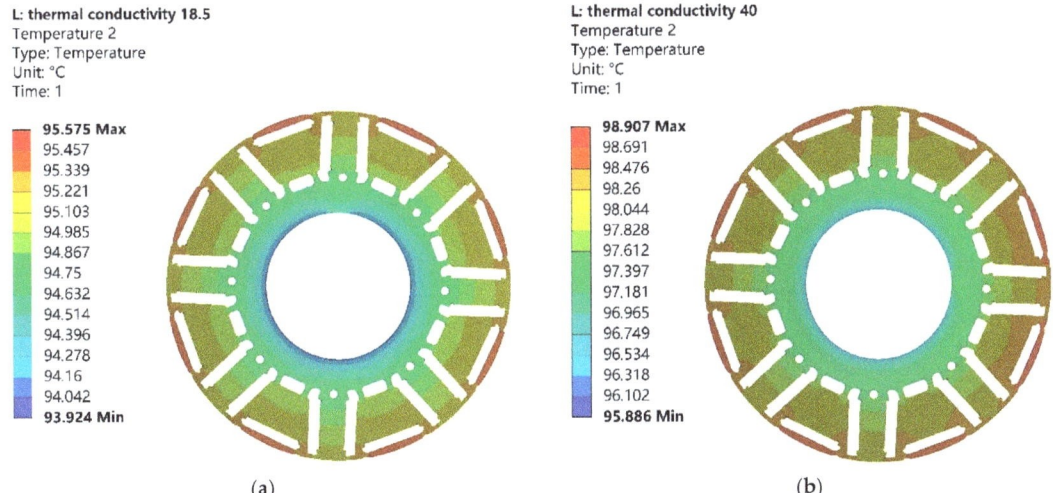

Figure 15. Temperature distribution of motor rotor cores of two materials at 19,000 r/min. (**a**) 10JNEX900. (**b**) 20SW1200H.

4. Research of Slotting Method

The magnetization properties of its teeth and yoke are particularly sensitive to the processing stresses generated during the machining process, so there are significant differences between the actual motor performance that is finally exhibited and the calculated values after simulation through the original material data.

Since high silicon steel cores are usually used in stators to reduce core losses, the effect of processing on the mechanical properties of the material is not yet clear. The commonly used core cutting methods are punching, laser cutting, wire electrical discharge machining (WEDM), and water jet cutting [24–26]. The deterioration of the magnetic properties by punching is more severe and due to the brittle material of high silicon steel the formability is poor. The process of water jet cutting for a single piece of silicon steel is complicated and its cutting accuracy for the whole core is low [27]. Due to the thinness and brittleness of high silicon steel sheets, they are prone to cracking. Moreover, combined with the complex structure of the motor rotor and its high-speed characteristics, a small defect in the core can lead to rotor cracking at high speeds. To ensure the success of the prototype, experimental studies on the cutting quality of WEDM and laser cutting methods and their effects on magnetic properties were conducted in this paper, and the WEDM method was finally chosen to manufacture the prototype.

4.1. Wire Electrical Discharge Machining

The WEDM equipment used in this study is the DK7732 from SUZHOU BAOMA, China. WEDM is the most commonly used core cutting process for prototyping. It has good universality and good cutting quality and is suitable for cutting cores of various new materials and complex shapes. However, the cutting efficiency of WEDM is low, and if the cutting area is large and the structure is complicated, it will lead to longer cutting time. Figure 16 shows a sample of the experimental ring cut by WEDM. It can be seen that the sample surface quality material is smooth, the burr is small, and there is no shattering. The grain between the slices is still clearly visible, and there is no conduction between the stacks. The maximum error of dimensional measurement after CMM is controlled at ±0.02 mm with high accuracy. Because this paper only considers the production of the principle prototype, after the above comprehensive analysis, WEDM is suitable for the production of the prototype in this paper.

Figure 16. Ring test specimen machined by WEDM.

4.2. Laser Cutting

For ordinary silicon steel sheets, laser cutting is a better choice of grooving method. The GF6025Plus laser cutting machine from HGTECH, China, was used for this study. Laser cutting has high cutting efficiency, its cutting speed can reach 12 m/min, and the cutting accuracy can also be controlled within ±0.03 mm. However, this method is only suitable for cutting single silicon steel sheet. If the stacked cores are cut by laser, it will cause the adhesive and coating between the stacked sheets to be melted at high temperature, resulting in a short circuit on the cutting surface and increasing the iron loss of the sample. Three problems arise when slotting high-silicon steel using this method: (1) Due to the high silicon steel sheet its brittleness is high, single piece cutting process if the operation is not standardized will easily appear single piece shattering situation. (2) Because the high silicon steel sheet is very thin, using laser cutting will cause the whole cutting and stacking process to become complicated. (3) The edge part of high silicon steel sheet processed by laser cutting will produce large burrs, which will lead to the reduction of the stacking coefficient of the iron core and eventually affect the torque performance of the motor. Figure 17 shows a test specimen of the laser cutting process molding. Therefore, it can be concluded from the analysis that high silicon steel is not suitable for grooving by this method.

4.3. The Effect of Slotting on the Magnetic Properties of the Material

Figure 18 shows the comparison of the magnetic performance test data of 10JNEX900 at 50 Hz and 400 Hz under two different cutting methods. As seen in Figure 18a, the saturation flux density and permeability of the laser cutting sample are significantly lower than those of the WEDM sample during the whole magnetization process at a frequency of 50 Hz. The magnetic flux density of the WEDM sample was 3.74% higher than that of the laser cutting sample when the magnetic field strength was 6000 A/m. From Figure 18b, it can be seen that the iron consumption of the laser cutting sample is significantly higher than that of the WEDM sample at 400 Hz. When the magnetic flux density was 1 T, the iron consumption of the laser cutting sample was 24.2% higher than that of the WEDM sample.

Figure 17. Laser cutting shaped ring test specimen.

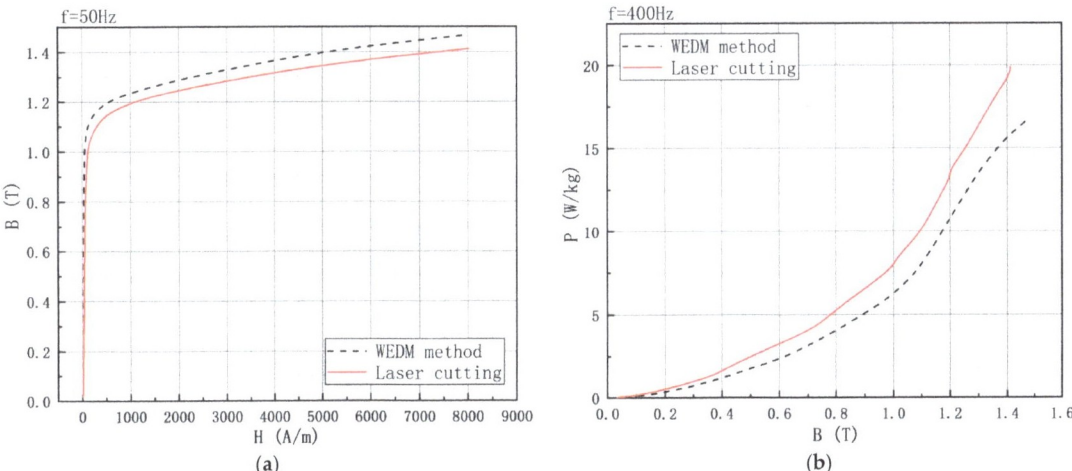

Figure 18. Comparison of magnetic properties of high silicon steel with two different processing methods. (**a**) B-H curve. (**b**) B-P curve.

Figure 19 illustrates the organization of the specimen edge cross-section after two different cutting methods. As seen in Figure 19b, the edge of the specimen after WEDM also has no obvious plastic deformation and is in the form of an arc-shaped oblique strip, but the edge area has been melted by high temperature and a small amount of burr has appeared. Due to the rapid high temperature and cooling process that the sample undergoes during cutting, a small thermal stress is generated inside it. As seen in Figure 19b, there is an obvious burning and melting phenomenon on the edge of the specimen after laser cutting, which is due to the thermal stress generated at the laser cutting temperature of more than 1000 degrees. After cooling, a large area of internal cooling stress is generated in its edge part. This internal stress is detrimental to the magnetic properties of the material and results in laser cutting samples with lower saturation flux density and higher iron losses. Therefore, from the point of view of the effect on the magnetic properties of the material, the magnetic properties of the material after WEDM show a better performance.

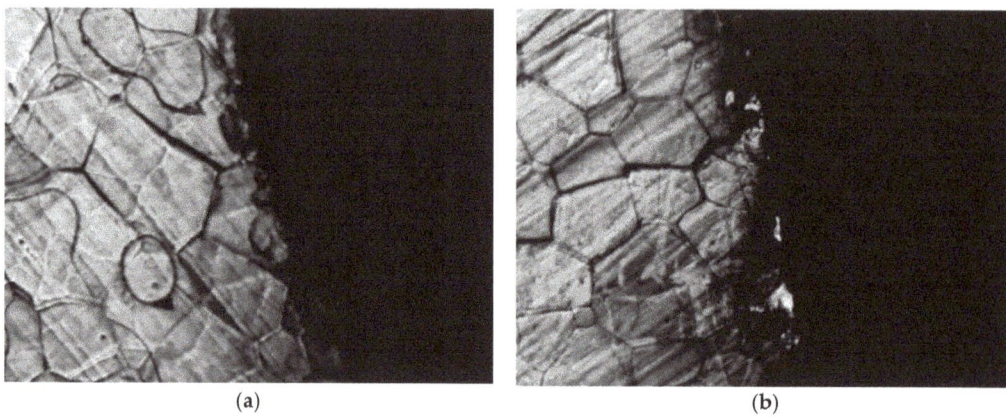

Figure 19. Changes in grain structure at the edges of high silicon steel after grooving in two different ways. (**a**) WEDM. (**b**) Laser cutting.

After the above analysis, the stator-rotor core of the principle prototype was machined by WEDM, as shown in Figure 20. It can be seen that the laminated core shows a good appearance after cutting, which verifies the thesis that WEDM is more suitable for high silicon steel cutting in the paper.

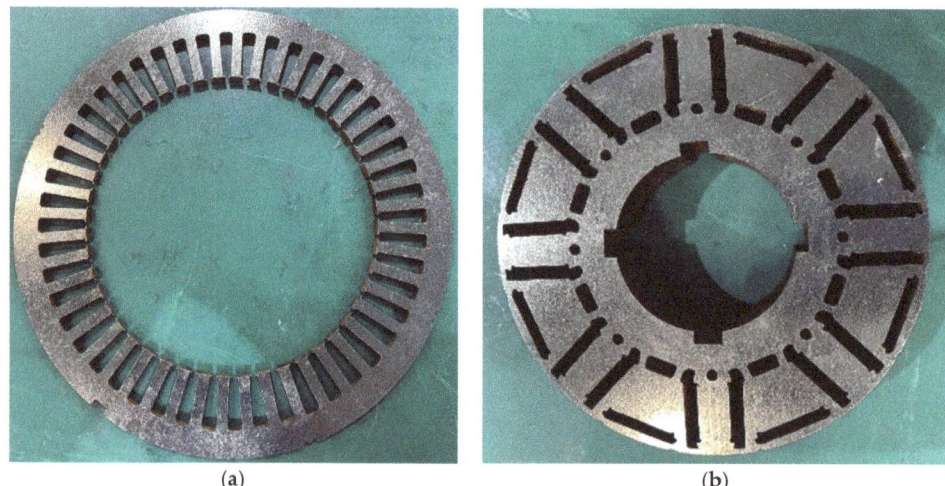

Figure 20. WEDM core of a motor. (**a**) Stator. (**b**) Rotor.

5. Conclusions

This paper presents an interior rotor structure of a high-speed permanent magnet synchronous motor with high silicon steel (10JNEX900) as the rotor material. To further determine the idea that high silicon steel is suitable for high-speed motor rotors, the magnetic properties of 10JNEX900 were measured in this paper. The results show that 10JNEX900 has better magnetic properties than the conventional high-strength steel 20SW1200H in both the normal state and the multi-physical field coupled state.

Meanwhile, considering the actual operating conditions of the high-speed motor rotor, the proposed rotor is analyzed in terms of electromagnetic performance, mechanical performance, and thermal performance, and the analysis results of the 10JNEX900 rotor and the 20SW1200H rotor are compared and analyzed in this paper. The results show that motors with high-silicon steel rotors have higher efficiency, better performance and lower temperature rise than rotors with conventional high-strength steel.

In order to solve the problem of difficult grooving due to the high brittleness of high silicon steel, this paper compares the two most commonly used methods for cutting high silicon steel cores. The results show that WEDM has better cut quality than laser cutting and its effect on the magnetic properties of the material is relatively small. The WEDM method is more suitable for slotting high-silicon steel cores. After the above comprehensive analysis, high silicon steel can be considered as a rotor core material for high-speed motors, which provides a solution to the problem of centrifugal force and loss in the rotor that exists during high speed operation of the motor.

In the future, we will continue the fabrication of the principle prototype and complete the performance test analysis of the whole machine. Further verification of the suitability of high-silicon steel as a rotor for high-speed motors will be performed, and problems arising from the prototype production will be fed back into the design phase for a more in-depth study of this direction.

Author Contributions: Conceptualization, D.M.; methodology, D.M. and B.T.; software, X.Z. and Y.L.; validation, D.M., B.T. and S.X.; formal analysis, D.M.; investigation, S.X.; resources, R.P.; data curation, D.M.; writing—original draft preparation, D.M. and S.X.; writing—review and editing, D.M. and R.P.; visualization, D.M.; supervision, R.P.; project administration, D.M.; funding acquisition, R.P. All authors have read and agreed to the published version of the manuscript.

Funding: This research was funded by programme of Scholars of the Xingliao Plan (No. XLYC2002113) and Shenyang University of Technology Interdisciplinary Team Project (No. 100600453).

Institutional Review Board Statement: Not applicable.

Informed Consent Statement: Not applicable.

Data Availability Statement: Not applicable.

Conflicts of Interest: The authors declare no conflict of interest.

References

1. Malinowski, J.; Hoyt, W.; Zwanziger, P.; Finley, B. Motor and Drive-System Efficiency Regulations: Review of Regulations in the United States and Europe. *IEEE Ind. Appl. Mag.* **2016**, *23*, 34–41. [CrossRef]
2. Li, Y.; Wang, J.; Zhang, H.; Pei, R. New design of electric vehicle motor based on high-strength soft magnetic materials. *COMPEL-Int. J. Comput. Math. Electr. Electron. Eng.* **2022**; ahead-of-print. [CrossRef]
3. Schneider, T.; Binder, A. Design and Evaluation of a 60,000 rpm Permanent Magnet Bearingless High Speed Motor. In Proceedings of the 2007 7th International Conference on Power Electronics and Drive Systems, Bangkok, Thailand, 27–30 November 2007; pp. 1–8.
4. Jumayev, S.; Merdzan, M.; Boynov, K.O.; Paulides, J.J.H.; Pyrhönen, J.; Lomonova, E.A. The effect of PWM on rotor ed-dy-current losses in high-speed permanent magnet machines. *IEEE Trans. Magn.* **2015**, *51*, 1–4. [CrossRef]
5. Tuysuz, A.; Steichen, M.; Zwyssig, C.; Kolar, J.W. Advanced cooling concepts for ultra-high-speed machines. In Proceedings of the 2015 9th International Conference on Power Electronics and ECCE Asia (ICPE-ECCE Asia), Seoul, Republic of Korea, 1–5 June 2015; pp. 2194–2202.
6. Hong, D.-K.; Woo, B.-C.; Lee, J.-Y.; Koo, D.-H. Ultra High Speed Motor Supported by Air Foil Bearings for Air Blower Cooling Fuel Cells. *IEEE Trans. Magn.* **2012**, *48*, 871–874. [CrossRef]
7. Ou, J.; Liu, Y.; Breining, P.; Gietzelt, T.; Wunsch, T.; Doppelbauer, M. Experimental Characterization and Feasibility Study on High Mechanical Strength Electrical Steels for High-Speed Motors Application. *IEEE Trans. Ind. Appl.* **2020**, *57*, 284–293. [CrossRef]
8. Zhang, H.; Zeng, L.; An, D.; Pei, R. Magnetic Performance Improvement caused by Tensile Stress in Equivalent Iron Core fabricated by High-Strength Non-Oriented Electrical Steel. *IEEE Trans. Magn.* **2021**, *58*, 1–5. [CrossRef]
9. Kasai, S.; Namikawa, M.; Hiratani, T. Recent progress of high silicon electrical steel in JFE steel. *JFE Tech. Rep.* **2016**, *21*, 14–19.
10. Ou, J.; Liu, Y.; Breining, P.; Gietzelt, T.; Wunsch, T.; Doppelbauer, M. Study of the Electromagnetic and Mechanical Properties of a High-silicon Steel for a High-speed Interior PM Rotor. In Proceedings of the 2019 22nd International Conference on Electrical Machines and Systems (ICEMS), Harbin, China, 11–14 August 2019.
11. Liu, G.; Liu, M.; Zhang, Y.; Wang, H.; Gerada, C. High-Speed Permanent Magnet Synchronous Motor Iron Loss Calculation Method Considering Multi-Physics Factors. *IEEE Trans. Ind. Electron.* **2019**, *67*, 5360–5368. [CrossRef]

12. Krings, A.; Mousavi, S.A.; Wallmark, O.; Soulard, J. Temperature Influence of NiFe Steel Laminations on the Characteristics of Small Slotless Permanent Magnet Machines. *IEEE Trans. Magn.* **2013**, *49*, 4064–4067. [CrossRef]
13. Chen, J.Q.; Wang, D.; Cheng, S.W. Modeling of temperature effects on magnetic property of non-oriented silicon steel lamination. *IEEE Trans. Magn.* **2015**, *51*, 1–4.
14. Xue, S.S.; Feng, J.H.; Guo, S.Y. A new iron loss model for temperature dependencies of hysteresis and eddy current losses in electrical machines. *IEEE Trans. Magn.* **2018**, *54*, 1–10. [CrossRef]
15. Xue, S.; Feng, J.; Guo, S.; Chen, Z.; Peng, J.; Chu, W.Q.; Huang, L.R.; Zhu, Z.Q. Iron Loss Model Under DC Bias Flux Density Considering Temperature Influence. *IEEE Trans. Magn.* **2017**, *53*, 1–4. [CrossRef]
16. Kurosaki, Y.; Mogi, H.; Fujii, H.; Kubota, T.; Shiozaki, M. Importance of punching and workability in non-oriented electrical steel sheets. *J. Magn. Magn. Mater.* **2008**, *320*, 2474–2480. [CrossRef]
17. Balyts' kyi, O.I.; Mascalzi, G. Selection of materials for high-speed motor rotors. *Mater. Sci.* **2002**, *38*, 293–303. [CrossRef]
18. Liu, Y.; Ou, J.; Schiefer, M.; Breining, P.; Grilli, F.; Doppelbauer, M. Application of an Amorphous Core to an Ultra-High-Speed Sleeve-Free Interior Perma-nent-Magnet Rotor. *IEEE Trans. Ind. Electron.* **2018**, *65*, 8498–8509. [CrossRef]
19. Saleem, A.; Alatawneh, N.; Chromik, R.R.; Lowther, D.A. Effect of Shear Cutting on Microstructure and Magnetic Properties of Non-Oriented Electrical Steel. *IEEE Trans. Magn.* **2015**, *52*, 1–4. [CrossRef]
20. Belhadj, A.; Baudouin, P.; Breaban, F.; Deffontaine, A.; Dewulf, M.; Houbaert, Y. Effect of laser cutting on microstructure and on magnetic properties of grain non-oriented electrical steels. *J. Magn. Magn. Mater.* **2003**, *256*, 20–31. [CrossRef]
21. Emura, M.; Landgraf, F.; Ross, W.; Barreta, J. The influence of cutting technique on the magnetic properties of electrical steels. *J. Magn. Magn. Mater.* **2003**, *254–255*, 358–360. [CrossRef]
22. Ma, D.; Li, J.; Tian, B.; Zhang, H.; Li, M.; Pei, R. Studies on Loss of a Motor Stator Iron Core with High Silicon Electrical Steel Considering Temperature and Compressive Stress Factors. In Proceedings of the 2022 IEEE 5th International Electrical and Energy Conference (CIEEC), Nangjing, China, 27–29 May 2022; pp. 4243–4248.
23. Xiao, L.; Yu, G.; Zou, J.; Xu, Y.; Liang, W. Experimental analysis of magnetic properties of electrical steel sheets under temperature and pressure coupling environment. *J. Magn. Magn. Mater.* **2019**, *475*, 282–289. [CrossRef]
24. Liew, G.S.; Soong, W.L.; Ertugrul, N.; Gayler, J. Analysis and performance investigation of an axial-field PM motor utilising cut amorphous magnetic material. In Proceedings of the 2010 20th Australasian Universities Power Engineering Conference, Christchurch, New Zealand, 5–8 December 2010; pp. 1–6.
25. Celie, J.; Stie, M.; Rens, J.; Sergeant, P. Effects of cutting and annealing of amorphous materials for high speed permanent magnet machines. In Proceedings of the 2016 XXII International Conference on Electrical Machines (ICEM), Lausanne, Switzerland, 4–7 September 2016; pp. 1630–1635.
26. Bali, M.; Muetze, A. Influences of CO_2 laser, FKL laser, and mechanical cutting on the magnetic properties of elec-trical steel sheets. *IEEE Trans. Ind. Appl.* **2015**, *51*, 4446–4454. [CrossRef]
27. Paltanea, G.; Paltanea, V.M.; Stefanoiu, R.; Nemoianu, I.V.; Gavrila, H. Correlation between Magnetic Properties and Chemical Composition of Non-Oriented Electrical Steels Cut through Different Technologies. *Materials* **2020**, *13*, 1455. [CrossRef] [PubMed]

Article

A High-Performance Magnetic Shield with MnZn Ferrite and Mu-Metal Film Combination for Atomic Sensors

Xiujie Fang [1,2,3], Danyue Ma [1,2,3,*], Bowen Sun [2,3], Xueping Xu [2,3], Wei Quan [2,3], Zhisong Xiao [1] and Yueyang Zhai [2,3,*]

1. School of Physics, Beihang University, Beijing 100191, China
2. Zhejiang Provincial Key Laboratory of Ultra-Weak Magnetic-Field Space and Applied Technology, Hangzhou Innovation Institute of Beihang University, Hangzhou 310000, China
3. Key Laboratory of Ultra-Weak Magnetic Field Measurement Technology, Ministry of Education, School of Instrumentation and Optoelectronic Engineering, Beihang University, Beijing 100191, China
* Correspondence: madanyue0419@buaa.edu.cn (D.M.); yueyangzhai@buaa.edu.cn (Y.Z.)

Abstract: This study proposes a high-performance magnetic shielding structure composed of MnZn ferrite and mu-metal film. The use of the mu-metal film with a high magnetic permeability restrains the decrease in the magnetic shielding coefficient caused by the magnetic leakage between the gap of magnetic annuli. The 0.1–0.5 mm thickness of mu-metal film prevents the increase of magnetic noise of composite structure. The finite element simulation results show that the magnetic shielding coefficient and magnetic noise are almost unchanged with the increase in the gap width. Compared with conventional ferrite magnetic shields with multiple annuli structures under the gap width of 0.5 mm, the radial shielding coefficient increases by 13.2%, and the magnetic noise decreases by 21%. The axial shielding coefficient increases by 22.3 times. Experiments verify the simulation results of the shielding coefficient of the combined magnetic shield. The shielding coefficient of the combined magnetic shield is 16.5%. It is 91.3% higher than the conventional ferrite magnetic shield. The main difference is observed between the actual and simulated relative permeability of mu-metal films. The combined magnetic shielding proposed in this study is of great significance to further promote the performance of atomic sensors sensitive to magnetic field.

Keywords: MnZn ferrite; mu-metal film; magnetic shield; magnetic noise; atomic sensors

Citation: Fang, X.; Ma, D.; Sun, B.; Xu, X.; Quan, W.; Xiao, Z.; Zhai, Y. A High-Performance Magnetic Shield with MnZn Ferrite and Mu-Metal Film Combination for Atomic Sensors. *Materials* **2022**, *15*, 6680. https://doi.org/10.3390/ma15196680

Academic Editor: Michael R. Koblischka

Received: 3 September 2022
Accepted: 19 September 2022
Published: 26 September 2022

Publisher's Note: MDPI stays neutral with regard to jurisdictional claims in published maps and institutional affiliations.

Copyright: © 2022 by the authors. Licensee MDPI, Basel, Switzerland. This article is an open access article distributed under the terms and conditions of the Creative Commons Attribution (CC BY) license (https://creativecommons.org/licenses/by/4.0/).

1. Introduction

Numerous ultra-high sensitivity atomic sensors that use quantum effects to detect physical quantities, such as atomic magnetometer [1,2], atomic gyroscope [3,4], atomic clock [5], and superconducting quantum interferometer [6,7], are magnetic-field sensitive. The stability of the environmental magnetic field has a direct effect on the measurement performance of sensors. Consequently, passive magnetic shielding and active magnetic compensation are widely used in the aforementioned sectors [8–10].

The accuracy of the magnetometer limits the accuracy of the active magnetic compensation. Moreover, its high-power consumption requires an external current source [9,10]. The passive magnetic shield uses the characteristics of higher permeability of magnetic material compared with air to cancel the magnetic flux density around the material. Utilizing a magnetic material with a higher permeability sufficiently reduces the magnetic flux density within the magnetic shield. A common magnetic shield that achieves a high shielding coefficient involves using a mu-metal multilayer shielding [11]. This type of magnetic shield successfully suppresses the influence of the fluctuating magnetic field in the environment on the sensors inside the magnetic shield. Nevertheless, it possesses a high conductivity. The Johnson current generated by the material results in a magnetic noise level of ~10 fT/Hz$^{1/2}$. This prevents the atomic sensors from being made more sensitive. A high-performance magnetic shield must have a high shielding coefficient and a low magnetic noise [12,13].

Reducing the magnetic noise of magnetic shielding systems has become a research priority in recent years. Kornack and Lee investigated the analytical formula for magnetic shielding noise [14,15]. The magnetic noise was computed using the power loss based on the fluctuation dissipation theory [16]. It was discovered that the magnetic noise was mostly caused by the complex permeability and conductivity of the material. The conductivity of MnZn ferrite material was ~10^{-6} S/m less than that of mu-metal material [17,18]. Through theoretical calculation and experimental research, the noise of the ferrite shield has been reduced by 25 times to 0.75 $fT/Hz^{1/2}$ compared with the noise of the mu-metal shield, which is above 40 Hz. Bevan employed square ferrite magnetic shielding in 2018 to reduce the magnetic noise level of the NMR gyroscope [19]. In addition, low-noise ferrite shielding is utilized in several high-sensitivity atomic sensors and measurement techniques, such as co-magnetometer [20,21], demonstrating its superior performance.

As observed in the previous research, the practical application of a ferrite magnetic shield presents several challenges. In lieu of being integrally created, numerous magnetic annuli are spliced together to produce the ferrite magnetic shield to meet the requirements of a large-scale magnetic shielding system and limit the influence of magnetic noise [22,23]. Due to the surface roughness and non-uniformity of the magnetic annuli, there will be an air gap, approximately 0.1–0.5 mm, in the ferrite shield when magnetic annuli are spliced [22]. The air gap between magnetic annuli diminishes the shielding coefficient and increases the radial magnetic noise. The shielding coefficient of ferrite magnetic shielding deteriorates. This decreases the environmental magnetic noise reduction and increases self-induced magnetic noise. Its effectiveness is impacted by the superposition of the two forms of magnetic noise. The research on the air gap of ferrite magnetic annuli focuses solely on its effect on the shielding coefficient and magnetic noise. There is currently no solution for the air gap. The air gap cannot be eliminated by finishing the contact surface of magnetic annuli.

This research quantitatively evaluates how the air gap of ferrite annuli affects the axial and radial shielding coefficient and magnetic noise. To reduce the effect of the gap on the shielding coefficient and magnetic noise, a magnetic shielding structure comprised of ferrite and mu-metal film is proposed. This configuration can use the mu-metal material to suppress the magnetic leakage field at the ferrite air gap. Outside the ferrite magnetic shielding, a mu-metal film is adhered to lessen the magnetic shielding noise of the mu-metal film. The shielding performance of ferrite and mu-metal film combination magnetic shield (FMCS) is quantified using the finite element method (FEM). This configuration has a better shielding coefficient than conventional ferrite magnetic shields. The magnetic noise study results indicated that this design reduced the effect of the air gap on magnetic noise. In addition, the configuration accounted for the occurrence of an air gap between the ferrite magnetic shield and the mu-metal film when they adhered. In this case, the shielding coefficient and magnetic noise were also explored in this work. Finally, the magnetic shielding coefficient measuring platform was constructed, and the experiment verified the accuracy of the calculation.

2. Methods

2.1. The Magnetic Shielding Structure Composed of Ferrite and Mu-Metal Film

The FMCS schematic comprises a ferrite shield and a mu-metal film. In Figure 1, the x-direction is the radial, while the z-direction is the axial. The size of the magnetic shielding model established by simulation and experiment is identical. Five magnetic annuli and two end caps constitute the ferrite shield. Each annulus has a height of 45 mm, an inner diameter of 114 mm, and a wall thickness of 13 mm. The thickness of the two end caps is 10 mm. In practical application, there are four 22 mm diameter access holes in the radial direction and two 28 mm diameter access holes in the axial direction. This study exclusively examines the effect of the air gap between ferrite magnetic annuli to prevent the access hole from influencing the results, and the access hole is not accounted for in the simulation. The ferrite magnetic shield material is the soft MnZn ferrite with a spinel structure. Figure 2 depicts the morphology of the MnZn ferrites, consisting of aggregated crystallites with homogeneous grains. Low porosity is present in the MnZn ferrite sample, and no air gaps emerge at the grain boundary. Owing to the discontinuous grain development during the sintering process, a few vacancies are observed within the grain. This intra-granular vacancy may impede the mobility of the domain wall, resulting in an adverse effect on the magnetic permeability. The composition of the MnZn ferrite is analyzed using an inductively coupled plasma emission spectrometer (Agilent, Palo Alto, CA, USA). The main elements, Fe, Mn, and Zn are shown in Table 1, in which other elements such as Na, P, Si, and Ca are also included. The addition of trace elements can reduce the loss. However, the mass fraction (wt.%) of other elements should not exceed 0.15%. The complex permeability of the ferrite magnetic shield is also measured experimentally. The real component of the complex permeability is 10029, and the imaginary component is 150 [24].

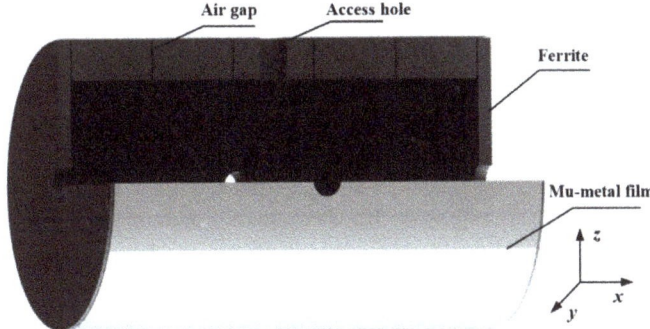

Figure 1. The schematic diagram of the FMCS structure.

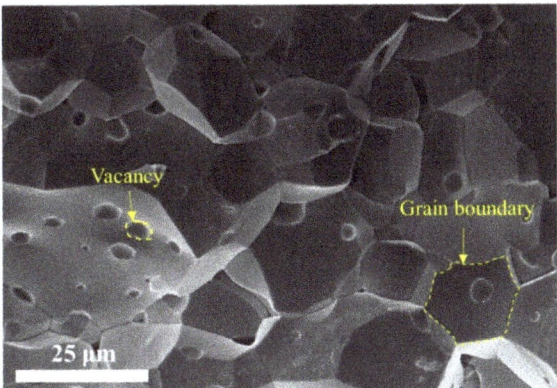

Figure 2. Morphology of high permeability MnZn ferrite. Grain boundary and air vacancy were indicated within the dashed lines.

Table 1. Composition analysis results of MnZn ferrite materials.

Element	Fe	Zn	Mn	Na	P	Si	Ca
wt.%	48.7568	13.4317	11.7299	0.0681	0.0676	0.0473	0.0288

A mu-metal film is adhered to the exterior of the ferrite magnetic shield. The thickness of the film ranges from 0.1 mm to 0.5 mm. The real component of the complex permeability of mu-metal film is 30,000, while the imaginary component is 1000. There are two methods for pasting. The first one involved tightly pasting the ferrite and mu-metal film, and the outside of the film was fixated using adhesive tape. The second one involved attaching the ferrite and mu-metal film using an insulating double-sided tape, creating a space between them. Both cases were analyzed in this work.

2.2. Analysis Method of Shielding Coefficient and Noise

The magnetic shielding coefficient is represented by the coefficient S. It is defined as the ratio of the magnetic flux density B_{shield} at the center point of the magnetic shield when the shield was present to the magnetic flux density B_0, with the shield at the same point without the shield. The equation is as follows [25]:

$$S = \frac{B_{shield}}{B_0} \quad (1)$$

The magnetic shielding coefficient for complex structures can be calculated using FEM and the commercial program ANSYS Electromagnetics Suite 19.2 (ANSYS Maxwell 3D, Canonsburg, PA, USA). To generate a highly uniform space magnetic field during the simulation of the static shielding coefficient, the magnetic field was simulated using the boundary condition approach. During the simulation, the magnetic flux density used was 25 µT, which was the actual measured magnetic flux density of the environment. Magnetic noise was an additional essential parameter for evaluating the performance of magnetic shielding. It consisted of the residual environmental magnetic noise after magnetic shielding and magnetic noise produced by magnetic shielding materials. The multilayer magnetic shield could attain a shielding coefficient of 10^4 to 10^6. The residual ambient magnetic noise was less than 0.1 fT/Hz$^{1/2}$, and the magnetic noise generated by magnetic shielding materials dominated.

Generally, reciprocal approaches were used to calculate the magnetic noise δB. The magnetic noise was obtained by calculating the power loss, P_s, in the material. P_s was caused by the magnetic field generated by the known excitation coil. The area was A, and the carry current is I [14,15].

$$\delta B = \frac{\sqrt{8k_B T}\sqrt{P_s}}{\omega A I} \quad (2)$$

where the Boltzmann constant is k_B, angular frequency is ω, and the Kelvin temperature is T. The power loss of low-frequency materials was mainly composed of eddy current power loss $P_e = \int_V \frac{1}{2}\sigma E^2 dV$ and hysteresis power loss $P_h = \int_V \frac{1}{2}\mu'' H_m^2 dV$. The magnetic noise of FMCS is:

$$\delta B_{FBCS} = \frac{\sqrt{8k_B T}\sqrt{P_e^{tot} + P_h^{tot} + P_e^{mu} + P_h^{mu}}}{\omega A I} \quad (3)$$

where P_e^{tot} and P_h^{tot} are the eddy current power loss and hysteresis power loss of ferrite shield, respectively. P_e^{mu} and P_h^{mu} are the eddy current power loss and hysteresis power loss of mu-metal film, respectively.

3. Experimental Calculation Results and Discussion

3.1. Magnetic Shielding Coefficient

A Ferrite magnetic shield is difficult to manufacture and process, especially when a large size is required. For this reason, a cylindrical ferrite shield is usually made of several short ferrite annuli pasted with ceramic adhesive. The thickness of the ceramic adhesive and the unevenness of the contact surface led to air gaps [23]. The air gap between magnetic annuli causes magnetic leakage. The cloud diagram of magnetic flux density when the air gap width was 0.5 mm is shown in Figure 3a, and Figure 3b demonstrates that the FMCS structure effectively prevented magnetic leakage at the air gap of the ferrite magnetic shield. The thickness of mu-metal film used for simulation was 0.5 mm.

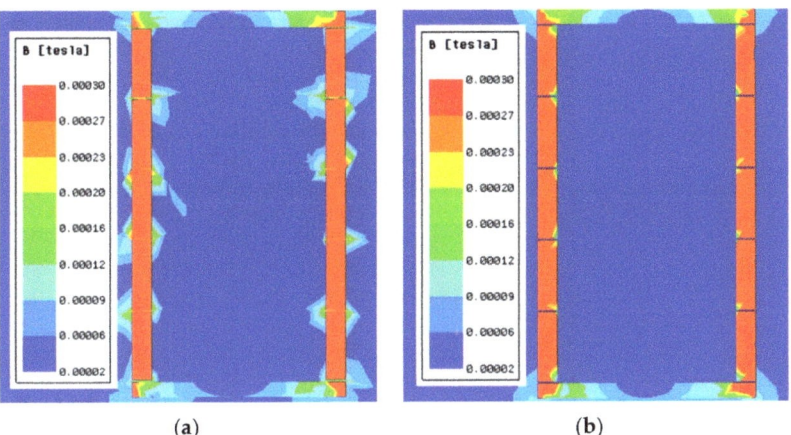

Figure 3. Magnetic flux density maps along the x–z plane; (**a**) conventional ferrite shield and (**b**) FMCS.

The relationship between mu-metal film thickness and magnetic shielding coefficient was investigated to quantify the inhibitory effect of FMCS structure on the air gap of the ferrite shield and to enhance the shielding coefficient. The conventional ferrite magnetic shielding coefficient variation with the gap width was first calculated. As shown in Figure 4, the x- and z-direction shielding coefficients decreased by 16.2% and 98.1%, respectively, when the gap width changed from 0 to 0.5 mm.

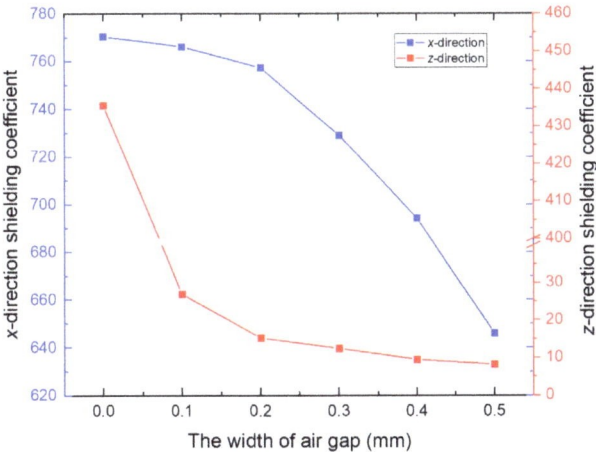

Figure 4. Relationship between magnetic shielding coefficient of the conventional ferrite and the gap width. When the gap width increases from 0 mm to 0.5 mm, the shielding coefficient for the x- and z-direction gradually decreases. The blue and red lines denote the x- and z-direction shielding coefficients, respectively.

Figure 5 illustrates the relationship between the magnetic shielding coefficient of the FMCS structure and the gap width for different film thicknesses. As shown in Figure 5a, the x-direction shielding coefficient of FMCS structure with varying film thicknesses decreases gradually as the gap width increases. When gap width increased from 0 to 0.5 mm, the shielding coefficient decreased by approximately 3.8%. Shielding coefficient increased as layer thickness increased. When the gap width was equivalent to 0.5 mm, the shielding coefficient of film (0.1 mm and 0.5 mm) increased by 9.8% and 13.2%, respectively, compared with conventional ferrite shielding. FMCS structure effectively suppressed the influence of the air gap on the x-direction shielding coefficient of conventional ferrite magnetic shielding. Figure 5b shows the relationship between the z-direction magnetic shielding coefficient of the FMCS structure and the width of the gap for different film thicknesses. Although the z-direction shielding coefficient decreased in the presence of an air gap, the width of the gap did not affect the shielding coefficient. In addition, the shielding coefficient improved with an increase in the layer thickness. When the gap width was 0.5 mm and film thickness was 0.1 mm, the z-direction shielding coefficient of the FMCS structure was 350, whereas the traditional ferrite magnetic shielding coefficient was only 15. The shielding coefficient increased by a factor of 22.3.

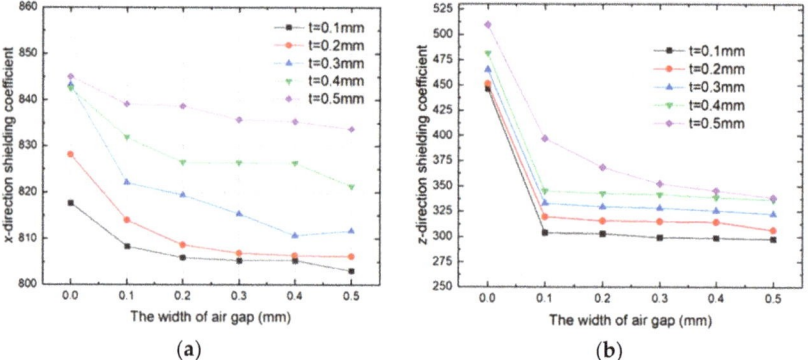

Figure 5. Relationship between the FMCS structure magnetic shielding coefficient and the gap width for different film thickness; (**a**) x-direction magnetic shielding coefficient, (**b**) z-direction magnetic shielding coefficient.

In addition, the ferrite and mu-metal film were attached to the FMCS framework with insulating double-sided tape, resulting in a gap between them. This configuration was also examined. Figure 6 depicts the results of the shielding coefficient calculation. When there was an air gap (width = 0.1 mm) between ferrite and mu-metal film (thickness = 0.5 mm), the shielding coefficient decreased by approximately 4.1% and 76.2%, respectively, compared to that in close contact. In practical application, ferrite and mu-metal films must be firmly pasted.

3.2. Magnetic Shield Noise

Magnetic noise is another important parameter for evaluating the shielding performance. Magnetic noise limits the enhancement of the sensitivity index, particularly in ultra-high-sensitivity magnetometers [26]. The sensitive axis of the magnetic field is generally oriented in the radial direction with an excellent magnetic shielding coefficient [2,4,27]. This study analyzed the influence of air gap on the shielding coefficient of conventional ferrite magnetic shields and FMCS structures. As shown in Figure 7, the air gap width increased from 0 to 0.5 mm, the thermal magnetization noise increased by 34.3%, and the eddy current noise decreased. However, the eddy current noise accounted for less than 1% of the total noise, which can be ignored.

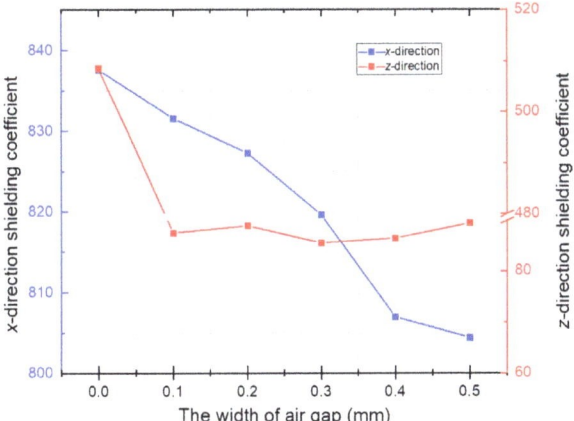

Figure 6. Shielding coefficient with an air gap between ferrite and mu-metal film. The ferrite and mu-metal film are fixed with insulating double-sided tape. The gap width caused by the isolation of double-sided adhesive tape is 0.1 mm.

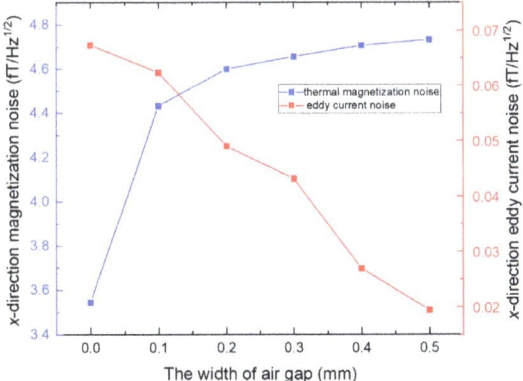

Figure 7. Magnetic noise of conventional ferrite magnetic shield in the presence of an air gap between ferrite annuli.

The magnetic noise calculation results of the FMCS structure are shown in Figure 8. As illustrated in Figure 8a, when mu-metal film was close to the ferrite magnetic shield, the FMCS structure demonstrated a noticeable effect on suppressing the air gap of the ferrite magnetic annuli. The magnetic noise increased by 4.5% when the air gap width was extended from 0 to 0.5 mm, but the magnetic noise did not increase with the increase of air gap width. When the air gap width was 0.5 mm and the film thickness was 0.1 mm, the magnetic noise of the FMCS structure was 21% less than that of the typical ferrite magnetic shield. The magnetic noise could be reduced further by increasing the film thickness. Figure 8b shows the magnetic noise of the FMCS structure when the ferrite shield and the film were adhered with double-sided tape and there was a gap between them. In this case, despite being able to maintain the air gap, the magnetic noise reduction was only 6.3% less than with standard ferrite magnetic shielding. Increasing the thickness of the film could also reduce the magnetic noise, but the effect was minimal.

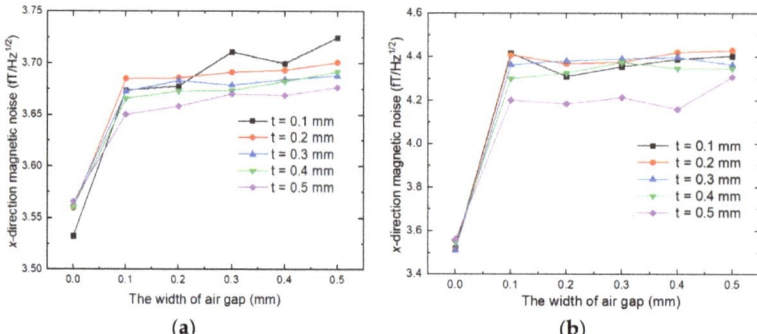

Figure 8. The magnetic noise calculation results of FMCS structure: (**a**) magnetic noise when there is no air gap between ferrite and mu-metal film and (**b**) magnetic noise when there is an air gap between ferrite and mu-metal film. The ferrite and mu-tal film are fixed with insulating double-sided tape.

4. Experimental Setup and Results

As the current commercial atomic magnetometer has a sensitivity of tens of fT, it is difficult to detect the magnetic noise of ferrite magnetic shielding. The noise may be measured by building an ultra-high sensitive magnetic shielding measurement device based on the spin-exchange relaxation-free (SERF) effect. The device is complicated and is easily affected by optical or electrical noise detection. It is challenging to isolate the magnetic noise. Therefore, this study only measured the shielding coefficient to verify the advantages of the proposed FMCS structure and the simulation results. The schematic illustration of the magnetic shielding coefficient measurement platform is shown in Figure 9. The magnetic shield composed of one layer of aluminum magnetic shield and four layers of mu-metal magnetic shields suppressed the external magnetic field to avoid the fluctuation of the external magnetic field from distorting the measurement results. The combined magnetic shield had a shielding coefficient of approximately 10^5 and a residual magnetic field of approximately 0.5 nT. The triaxial coil generated a known uniformly stable magnetic field. The magnetic field generated by the coil was enhanced under the impact of the ferromagnetic boundary of the outer combined magnetic shield. The three-axis coil was calibrated within the combined magnetic shield. The radial and axial coils were 25 nT/mA and 35 nT/mA, respectively. A commercial atomic magnetometer was used to measure the magnetic field in the FCMS structure.

For MnZn ferrite materials, $\sigma \approx 1\ \Omega^{-1}\mathrm{m}^{-1}$, correspondingly, the threshold frequency was higher than kHz, and below the threshold frequency, the shielding coefficient of ferrite magnetic shield did not change with the frequency, while the application frequency of SERF magnetometer was below 100 Hz. We also measured the frequency dependence of the shielding coefficient of the FMCS structure from DC to 100 Hz. The change of the shielding coefficient with frequency was only 2%. Figure 10 depicts the residual magnetic field in the FMCS structure and the conventional magnetic shield when the known DC external magnetic field was applied. Due to the increase in magnetic permeability, the x-direction shielding coefficient (a) of the conventional ferrite magnetic shield increased with the external magnetic field. When the external magnetic field was between 400 and 1800 nT, the average shielding coefficient was 703.26, which was close to the simulation value, proving that the analysis was accurate. The external magnetic field applied when measuring the z-direction magnetic shielding coefficient of conventional magnetic shielding was between 50 and 120 nT. The reason why the magnetic field applied in the z-direction was smaller than that in the x-direction was to prevent the residual magnetic field inside the shield from exceeding the measuring range of the atomic magnetometer due to the smaller measuring range of the atomic magnetometer and the smaller shielding coefficient in the z-direction. The average value of the z-direction shielding coefficient was 36.3, which confirmed that the air gap of a conventional ferrite magnetic shield with multiple annuli significantly impacted the axial shielding coefficient. Our experiments proved that the FMCS structures were superior to the conventional ferrite magnetic shielding. The average shielding coefficients of the FMCS structure (c) in the x and z directions were 819.22 and 383.6, respectively, which was an increase of 16.5% and 91.3% over the conventional ferrite magnetic shielding. The difference between the measured value and the simulated value was mainly due to the air gap during the film pasting process and the permeability of the material.

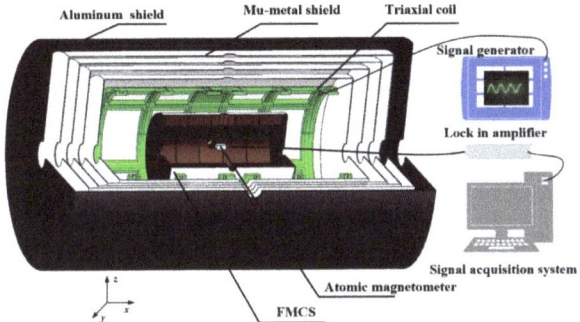

Figure 9. Schematic illustration of the magnetic shielding coefficient measurement platform. The ferrite shield size is the same as the structure used in the simulation, and the thickness of the film is 0.5 mm.

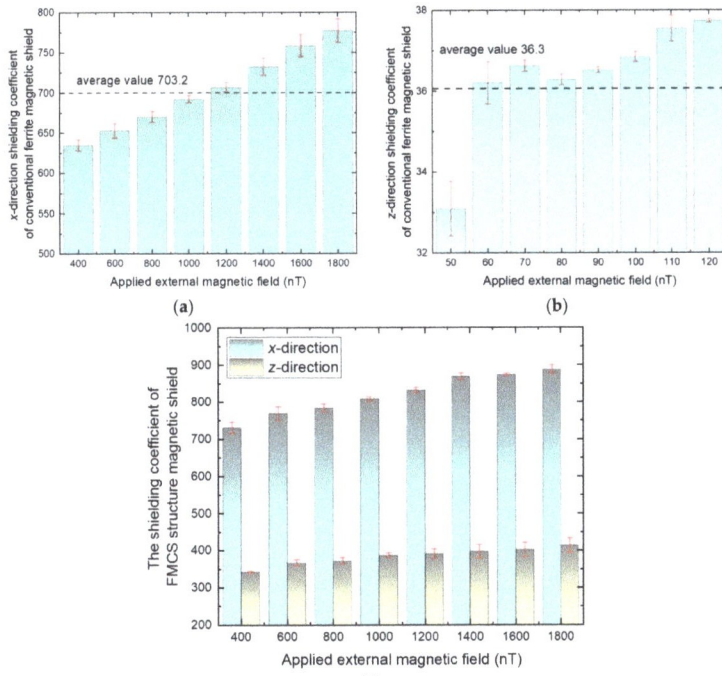

Figure 10. The (**a**) x-direction and (**b**) z-direction shielding coefficient of conventional ferrite magnetic shield versus applied external magnetic field; (**c**) Shielding coefficient of the FMCS when the external magnetic field is applied.

5. Conclusions

In this study, a high-performance magnetic shielding structure composed of ferrite and mu-metal film is proposed. Mu-metal film with a high relative permeability was used to suppress the reduction of magnetic shielding coefficient resulting from magnetic leakage in the magnetic annuli gap. The thickness of the mu-metal film, between 0.1 and 0.5 mm, prevented the increase in combined magnetic noise. FEM was used to calculate the shielding coefficient and magnetic noise of ferrite and mu-metal film combined shielding layer under varying gap widths of magnetic annuli. The results showed that the magnetic shielding coefficient and magnetic noise of the FMCS structures barely changed as the gap width increased. Under the gap width of 0.5 mm, the x-direction shielding coefficient increased by 13.2%, whereas the magnetic noise decreased by 21% compared to the conventional ferrite magnetic shielding with a gap. The shielding coefficient in the z-direction increased by 22.3 times. In addition, the influence of different assembly methods of ferrite and mu-metal film on the magnetic shielding performance was analyzed. The shielding performance declined when there was a gap between ferrite and the double-sided adhesive tape-pasted film. The gap between the film and the ferrite magnetic shielding must be avoided when the FMCS structure was employed to increase the shielding coefficient and decrease the magnetic noise. Amorphous, nanocrystalline, and other thin film materials may be utilized apart from mu-metal materials.

The experiments validated the shielding coefficient simulation results for the combined magnetic shield. The experimental results showed that the shielding coefficient of the composite magnetic shield was 16.5% and 91.3% greater than that of the conventional ferrite magnetic shield, respectively. The main difference was between the actual relative permeability of the mu-metal film and its simulated relative permeability. The high-performance combined magnetic shielding proposed in this study is of great significance for enhancing the sensitivity of atomic sensors.

Author Contributions: X.F. and D.M.: Investigation, data curation and writing—original draft. B.S.: simulation analysis, resources, and visualization. X.X.: data curation and supervision. W.Q.: construction of measuring device and data acquisition. Z.X. and Y.Z.: funding acquisition, supervision, and project administration. All authors have read and agreed to the published version of the manuscript.

Funding: This work was supported by the China Postdoctoral Science Foundation under Grant No. 2022M710321, the National Natural Science Foundation of China under Grant No. 62203028, and the National Natural Science Foundation of China under Grant No. 4191002.

Data Availability Statement: Not applicable.

Conflicts of Interest: The authors declare no conflict of interest.

References

1. Kim, Y.J.; Savukov, I. Parallel high-frequency magnetic sensing with an array of flux transformers and multi-channel optically pumped magnetometer for hand MRI application. *J. Appl. Phys.* **2020**, *128*, 154503. [CrossRef]
2. Zhang, R.; Mhaskar, R.; Smith, K.; Balasubramaniam, E.; Prouty, M. Vector measurements using all optical scalar atomic magnetometers. *J. Appl. Phys.* **2021**, *129*, 044502. [CrossRef]
3. Chalkov, V.V.; Shevchenko, A.N. Studying the atomic gyroscope magnetic shield residual magnetization by comparing parametric resonance signals. *J. Phys. Conf. Ser.* **2020**, *1536*, 012013. [CrossRef]
4. Chen, Y.; Zhao, L.; Zhang, N.; Yu, M.; Ma, Y.; Han, X.; Zhao, M.; Lin, Q.; Yang, P.; Jiang, Z. Single beam Cs-Ne SERF atomic magnetometer with the laser power differential method. *Opt. Express* **2022**, *30*, 16541–16552. [CrossRef]
5. Kitching, J. Chip-scale atomic devices. *Appl. Phys. Rev.* **2018**, *5*, 031302. [CrossRef]
6. Rea, M.; Holmes, N.; Hill, R.M.; Boto, E.; Leggett, J.; Edwards, L.J.; Woolger, D.; Dawson, E.; Shah, V.; Osborne, J.; et al. Precision magnetic field modelling and control for wearable magnetoencephalography. *NeuroImage* **2021**, *241*, 118401. [CrossRef]
7. Forgan, E.M.; Muirhead, C.M.; Rae, A.I.M.; Speake, C.C. A proposal for detection of absolute rotation using superconductors and large voltages. *Phys. Lett. A* **2020**, *386*, 126994. [CrossRef]
8. Pan, D.; Lin, S.; Li, L.; Li, J.; Jin, Y.; Sun, Z.; Liu, T. Research on the design method of uniform magnetic field coil based on the MSR. *IEEE Trans. Ind. Electron.* **2020**, *67*, 1348–1356. [CrossRef]
9. He, K.; Wan, S.; Sheng, J.; Liu, D.; Gao, J.H. A high-performance compact magnetic shield for optically pumped magnetometer-based magnetoencephalography. *Rev. Sci. Instrum.* **2019**, *90*, 064102. [CrossRef]

10. Fan, W.; Quan, W.; Liu, F.; Xing, L.; Liu, G. Suppression of the bias error induced by magnetic noise in a spin-exchange relaxation-free gyroscope. *IEEE Sens. J.* **2019**, *19*, 9712–9721. [CrossRef]
11. Li, J.; Wei, Q.; Han, B.; Liu, F.; Xing, L.; Liu, G. Multilayer cylindrical magnetic shield for SERF atomic Co-magnetometer application. *IEEE Sens. J.* **2019**, *19*, 2916–2923. [CrossRef]
12. Savukov, I.; Kim, Y.J. Investigation of magnetic noise from conductive shields in the 10–300 kHz frequency range. *J. Appl. Phys.* **2020**, *128*, 234501. [CrossRef]
13. Eshraghi, M.J.; Kim, J.M. Low-frequency magnetic noise reduction using superconducting lead sheet magnetic shield for a two-stage pulse-tube cryocooler. *Annu. Rep. Riss* **2009**, *6*, 77–82.
14. Kornack, T.W.; Smullin, S.J.; Lee, S.K.; Romalis, M.V. A low-noise ferrite magnetic shield. *Appl. Phys. Lett.* **2007**, *90*, 223501. [CrossRef]
15. Lee, S.K.; Romalis, M.V. Calculation of magnetic field noise from high-permeability magnetic shields and conducting objects with simple geometry. *J. Appl. Phys.* **2008**, *103*, 084904. [CrossRef]
16. Smythe, W.R. *Static and Dynamic Electricity*; McGraw-Hill: New York, NY, USA, 1968; pp. 188–190.
17. Petrescu, L.-G.; Petrescu, M.C.; Ionita, V.; Cazacu, E.; Constantinescu, C.-T. Magnetic Properties of Manganese-Zinc Soft Ferrite Ceramic for High Frequency Applications. *Materials* **2019**, *12*, 3173. [CrossRef]
18. Jin, X.Q.; Zhang, Z.S.; Chen, W. High frequency, wide temperature range, high DC superposition characteristics, high permeability MnZn ferrite material. *J. Magn. Mater. Devices* **2005**, *36*, 54–55.
19. Bevan, D.; Bulatowitz, M.; Clark, P.; Flicker, J.; Griffith, R.; Larsen, M.; Luengo-Kovac, M.; Pavell, J.; Rothballer, A.; Sakaida, D.; et al. Nuclear magnetic resonance gyroscope: Developing a primary rotation sensor. In Proceedings of the IEEE International Symposium on Inertial Sensors and Systems, Moltrasio, Italy, 26–29 March 2018; pp. 1–2.
20. Fu, Y.; Sun, J.; Ruan, J.; Quan, W. A nanocrystalline shield for high precision co-magnetometer operated in spin-exchange relaxation-free regime. *Sens. Actuators A Phys.* **2022**, *339*, 113487. [CrossRef]
21. Walker, T.G.; Larsen, M.S. Spin-exchange-pumped NMR gyros. *Adv. Atom. Mol. Opt. Phys.* **2016**, *65*, 373–401. [CrossRef]
22. Lu, J.; Sun, C.; Ma, D.; Yang, K.; Zhao, J.; Han, B.; Quan, W.; Zhang, N.; Ding, M. Effect of gaps on magnetic noise of cylindrical ferrite shield. *J. Phys. D Appl. Phys.* **2021**, *54*, 255002. [CrossRef]
23. Lu, J.; Ma, D.; Yang, K.; Quan, W.; Zhao, J.; Xing, B.; Han, B.; Ding, M. Study of magnetic noise of a multi-annular ferrite shield. *IEEE Access* **2020**, *8*, 40918–40924. [CrossRef]
24. Ma, D.; Lu, J.; Fang, X.; Yang, K.; Wang, K.; Zhang, N.; Han, B.; Ding, M. Parameter modeling analysis of a cylindrical ferrite magnetic shield to reduce magnetic noise. *IEEE Trans. Ind. Electron.* **2022**, *69*, 991–998. [CrossRef]
25. Zhang, H.; Zou, S.; Chen, X.Y. Parameter modeling analysis and experimental verification on magnetic shielding cylinder of all-optical atomic spin magnetometer. *J. Sens.* **2014**, *44*, 1177–1180.
26. Dang, H.B.; Maloof, A.C.; Romalis, M.V. Ultrahigh sensitivity magnetic field and magnetization measurements with an atomic magnetometer. *Appl. Phys. Lett.* **2010**, *97*, 151110. [CrossRef]
27. Zhang, Y.; Huang, S.; Xu, F.; Hu, Z.; Lin, Q. Multi-channel spin exchange relaxation free magnetometer towards two-dimensional vector, magnetoencephalography. *Opt. Express* **2019**, *27*, 597–607. [CrossRef]

Article

Analysis and Suppression of Thermal Magnetic Noise of Ferrite in the SERF Co-Magnetometer

Haoying Pang [1,2], Feng Liu [1,2,*], Wengfeng Fan [1,2], Jiaqi Wu [1,2], Qi Yuan [1,2], Zhihong Wu [1,2] and Wei Quan [2,3,*]

1. School of Instrumentation and Optoelectronic Engineering, Beihang University, Beijing 100191, China
2. Zhejiang Provincial Key Laboratory of Ultra-Weak Magnetic-Field Space and Applied Technology, Hangzhou Innovation Institute, Beihang University, Hangzhou 310051, China
3. Innovative Research Institute of Frontier Science, Beihang University, Beijing 100191, China
* Correspondence: liufeng1991@buaa.edu.cn (F.L.); quanwei@buaa.edu.cn (W.Q.)

Abstract: The ferrite magnetic shield is widely used in ultra-high-sensitivity atomic sensors because of its low noise characteristics. However, its noise level varies with temperature and affects the performance of the spin-exchange relaxation-free (SERF) co-magnetometer. Therefore, it is necessary to analyze and suppress the thermal magnetic noise. In this paper, the thermal magnetic noise model of a ferrite magnetic shield is established, and the thermal magnetic noise of ferrite is calculated more accurately by testing the low-frequency complex permeability at different temperatures. A temperature suppression method based on the improved heat dissipation efficiency of the ferrite magnetic shield is also proposed. The magnetic noise of the ferrite is reduced by 46.7%. The experiment is basically consistent with the theory. The sensitivity of the co-magnetometer is decreased significantly, from $1.21 \times 10^{-5} \circ/s/Hz^{1/2}$ to $7.02 \times 10^{-6} \circ/s/Hz^{1/2}$ at 1 Hz. The experimental results demonstrate the effectiveness of the proposed method. In addition, the study is also helpful for evaluating the thermal magnetic noise of other materials.

Keywords: ferrite complex permeability; low frequency; SERF co-magnetometer; thermal magnetic noise

1. Introduction

Atomic magnetometers and co-magnetometers based on a spin-exchange relaxation-free (SERF) regime have been widely used in frontier science research [1,2], ultra-weak magnetic field measurement [3–5], inertial navigation [6,7] and magnetoencephalography [8,9] because of their ultra-high-sensitivity magnetic field and inertial measurement capability. In particular, the SERF co-magnetometer has the potential for high accuracy, miniaturization and low cost, and has been seen as the development direction for the next generation of rotational sensors [10–12]. The passive magnetic shield is a crucial part of the co-magnetometer [13]. It is usually made of μ-metal material with high magnetic permeability, which shields the external environment from electromagnetic interference and provides the necessary low magnetic field environment for the SERF co-magnetometer [14,15].

Although passive magnetic shields can provide adequate shielding against external interference, the shielding material generates magnetic noise due to Johnson currents, which becomes a limiting factor in improving the sensitivity of SERF co-magnetometers [16,17]. In recent years, ferrite materials have been used as inner shielding materials for SERF co-magnetometers and magnetometers due to their high resistivity and low loss [18]. Kornack et al. first used ferrite in the SERF magnetometer. The noise level reached 0.75 fT/Hz$^{1/2}$, 25 times lower than the μ-metal material [19]. Since then, ferrite has been widely used in various high-sensitivity atomic sensors, such as atomic magnetometers [5], nuclear magnetic resonance gyroscopes [20], etc. In 2019, Fan et al. used a ferrite magnetic shield in a SERF co-magnetometer and noted that ferrite with low magnetic noise and high temperature stability is the superior inner shielding material [16].

Many researchers have studied the noise characteristics of ferrite to reduce magnetic noise further [21,22]. The magnetic noise of cylindrical ferrite was calculated using the fluctuation–dissipation theorem created by Kornack [19]. They pointed out that the magnetic noise of ferrite is mainly related to its complex permeability, temperature and size. Subsequently, Lee et al. performed calculations for shields with simple geometries [23]. In addition, Ma et al. developed a more accurate model to analyze the longitudinal and transverse magnetic noise of the ferrite [24]. The magnetic noise is also effectively reduced by optimizing the dimensional parameters of the cylindrical ferrite shield. Yang proposed a more accurate method for measuring the complex permeability of ferrite [25]. In addition, for large-size ferrite, Lu studied the magnetic noise of the multi-annular ferrite shield and pointed out that the effect of the gap is negligible when the gap width is less than 0.01 mm [26]. However, the temperature properties of the material have not been studied in the literature. The characteristics of high-temperature superconducting materials [27] and soft magnetic materials are affected by temperature. Chrobak et al. obtained the temperature dependence of the critical current of the film from the temperature dependence of the imaginary part of the AC susceptibility of the high-temperature superconducting material [28]. Additionally, Fiorillo et al. analyzed the loss and permeability versus temperature in soft magnetic ferrites [29]. This does not apply to the low-frequency range required by the SERF co-magnetometer. Therefore, our work needs to consider the effect of temperature on magnetic noise at low frequencies and the variation in ferrite complex permeability at different temperatures.

In this paper, the thermal magnetic noise of ferrite is analyzed theoretically, and a temperature suppression method based on improving the heat dissipation efficiency is proposed. First, we tested the complex permeability of the ferrite at different temperatures and calculated the variation in magnetic noise according to temperature. After that, a SERF magnetometer was used to compare the magnetic noise before and after the ferrite temperature suppression. The theory's accuracy was verified by the basic agreement between theory and experiment. Finally, the sensitivity of the co-magnetometer system was tested, and it was experimentally demonstrated that the sensitivity performance of the system could be significantly improved by reducing the ferrite temperature.

2. Methods

According to the generalized Nyquist theory, the random motion of the electric charge caused by the thermal motion of atoms in a conductor will produce an arbitrary change in current, and the change in current will produce a magnetic field [23,30]. Therefore, atomic thermal motion inside the ferrite generates magnetic interference in the shielded space, and this interference is the primary source of magnetic noise in the SERF co-magnetometer. The fluctuation–dissipation theorem is used to calculate the magnetic noise. Assuming that an excitation coil with an applied oscillating current is placed in a ferrite shield and another pickup coil is placed in the same position, the total power loss in the shielded cylinder is defined as $P(f)$. The relationship between the magnetic noise $\delta B(f)$ and $P(f)$ is as follows:

$$\delta B(f) = \frac{\sqrt{4k_B T}\sqrt{2P(f)}}{2\pi f A I}, \qquad (1)$$

where k_B is the Boltzmann constant, T is the absolute temperature of the shields, A is the area of the excitation coil and I and f are the current and frequency of the excitation coil. $P(f)$ contains eddy-current loss, hysteresis loss and excess loss. For the ferrite shield, its power loss at low frequency is mainly hysteresis loss, so the expression of $P(f)$ is

$$P(f) \approx P_{hyst}(f) = \int_V \pi f \mu'' H^2 dV, \qquad (2)$$

where $P_{hyst}(f)$ is the hysteresis loss and V is the volume of the magnetic shield. $\mu = \mu' - i\mu''$ is the complex permeability of the magnetic material; μ' and μ'' are the real and imaginary parts of the complex magnetic permeability, respectively. H is the magnetic field strength. The above equation represents the volume integral of the magnetic shield in an oscillating magnetic field with an amplitude of H.

Considering the ferrite magnetic shield as a long and hollow cylinder, the longitudinal magnetic field noise in the shield is [19]

$$\delta B(f) = \frac{0.26\mu_0}{r\sqrt{t}}\sqrt{\frac{2k_B T \mu''}{\pi f \mu'^2}}, \qquad (3)$$

μ_0 is the vacuum permeability, and r and t are the inner diameter and thickness of the shield. The noise in this equation is directly related to the temperature. In addition, the complex permeability of the ferrite increases with increasing temperature. Therefore, the analysis of the thermal magnetic noise of ferrite requires accurate testing of the low-frequency complex permeability at different temperatures.

The coils are wound uniformly around the ferrite sample, corresponding to an equivalent resistive and inductive circuit. The complex permeability of the ferrite can be measured by measuring the inductance and resistance parameters of the circuit. The relationship between complex permeability, inductance and resistance is as follows (see Appendix A for details):

$$L_s = \frac{\mu_0 A_e N^2}{l_e}\mu', \qquad (4)$$

$$R_y = \frac{2\pi f_c \mu_0 A_e N^2}{l_e}\mu'' + R_w, \qquad (5)$$

where L_s is the inductance of the sample, l_e is the effective magnetic circuit length, A_e is the effective cross-sectional area, N is the number of turns of the winding, R_y is the resistance value containing the resistance of the test coil, R_w is the resistance value of the measurement coil and f is the test frequency.

The complex magnetic permeability at a specific magnetic field strength was obtained using the above method. However, the complex permeability at zero magnetic field conditions is required for theoretical calculations. Under small magnetic field excitation, the complex permeability is related to the magnetic field strength as $\mu_h = \mu_a + \eta H$. Here, μ_h is the amplitude permeability, μ_a is the permeability at zero fields and η is the Rayleigh constant. Therefore, in this paper, the complex permeability is tested at different magnetic field strengths, and then the complex permeability is obtained at zero field conditions based on the above relationship.

3. Experimental Setup

The experimental setup in this section is divided into two parts. First, a complex permeability test system is built to test the complex permeability of ferrite at different temperatures and quantitatively analyze the relationship between magnetic noise and temperature. Then, the K-Rb-[21]Ne co-magnetometer system is built to test the effectiveness of the thermal and magnetic noise suppression method.

The ferrite complex permeability test system is shown in Figure 1. The magnetic shielding system is used to shield the external environment from magnetic field interference and contains an oven, a sample of ferrite wound with demagnetizing and enameled wires. The demagnetizing wire is used to demagnetize the sample, and the enameled wire is used as the test coil. The inner diameter of the ferrite sample is 15 mm, the outer diameter is 25 mm, the height is 7.5 mm and the permeability is 2000. A self-made temperature control system is used to control the temperature of the sample, and the temperature control accuracy is 10 mK. The Omicron-lab BODE 100 impedance analyzer is used to apply different excitation intensities to the sample and to measure the inductance and resistance. The range of magnetic excitation applied to the sample is 0.3–0.8 A/m. To reduce the influence of remanence, the shielding system and the sample are demagnetized, respectively. Furthermore, the temperature is changed and left for at least half an hour to allow the sample to equilibrate with the ambient temperature. The measurements are repeated ten times for each set of conditions to ensure the precision of the test.

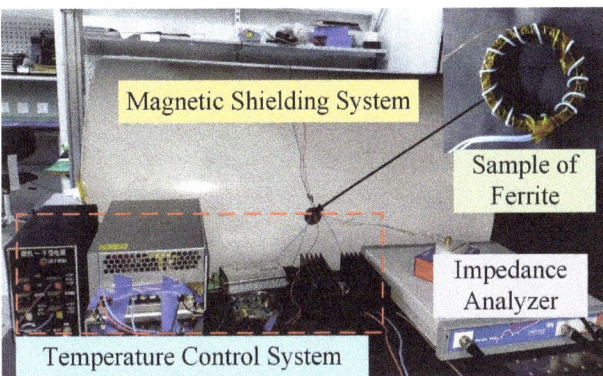

Figure 1. Experimental test system for temperature characteristics of ferrite complex permeability.

The schematic diagram of the K-Rb-^{21}Ne co-magnetometer system is shown in Figure 2. The magnetic shielding system consists of the outer layer of the Permalloy shield and the inner layer of the ferrite shield. Their thicknesses are 2 mm and 6 mm, respectively. The outer shield has a permeability of about 20,000. It is used to shield the external ambient magnetic field, but its large eddy current loss results in a relatively high level of magnetic noise in the system. The inner ferrite shield is used to shield the magnetic noise of Permalloy. Therefore, the magnetic noise of the system is mainly the magnetic noise of the ferrite. This is the prominent noise studied in this paper. Inside the magnetic shielding system are coils, an oven, and an alkali metal vapor cell. Coils are used to compensate for the residual magnetic field inside the shield. The oven heats an 8 mm diameter GE180 aluminosilicate vapor cell to an operating temperature of 180 °C. External to the magnetic shielding system is the optical system, which provides the pump and probe lights necessary for the operation of the co-magnetometer. When the system is operated in SERF magnetometer mode, the vapor cell contains a mixture of K and Rb alkali metals, 50 Torr of N_2 quenching gas and 2 atm of ^{4}He buffer gas. The SERF magnetometer is used to test the magnetic noise level of the system. When the co-magnetometer system operates in gyroscope mode, only the vapor cell needs to be replaced. At this time, the vapor cell contains a mixture of K and Rb alkali metals, 2 atm of ^{21}Ne and 50 Torr of N_2. The co-magnetometer system is used to verify the effectiveness of the thermal magnetic noise analysis and suppression method proposed in this paper.

The temperature near the vapor cell of the co-magnetometer system is as high as 180 °C, while the whole system is at an ambient temperature of 25 °C. The temperature is reduced from high to low temperature by heat conduction and heat convection. The ferrite shield is part of the heat dissipation path, and its temperature is higher than the ambient temperature. The closer it is to the vapor cell, the higher its temperature. Because the performance of the co-magnetometer system is related to the volume of the magnetic shield and the vapor cell temperature, the temperature of the ferrite shield cannot be reduced by changing the cell temperature and the volume of the magnetic shield. Therefore, this paper uses the change in heat dissipation efficiency to reduce the temperature of the ferrite and suppress the thermal magnetic coupling noise. The exact method is shown in Figure 3. On the one hand, a high thermal conductivity silicon pad is wrapped around the outside of the magnetic shield to improve the heat dissipation efficiency of the shield. The thermal silicone pad has a thermal conductivity of 4 W/(m·K). On the other hand, the aerogel insulation blanket is wrapped around the outside of the oven, and the thermal conductivity of the aerogel insulation blanket is 0.02 W/(m·K), which reduces the heat transfer from the cell to the shield. This method utilizes the gaps between the magnetic shields and does not negatively affect the original system. Solutions filled with thermally conductive or insulating materials have a wide range of engineering applications, such as nuclear magnetic resonance gyroscopes.

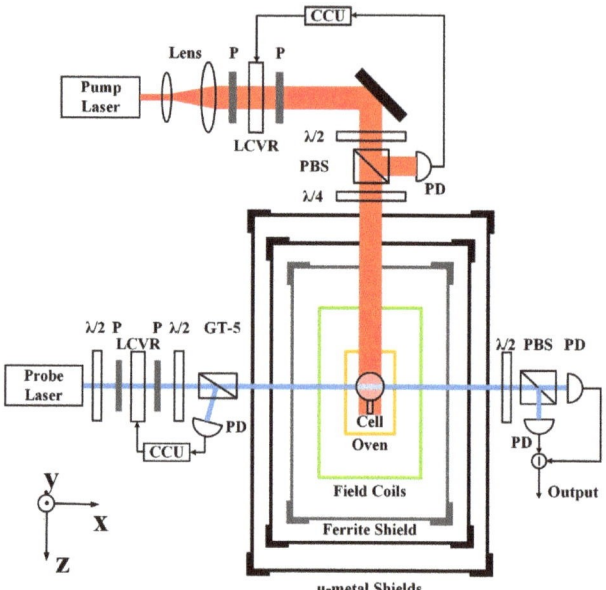

Figure 2. Schematic diagram of the co-magnetometer system. P: polarizer; LCVR: liquid crystal variable retarder; CCU: circuit control system; $\lambda/2$: half wave plate; PBS: polarizing beam splitter; $\lambda/4$: quarter wave plate; PD: photodiode; GT-5: Glan–Taylor polarizer.

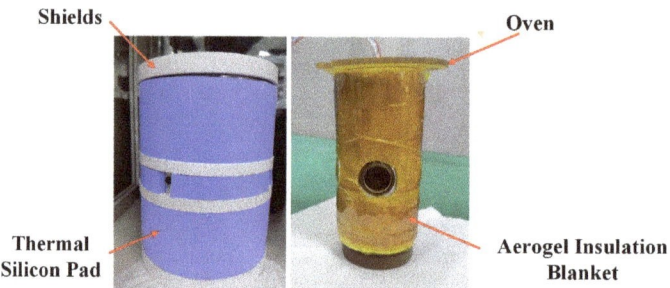

Figure 3. Temperature suppression method for the magnetic shield.

4. Results and Analysis

4.1. Measurement of Ferrite Complex Permeability at Different Temperatures

First, the BODE 100 impedance analyzer is used to apply a current I = 1.13 mA to the test coil, producing the corresponding magnetic field strength H = 0.3 A/m. BODE 100 is also used to measure resistance R_y and inductance L_e in the frequency range of 20–200 Hz. According to Equation (4), the real part of the magnetic permeability can be calculated. The imaginary part of the permeability μ'' and the resistance of the wire R_w can be obtained by linear fitting the test data according to Equation (5). After that, to get the complex permeability under zero magnetic field excitation, the complex permeability under different excitations are measured. The range of current I applied to the test coil is 1.13 mA to 3.76 mA, and the corresponding magnetic field strength is 0.3 to 1 A/m. By linear fitting the magnetic field strength H and the permeability, the permeability at zero magnetic field strength can be obtained. The complex permeability is tested at 30 °C, 50 °C, 70 °C and 90 °C for different intensity excitations, respectively. As shown in Figure 4, the different marks indicate the actual measured values of complex permeability at different temperatures, the lines indicate the fitted curves of the actual measured values and the error bars are the standard deviations of 10 times test at different temperatures. The complex permeability at zero excitation field at different temperatures is obtained from the fitted curves and is summarized in Table 1. Equation (3) is used to calculate the magnetic noise according to the size of the ferrite in the co-magnetometer. The inner and outer diameters of the ferrite magnetic shield are 72 mm and 84 mm, respectively. The theoretical magnetic noise at 1 Hz is shown in Table 1. When the ferrite temperature is reduced from 90 °C to 30 °C, the magnetic noise at 1 Hz is reduced from 12.93 fT/Hz$^{1/2}$ to 7.45 fT/Hz$^{1/2}$.

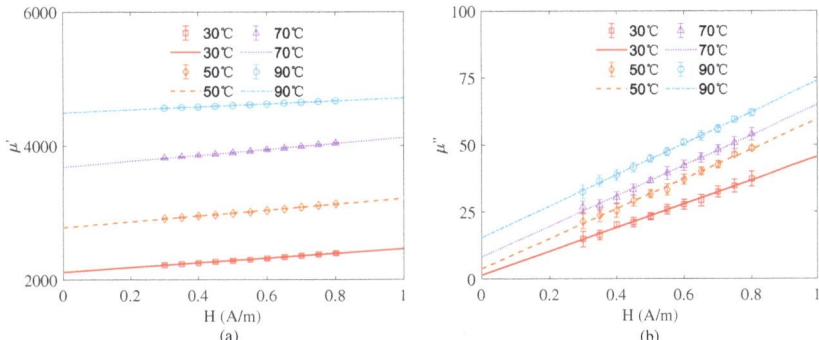

Figure 4. (**a**) Real part of permeability at different temperatures. (**b**) Imaginary part of permeability at different temperatures.

Table 1. Variation in ferrite complex permeability with temperature.

Temperature (°C)	μ'	μ''	$\delta B(f)$ at 1 Hz (fT/Hz$^{1/2}$)
30	2113.4	1.37	7.45
50	2777.3	3.77	9.77
70	3679.0	8.23	11.22
90	4494.6	15.42	12.93

4.2. Magnetic Noise Measurement

In this paper, we reduce the temperature of the magnetic shield by changing the heat transfer in the co-magnetometer, increasing the insulation of the oven as well as improving the efficiency of heat dissipation. When the magnetic shield temperature is not suppressed, the shield temperature is 79.1 °C, and the shield temperature after suppression is 33.2 °C.

Due to the extremely low conductivity of ferrite, its eddy current loss below KHz can be disregarded, and the hysteresis loss magnetic noise becomes the dominant low-frequency magnetic noise in the SERF magnetometer, which can be regarded as 1/f noise. In this subsection, the magnetic noise of the system is tested using a SERF magnetometer. The magnetic noise is tested before and after the temperature suppression of the shield, respectively. Figure 5 indicates the magnetic noise curves of the magnetometer in both cases. For comparison, the probe background noise of the system (solid black line) is also tested. The probe background noise is significantly lower than the sensitivity curve of the magnetometer, indicating that the sensitivity of the magnetometer is primarily limited by the magnetic noise. Meanwhile, in the low-frequency band, the system sensitivity curve shows a significant 1/f noise, indicating that it is primarily due to the hysteresis loss magnetic noise of the ferrite. The solid blue line and the solid red line show the sensitivity curves before and after temperature suppression of the shield, respectively. The system sensitivity is 15.24 and 8.12 fT/Hz$^{1/2}$ at 1 Hz, respectively. The noise at 30 Hz is due to the application of a calibrated magnetic field with a peak-to-peak value of 0.5 nT, and the noise at 50 Hz is the power frequency noise of the circuit. As the temperature increases, the intense thermal motion of the atoms inside the sample prevents the magnetic moment of the atoms from varying with the applied magnetic field, leading to an increase in the Rayleigh constant. Thus, the hysteresis loss increases, leading to an increase in the magnetic noise of the system. The magnetic noise of the system is reduced by 46.7% by the shield temperature suppression method proposed in this paper. The experiments basically agree with the theory, and the magnetometer test results are slightly larger than the calculated results in the previous subsection. This difference may be due to the openings in the ferrite and the gap between the ferrite annulus.

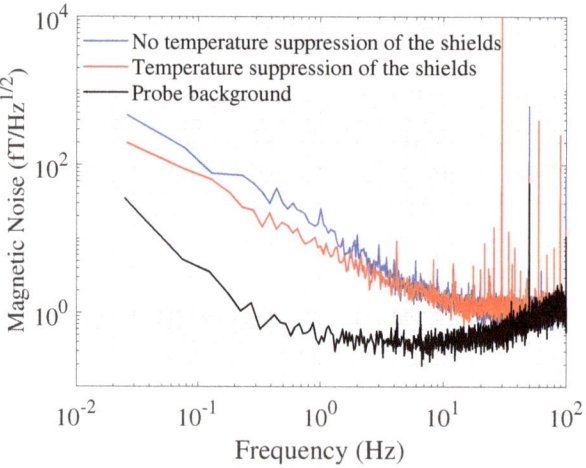

Figure 5. Magnetic noise measurement results of SERF magnetometer.

4.3. Sensitivity Measurement of SERF Co-Magnetometer

Finally, to demonstrate the accuracy of the theoretical analysis and the effectiveness of the suppression method, the sensitivity is tested in a SERF co-magnetometer.

The system signal error caused by magnetic noise can be expressed as

$$N_B(f) = \delta B(f)|G_B(f)|, \tag{6}$$

where $|G_B(f)|$ is the amplitude–frequency response of the system to the magnetic field [31]. When the magnetic noise of a co-magnetometer system is reduced, its inertial measurement sensitivity is reduced accordingly. The probe background noise of the system is tested, as shown by the solid black line in Figure 6, with a noise value of $2.21 \times 10^{-6\circ}/\text{s}/\text{Hz}^{1/2}$ at 1 Hz. The probe background noise is much lower than the system noise, indicating that the system is in a normal inertial measurement state, so the system noise reduction provided by the ferrite magnetic noise reduction will not be annihilated by the probe background noise. The blue and red solid lines in Figure 6 indicate the sensitivity curves of the co-magnetometer system before and after temperature suppression of the ferrite shield, respectively. Since the bandwidth of the SERF co-magnetometer is within 5 Hz, this work focuses on the sensitivity at low frequencies. After the ferrite temperature suppression, the system's noise is significantly reduced in the 5 Hz range. The system's sensitivity is reduced from $1.21 \times 10^{-5\circ}/\text{s}/\text{Hz}^{1/2}$ to $7.02 \times 10^{-6\circ}/\text{s}/\text{Hz}^{1/2}$ at 1 Hz. The sensitivity performance of the system is improved by 42.0%.

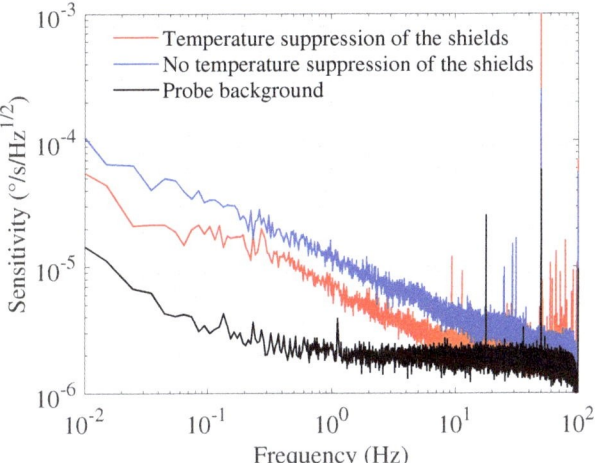

Figure 6. SERF magnetometer sensitivity measurement results.

5. Conclusions

An accurate method for analyzing and measuring the thermal magnetic noise of ferrite in the SERF co-magnetometer is proposed for the first time, and a suppression method is proposed accordingly, which significantly improves the measurement sensitivity of the co-magnetometer. Firstly, a thermal magnetic noise model of ferrite is established, and a low-frequency complex permeability test method at different temperatures is proposed. The complex permeability test system and the co-magnetometer system are built to test the thermal magnetic noise of ferrite, respectively. The results show that the measured and theoretical results of magnetic noise at different temperatures are basically consistent. The magnetic noise of the ferrite at 1 Hz is reduced from 15.24 $fT/Hz^{1/2}$ to 8.12 $fT/Hz^{1/2}$ by the proposed temperature suppression method. This study can not only evaluate the thermal magnetic noise of other low-noise magnetic materials, but is also essential to improve the sensitivity of SERF co-magnetometer and magnetometer.

Author Contributions: Conceptualization, H.P. and F.L.; methodology, H.P. and W.F.; validation, J.W., Q.Y. and Z.W.; formal analysis, H.P. and W.Q.; investigation, F.L. and W.F.; data curation, F.L.; writing—original draft preparation, H.P.; writing—review and editing, F.L., W.F. and W.Q.; project administration, W.Q. All authors have read and agreed to the published version of the manuscript.

Funding: This research was supported in part by the National Science Fund for Distinguished Young Scholars Nos.61925301, in part by the National Natural Science Foundation of China under Grant 62103026, Grant 62003022, and Grant 62003024.

Institutional Review Board Statement: Not applicable.

Informed Consent Statement: Not applicable.

Data Availability Statement: Not applicable.

Conflicts of Interest: The authors declare no conflict of interest.

Sample Availability: The data presented in this study are available on request from the corresponding author.

Abbreviations

SERF   Spin-Exchange Relaxation Free

Appendix A

Principle of Complex Magnetic Permeability Measurement

According to the measurement principle of the coil method, the relationship between the complex permeability of the ferrite and the inductance and resistance in the equivalent circuit is as follows:

$$\mu' = \frac{l_e L_s}{\mu_0 A_e N^2}, \tag{A1}$$

$$\mu'' = \frac{l_e (R_y - R_w)}{2\pi f_c \mu_0 A_e N^2}. \tag{A2}$$

The definition of each variable in the formula is shown in the following table:

Table A1. Definition of variables.

Trem	Description
L_s	inductance of the sample
l_e	effective magnetic circuit length
A_e	effective cross-sectional area
N	number of turns of the winding
R_y	resistance value containing the resistance of the test coil
R_w	resistance value of the measurement coil
f	test frequency

The expressions of l_e and A_e are:

$$l_e = \frac{\pi \ln(D_y/d_y)}{d_y^{-1} - D_y^{-1}},$$
$$A_e = \frac{h_y \ln^2(D_y/d_y)}{2(d_y^{-1} - D_y^{-1})}, \tag{A3}$$

where D_y is the outer diameter of the sample, d_y is the inner diameter, and h_y is the height of the sample.

The real part of the magnetic permeability can be calculated directly from the inductance value. The imaginary part contains the resistance value of the measurement coil, and Equation (A2) needs to be rewritten as

$$R_y = \frac{2\pi f_c \mu_0 A_e N^2}{l_e}\mu'' + R_w. \tag{A4}$$

By fitting the above equation to f_c and R_y, the slope and intercept are the imaginary part of the magnetic permeability and the resistance of the measurement coil, respectively.

References

1. Bear, D.; Stoner, R.E.; Walsworth, R.L.; Kostelecky, V.A.; Lane, C.D. Limit on Lorentz and CPT Violation of the Neutron Using a Two-Species Noble-Gas Maser. *Phys. Rev. Lett.* **2002**, *89*, 209902. [CrossRef]
2. Brown, I.M. A New Limit on Lorentz- and CPT -Violating Neutron Spin Interactions Using a K-³He Comagnetometer. Ph.D. Thesis, Princeton University, Princeton, NJ, USA, 2011.
3. Allred, J.C.; Lyman, R.N.; Kornack, T.W.; Romalis, M.V. High-Sensitivity Atomic Magnetometer Unaffected by Spin-Exchange Relaxation. *Phys. Rev. Lett.* **2002**, *89*, 130801. [CrossRef] [PubMed]
4. Kominis, I.K.; Kornack, T.W.; Allred, J.C.; Romalis, M.V. A Subfemtotesla Multichannel Atomic Magnetometer. *Nature* **2003**, *422*, 596–599. [CrossRef] [PubMed]
5. Dang, H.B.; Maloof, A.C.; Romalis, M.V. Ultrahigh sensitivity magnetic field and magnetization measurements with an atomic magnetometer. *Appl. Phys. Lett.* **2010**, *97*, 151110. [CrossRef]
6. Kornack, T.W.; Romalis, M.V. Dynamics of Two Overlapping Spin Ensembles Interacting by Spin Exchange. *Phys. Rev. Lett.* **2002**, *89*, 253002. [CrossRef] [PubMed]
7. Kornack, T.W.; Ghosh, R.K.; Romalis, M.V. Nuclear Spin Gyroscope Based on an Atomic Comagnetometer. *Phys. Rev. Lett.* **2005**, *95*, 230801. [CrossRef] [PubMed]

8. Boto, E.; Holmes, N.; Leggett, J.; Roberts, G.; Shah, V.; Meyer, S.S.; Munoz, L.D.; Mullinger, K.J.; Tierney, T.M.; Bestmann, S.; et al. Moving magnetoencephalography towards real-world applications with a wearable system. *Nature* **2018**, *555*, 657–661. [CrossRef] [PubMed]
9. Savukov, I.; Kim, Y.J.; Shah, V.; Boshier, M.G. High-sensitivity operation of single-beam optically pumped magnetometer in a kHz frequency range. *Meas. Sci. Technol.* **2017**, *28*, 035104. [CrossRef]
10. Fan, W.; Quan, W.; Liu, F.; Xing, L.; Liu, G. Suppression of the Bias Error Induced by Magnetic Noise in a Spin-Exchange Relaxation-Free Gyroscope. *IEEE Sens. J.* **2019**, *19*, 9712–9721. [CrossRef]
11. Pang, H.; Duan, L.; Quan, W.; Wang, J.; Wu, W.; Fan, W.; Liu, F. Design of Highly Uniform Three Dimensional Spherical Magnetic Field Coils for Atomic Sensors. *IEEE Sens. J.* **2020**, *20*, 11229–11236. [CrossRef]
12. Pang, H.; Fan, W.; Huang, J.; Liu, F.; Liu, S.; Quan, W. A Highly Sensitive In-Situ Magnetic Field Fluctuation Measurement Method Based on Nuclear-Spin Depolarization in an Atomic Comagnetometer. *IEEE Trans. Instrum. Meas.* **2022**, *71*, 9505408. [CrossRef]
13. Ma, D.; Ding, M.; Lu, J.; Yao, H.; Zhao, J.; Yang, K.; Cai, J.; Han, B. Study of Shielding Ratio of Cylindrical Ferrite Enclosure with Gaps and Holes. *IEEE Sens. J.* **2019**, *19*, 6085–6092. [CrossRef]
14. Fu, Y.; Sun, J.; Ruan, J.; Quan, W. A Nanocrystalline Shield for High Precision Co-magnetometer Operated in Spin-Exchange Relaxation-Free Regime. *Sens. Actuators A-Phys.* **2022**, *339*, 113487. [CrossRef]
15. Paperno, E.; Romalis, M.V.; Noam, Y. Optimization of five-shell axial magnetic shields having openings in the end-caps. *IEEE Trans. Magn.* **2004**, *40*, 2170–2172. [CrossRef]
16. Fan, W.; Quan, W.; Liu, F.; Pang, H.; Xing, L.; Liu, G. Performance of Low-Noise Ferrite Shield in a K-Rb-^{21}Ne Co-Magnetometer. *IEEE Sens. J.* **2019**, *20*, 2543–2549. [CrossRef]
17. Munger, C.T. Magnetic Johnson noise constraints on electron electric dipole moment experiments. *Phys. Rev. A* **2005**, *72*, 012506. [CrossRef]
18. Silveyra, J.M.; Ferrara, E.; Huber, D.L.; Monson, T.C. Soft magnetic materials for a sustainable and electrified world. *Science* **2018**, *362*, 418. [CrossRef] [PubMed]
19. Kornack, T.W.; Smullin, S.J.; Lee, S.K.; Romalis, M.V. A Low-Noise Ferrite Magnetic Shield. *Appl. Phys. Lett.* **2007**, *90*, 223501. [CrossRef]
20. Jia, Y.C.; Liu, Z.C.; Zhou, B.Q.; Liang, X.Y.; Wu, W.F.; Peng, J.P.; Ding, M.; Zhai, Y.Y.; Fang, J.C. Pump Beam Influence on Spin Polarization Homogeneity in the Nuclear Magnetic Resonance Gyroscope. *J. Phys. D-Appl. Phys.* **2019**, *52*, 355001.
21. Liu, L.B.; Yan, Y.; Ngo, K.D.T.; Lu, G.Q. NiCuZn Ferrite Cores by Gelcasting: Processing and Properties. *IEEE Trans. Ind. Appl.* **2017**, *53*, 5728–5733. [CrossRef]
22. Dobak, S.; Beatrice, C.; Tsakaloudi, V.; Fiorillo, F. Magnetic Losses in Soft Ferrites. *Magnetochemistry* **2022**, *8*, 60. [CrossRef]
23. Lee, S.K.; Romalis, M.V. Calculation of Magnetic Field Noise From High-Permeability Magnetic Shields and Conducting Objects with Simple Geometry. *J. Appl. Phys.* **2008**, *103*, 084904. [CrossRef]
24. Ma, D.; Lu, J.; Fang, X.; Yang, K.; Wang, K.; Zhang, N.; Han, B.; Ding, M. Parameter Modeling Analysis of a Cylindrical Ferrite Magnetic Shield to Reduce Magnetic Noise. *IEEE Trans. Ind. Electron.* **2022**, *69*, 991–998. [CrossRef]
25. Yang, K.; Lu, J.; Ding, M.; Zhao, J.; Ma, D.; Li, Y.; Xing, B.; Han, B.; Fang, J. Improved Measurement of the Low-Frequency Complex Permeability of Ferrite Annulus for Low-Noise Magnetic Shielding. *IEEE Access* **2019**, *7*, 126059–126065. [CrossRef]
26. Lu, J.; Ma, D.; Yang, K.; Quan, W.; Zhao, J.; Xing, B.; Han, B.; Ding, M. Study of Magnetic Noise of a Multi-Annular Ferrite Shield. *IEEE Access* **2020**, *8*, 40918–40924. [CrossRef]
27. Pczkowski, P.; Zachariasz, P.; Kowalik, M.; Tokarz, W.; Naik, S.P.K.; Żukrowski, J.; Jastrzębski, C.; Dadiel, L.J.; Tabiś, W.; Gondek, Ł. Iron diffusivity into superconducting $YBa_2Cu_3O_{7-\delta}$ at oxygen-assisted sintering: Structural, magnetic, and transport properties. *J. Eur. Ceram. Soc.* **2021**, *41*, 7085–7097. [CrossRef]
28. Chrobak, M.; Woch, W.M.; Szwachta, G.; Zalecki, R.; Gondek, Ł; Kołodziejczyk, A.; Kusiński, J. Thermal Fluctuations in YBCO Thin Film on MgO Substrate. *ACTA Phys. Pol. A* **2014**, *126*, A88–A91. [CrossRef]
29. Fiorillo, F.; Beatrice, C.; Coisson, M.; Zhemchuzhna, L. Loss and Permeability Dependence on Temperature in Soft Ferrites. *IEEE Trans. Magn.* **2009**, *45*, 4242–4245. [CrossRef]
30. Sidles, J.A.; Garbini, J.L.; Dougherty, W.M.; Chao, S.H. The classical and quantum theory of thermal magnetic noise, with applications in spintronics and quantum microscopy. *Proc. IEEE* **2003**, *91*, 799–816. [CrossRef]
31. Fan, W.; Quan, W.; Zhang, W.; Xing, L.; Liu, G. Analysis on the Magnetic Field Response for Nuclear Spin Co-magnetometer Operated in Spin-Exchange Relaxation-Free Regime. *IEEE Access* **2019**, *7*, 28574–28580. [CrossRef]

Article

Study on the Magnetic Noise Characteristics of Amorphous and Nanocrystalline Inner Magnetic Shield Layers of SERF Co-Magnetometer

Ye Liu [1,2], Hang Gao [1,2], Longyan Ma [2,3], Jiale Quan [1,2], Wenfeng Fan [1,2], Xueping Xu [1,2], Yang Fu [1,2], Lihong Duan [1,2,*] and Wei Quan [1,2,*]

1. School of Instrumentation and Optoelectronic Engineering, Beihang University, Beijing 100191, China
2. Zhejiang Provincial Key Laboratory of Ultra-Weak Magnetic-Field Space and Applied Technology, Hangzhou Innovation Institute, Beihang University, Hangzhou 310051, China
3. Research institute for Frontier Science, Beihang University, Beijing 100191, China
* Correspondence: duanlihong@buaa.edu.cn (L.D.); quanwei@buaa.edu.cn (W.Q.)

Citation: Liu, Y.; Gao, H.; Ma, L.; Quan, J.; Fan, W.; Xu, X.; Fu, Y.; Duan, L.; Quan, W. Study on the Magnetic Noise Characteristics of Amorphous and Nanocrystalline Inner Magnetic Shield Layers of SERF Co-Magnetometer. *Materials* 2022, 15, 8267. https://doi.org/10.3390/ma15228267

Academic Editor: Simone Pisana

Received: 7 October 2022
Accepted: 16 November 2022
Published: 21 November 2022

Publisher's Note: MDPI stays neutral with regard to jurisdictional claims in published maps and institutional affiliations.

Copyright: © 2022 by the authors. Licensee MDPI, Basel, Switzerland. This article is an open access article distributed under the terms and conditions of the Creative Commons Attribution (CC BY) license (https://creativecommons.org/licenses/by/4.0/).

Abstract: With the widespread use of magneto-sensitive elements, magnetic shields are an important part of electronic equipment, ultra-sensitive atomic sensors, and in basic physics experiments. Particularly in Spin-exchange relaxation-free (SERF) co-magnetometers, the magnetic shield is an important component for maintaining the SERF state. However, the inherent noise of magnetic shield materials is an important factor limiting the measurement sensitivity and accuracy of SERF co-magnetometers. In this paper, both amorphous and nanocrystalline materials were designed and applied as the innermost magnetic shield of an SERF co-magnetometer. Magnetic noise characteristics of different amorphous and nanocrystalline materials used as the internal magnetic shielding layer of the magnetic shielding system were analyzed. In addition, the effects on magnetic noise due to adding aluminum to amorphous and nanocrystalline materials were studied. The experimental results show that compared with an amorphous material, a nanocrystalline material as the inner magnetic shield layer can effectively reduce the magnetic noise and improve the sensitivity and precision of the rotation measurement. Nanocrystalline material is very promising for inner shield composition in SERF co-magnetometers. Furthermore, its ultra-thin structure and low cost have significant application value in the miniaturization of SERF co-magnetometers.

Keywords: soft magnetic material; amorphous; nanocrystalline; SERF co-magnetometer; magnetic shield

1. Introduction

In recent years, a new generation of high-precision atomic sensors, SERF co-magnetometers, have been widely used in Lorentz and CPT violation testing [1,2], magnetic field measurement [3,4], and rotation measurement [5,6]. Their high sensitivity to rotation holds promise as a rotation sensor in next-generation inertial navigation systems [7,8]. High shield factor and low noise in the magnetic shield system is necessary to ensure the normal operation of SERF co-magnetometers [9,10]; however, the inherent noise of the material comprising the magnetic shield system is the main noise term restricting their sensitivity and accuracy [11,12].

Soft magnetic materials are commonly used materials for magnetic shields because of their high magnetic permeability and easy magnetization and demagnetization [13–16]. For example, μ-metal is often used as a magnetic shielding material because of its high magnetic permeability [17–19]. The shield coefficient of magnetic shields made of μ-metal can reach 10^6, which can effectively shield against the external environmental magnetic field and quasi-static magnetic field fluctuations [20]. However, a low-resistivity μ-metal shield also produces high intrinsic magnetic noise due to Johnson current noise, which is the main factor limiting the accuracy of SERF co-magnetometers [4,11,21].

In addition to μ-metal, ferrite materials have been increasingly used as magnetic shielding in recent years because they generate less noise than μ-metal [22–25]. Kornack et al. [26] used manganese–zinc ferrite material in the inner layer of a magnetic shield system, and found that the noise it generated was much lower than that of μ-metal. Subsequently, scholars conducted more in-depth research on the application of manganese–zinc ferrite in SERF co-magnetometers. For example, Fan et al. [27] reported the performance of low-noise ferrites in a K-Rb-^{21}Ne co-magnetometer. Ma et al. [28] studied in detail the magnetic noise model of ferrite and the effect of shielding barrel configuration on noise. However, MnZn ferrite still cannot meet the requirements of low noise, and the manufacture of manganese–zinc ferrite is difficult and expensive. In addition, MnZn ferrite has a low Curie temperature and poor temperature stability. The working temperature of the sensitive element of the co-magnetometer can reach up to 200 degrees, and MnZn ferrite as the innermost magnetic shield produces magnetic field drift. Consequently, it is still necessary to explore new inner shield materials.

Amorphous and nanocrystalline alloys are a new generation of soft magnetic materials with promising application prospects [29]. They possess excellent magnetic properties, such as high saturation magnetic induction, low coercivity, high permeability, and high resistivity [30]. In addition, their unique nanostructure and relatively thin lamination can suppress eddy current loss, and is widely used in high-frequency fields, such as manufacturing transformer cores, power-supply switching and power electronic components [31,32]. Among them, iron-based amorphous and nanocrystalline alloys can be used as materials for making magnetic shields because of their high magnetic permeability [33]. Fu et al. applied nanocrystals to an inner-layer magnetic shield, proving the application potential of nanocrystals in the magnetic shield of SERF co-magnetometers [34]. However, the application and performance of amorphous and nanocrystalline alloys in SERF co-magnetometer magnetic shields has not been deeply studied.

In this paper, inner magnetic shields made of iron-based amorphous and nanocrystalline alloys for high-precision SERF co-magnetometers are proposed and designed. Firstly, a more accurate magnetic noise-error response model for the SERF co-magnetometer was established according to the Bloch equation. Next, the magnetic noise of a magnetic shield comprising amorphous and nanocrystalline alloys was measured separately using a SERF co-magnetometer. Lateral comparison of the magnetic noise generated by amorphous and nanocrystalline inner layer magnetic shields was conducted. At the same time, the magnetic noise of the inner magnetic shield comprising different layers of amorphous material and layers of nanocrystalline material were compared longitudinally. The magnetic noise characteristics of an amorphous inner magnetic shield and a nanocrystalline inner magnetic shield were studied and analyzed. Finally, the magnetic noise characteristics of amorphous and nanocrystalline inner magnetic shields with added aluminum were studied. Our theory and experiments show that nanocrystalline shields have lower thermal magnetization noise and better temperature stability compared to amorphous shields, making them a more favorable choice for the innermost shield material in SERF co-magnetometers. This is of great significance for miniaturization and engineering of low-noise magnetic shield systems for SERF co-magnetometers in future.

2. Magnetic Shield Noise Error Model Analysis

The time evolution of the coupled spin ensemble in a SERF co-magnetometer can be solved analytically with a complete set of Bloch equations [35,36]:

$$\frac{\partial \mathbf{P}^e}{\partial t} = \frac{\gamma^e}{Q(P^e)}(\mathbf{B} + \lambda M^n \mathbf{P}^n + \mathbf{L}) \times \mathbf{P}^e - \mathbf{\Omega} \times \mathbf{P}^e + \frac{(R_p \mathbf{s}_p + R_m \mathbf{s}_m + R_{se}^{en} \mathbf{P}^n - R_{tot}^e \mathbf{P}^e)}{Q(P^e)}$$

$$\frac{\partial \mathbf{P}^n}{\partial t} = \gamma^n (\mathbf{B} + \lambda M^e \mathbf{P}^e) \times \mathbf{P}^n - \mathbf{\Omega} \times \mathbf{P}^n + R_{se}^{ne} \mathbf{P}^e - R_{tot}^n \mathbf{P}^n$$

(1)

Here, $\mathbf{P}^e$ and $\mathbf{P}^n$ are the electron spin polarization and nuclear spin polarization, respectively. γ^e and γ^n are the gyromagnetic ratios of the electron and the nuclear spins, respectively. $\mathbf{Q}$ is the electron slowing-down factor, $\mathbf{B}$ is the ambient magnetic field vector, $\mathbf{L}$ is AC-Stark shift, $\mathbf{\Omega}$ is the measured rotation vector. s_p and s_m are the photon spin vectors of the pump and probe beams, oriented along the z and x axes, respectively. R^e_{tot} is sum of transverse relaxation of electron spin, R^{en}_{se} is the spin-exchange rate from collisions with the nuclear spins by the electron spins, R^{ne}_{se} is the spin-exchange rate from collisions with the electron spins by the nuclear spins, and R^n_{tot} is the total relaxation rate of nuclear spins. λM^e and λM^n are the electron spin magnetic moment of alkali metal atoms and the nuclear spin magnetic moment of inert gas atoms respectively.

When the co-magnetometer works under steady-state conditions, The input and output of the system can be approximated in the following linear form

$$\dot{\mathbf{X}} = \mathbf{AX} + \mathbf{WU} \tag{2}$$

Here, the state vector $\mathbf{X}$ can be written as $\mathbf{X} = \left[P^e_x, P^e_y, P^n_x, P^n_y\right]^T$, and the input vector $\mathbf{U}$ can be written as $\mathbf{U} = \left[\Omega_x, \Omega_y, B_x, B_y\right]^T$. The matrix A can be written as:

$$\mathbf{A} = \begin{bmatrix} -\frac{R^e_{tot}}{Q} & \frac{\gamma^e B^e_z}{Q} & \frac{R^{en}_{se}}{Q} & \frac{\gamma^e B^n_z P^e_z}{QP^n_z} \\ -\frac{\gamma^e B^e_z}{Q} & -\frac{R^e_{tot}}{Q} & -\frac{\gamma^e B^n_z P^e_z}{QP^n_z} & \frac{R^{en}_{se}}{Q} \\ R^{ne}_{se} & \frac{\gamma^n P^n_z B^e_z}{P^e_z} & -R^n_{tot} & \gamma^n B^n_z \\ -\frac{\gamma^n B^e_z P^n_z}{P^e_z} & R^{ne}_{se} & -\gamma^n B^n_z & -R^n_{tot} \end{bmatrix} \tag{3}$$

The matrix $\mathbf{W}$ can be written as

$$\mathbf{W} = \begin{bmatrix} 0 & -P^e_z & 0 & \frac{P^e_z \gamma^e}{Q} \\ P^e_z & 0 & -\frac{P^e_z \gamma^e}{Q} & 0 \\ 0 & -P^n_z & 0 & P^n_z \gamma^n \\ P^n_z & 0 & -P^n_z \gamma^n & 0 \end{bmatrix} \tag{4}$$

Therefore, the transfer function that characterizes the input–output relationship in the system at steady state can be expressed as $\mathbf{G}(s) = (s\mathbf{I} - \mathbf{A})^{-1}\mathbf{W}$. After calculation and simplification, the transfer function between input B_x and output P^e_x can be expressed as [12]

$$G_{Bx}(s) = P^e_x(s)/B_x(s) = \frac{k_{Bx}(s - \omega_{Bx1})(s - \omega_{Bx2})}{\left[(s - \varphi_1)^2 + \omega_1^2\right]\left[(s - \varphi_2)^2 + \omega_2^2\right]} \tag{5}$$

Here, the zeros are

$$\omega_{Bx1} = -\frac{P^e_z R^n_{tot} \gamma^e + P^n_z R^{en}_{se} \gamma^n}{P^e_z \gamma^e} \tag{6}$$

$$\omega_{Bx2} = \frac{B^e_z P^e_z R^n_{tot} \gamma^e + B^n_z P^n_z R^{en}_{se} \gamma^n + B^n_z P^e_z R^e_{tot} \gamma^n}{P^e_z (B^e_z \gamma^e + B^n_z \gamma^n Q)} \tag{7}$$

There are two separate oscillations with different frequencies φ_1, φ_2, and decay rates ω_1, ω_2. B^n_z is the effective field for the z-direction nuclear magnetization and B^e_z is the effective field for the z-direction electron magnetization. The co-magnetometer is sensitive to B_x when they are sensitive to Ω_y, which is why the performance of gyroscopes based on SERF co-magnetometers is limited by magnetic noise. Therefore, low-noise shielding is very important for SERF co-magnetometers.

When the SERF co-magnetometer is sensitively rotated, the fluctuation of B_x will cause the SERF co-magnetometer to produce bias drift, and the navigation error will accumulate over time. Therefore, it is necessary to establish a magnetic noise error model to analyze the equivalent error generated by magnetic noise. To obtain the power spectrum of the rotating signal of the SERF co-magnetometer, it is necessary to obtain the power spectral density of the static test data and the scale factor of the rotation, and then divide the static power spectral density by the scale factor to obtain the final signal power spectrum. At this point, the stochastic process associated with the system output signal can be regarded as a stationary stochastic process. According to the stochastic process analysis theory, the bilateral power spectral density (PSD) of the output signal when a stationary stochastic process is superimposed on a linear system can be expressed as:

$$S_Y(\omega) = S_X(\omega)|H(\omega)|^2 \tag{8}$$

where $S_X(\omega)$ is the bilateral PSD of the input process, $S_Y(\omega)$ is the bilateral PSD of the output process, $|H(\omega)|^2$ is the power transfer function of the linear system, and $|H(\omega)|$ is the amplitude–frequency response of the system. The steady state rotation rate in the time domain can be expressed as:

$$\Omega_y(t) = K_{\Omega_y} P_x^e(t) \tag{9}$$

The Ω_y can be described in the Laplace transform domain as:

$$\Omega_y(s) = K_{\Omega_y} G_{Bx}(s) B_x(s) = \frac{K_{\Omega_y} K_{Bx}(s - z_{Bx1})(s - z_{Bx2}) B_x(s)}{\left[(s - \Gamma_1)^2 + \omega_1^2\right]\left[(s - \Gamma_2)^2 + \omega_2^2\right]}. \tag{10}$$

where the scale factor K_{Ω_y} can be expressed as:

$$K_{\Omega_y} = \frac{(\gamma^e B^e R_{tot}^n + \gamma^n B^n R_{tot}^e)^2 + (R_{tot}^n R_{tot}^e)^2}{P_z^e \gamma^e R_{tot}^e (B^n)^2 \gamma^n} \tag{11}$$

In the SERF co-magnetometer inertial measurement system, the hysteresis loss magnetic noise and eddy current loss magnetic noise of the innermost shielding cylinder are uncorrelated, therefore the total angular velocity PSD caused by the two magnetic noises can be expressed as:

$$S_{\Omega y}^T(\omega) = S_{\Omega y}^{magn}(\omega) + S_{\Omega y}^{eddy}(\omega) \tag{12}$$

Among them, $S_{\Omega y}^{magn}(\omega)$ is the angular velocity PSD corresponding to the hysteresis loss magnetic noise, and $S_{\Omega y}^{eddy}(\omega)$ is the angular velocity PSD corresponding to the eddy current loss magnetic noise. When the co-magnetometer is in a steady state, the steady-state PSD can be expressed as:

$$\begin{aligned} S_\Omega^{magn}(\omega) &= S^{magn}(\omega)|G_{Bx}(0)|^2 K_{\Omega_y}^2 \\ S_\Omega^{eddy}(\omega) &= S^{eddy}(\omega)|G_{Bx}(0)|^2 K_{\Omega_y}^2 \end{aligned} \tag{13}$$

where $S^{magn}(\omega) = C_{magn}^2/2\omega$ and $S^{eddy}(\omega) = C_{eddy}^2/2$ are the bilateral PSD of hysteresis loss noise and eddy current loss noise, respectively. C_{magn} and C_{eddy} are the noise figures of hysteresis loss noise and eddy current loss noise, respectively.

The relationship between the Allan variance and the bilateral PSD of the angular velocity is:

$$\sigma^2(T) = \frac{4}{\pi T} \int_0^\infty S_\Omega\left(\frac{u}{\pi T}\right) \frac{\sin^4 u}{u^2} du \tag{14}$$

Here $u = \pi \nu T$, where ν is frequency and T is the group time.

Bringing $S_{\Omega y}^{magn}(\omega)$ and $S_{\Omega y}^{eddy}(\omega)$ into Equation (14) produces their Allan variance expression:

$$\left[\sigma_\Omega^{magn}(T)\right]^2 = \frac{C_{magn}^2 |G_{Bx}(0)|^2}{\pi} \ln 2 (T \gg 1/\nu_0) \tag{15}$$

$$\left[\sigma_\Omega^{eddy}(T)\right]^2 = \frac{C_{eddy}^2 |G_{Bx}(0)|^2}{2T} \tag{16}$$

Here, ν_0 represents the corner frequency of the hysteresis loss magnetic noise. The analysis shows that the Allan variance of the low-frequency angular velocity corresponding to the hysteresis loss magnetic noise does not change with the correlation time. Eddy current loss noise is inversely proportional to T and has a $-1/2$ slope with respect to T in a log–log plot.

Therefore, according to the above analysis, it is not difficult to see that we can use the magnetic noise PSD, the scale coefficient and the amplitude–frequency response to Bx to evaluate the performance of the inner magnetic shield.

3. Experimental Setup

The K-Rb-^{21}Ne co-magnetometer was used to conduct the experiment, and the schematic diagram of the structure is shown in Figure 1. The constructed K-Rb-^{21}Ne co-magnetometer structural components are similar to previous work [12]. The magnetic shield system of the K-Rb-^{21}Ne co-magnetometer consists of three shield layers, of which the outer two layers are μ-metal. μ-metal has a high shielding coefficient and is used to shield the interference of geomagnetism and external magnetic fields. The inner shielding layer is to shield the magnetic field and reduce the magnetic noise in the shielding system. In order to ensure that the device works in a low magnetic field environment, the magnetic shield system must be degaussed before the device is started. After degaussing, the residual magnetic field in the shielded barrel is compensated with a high-precision three-axis magnetic field coil. The sensitive core is a 10 mm diameter cell made of GE180 aluminosilicate glass. The cell is placed in a boron nitride oven, which is heated using a heating coil connected to the oven, allowing it to operate at a maximum temperature exceeding 200 °C.

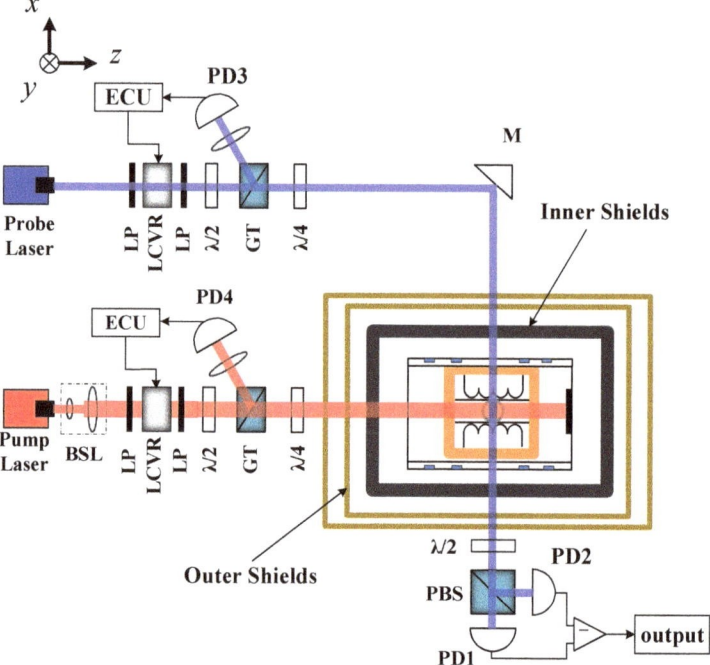

Figure 1. Schematic of K-Rb-^{21}Ne co-magnetometer. BSL: beam shaping lenses. LP: linear polarizer. LCVR: liquid crystal variable retarder. $\lambda/2$: half wave plate. GT: Glan–Taylor polarizer. $\lambda/4$: quarter wave plate. PBS: polarizing beam splitter. M: reflection mirror. PD: photodiode. ECU: electronic control unit. The magnetic shield system of common magnetic force consists of three shield layers. The outer shield consists of two 2 mm thick μ-metal shields with outer radii of 54 mm and 64 mm and lengths of 150 mm and 180 mm, respectively. The inner shield is the focus of our research.

The experimental setup shown in Figure 1 is a structural diagram of an experimental setup we built, which can be used to sense magnetic fields or to sense rotation. The difference between the two devices is that different atoms are used in the cell. When the device operates as a SERF magnetometer sensitive to magnetic field, the cell contains the natural-abundance K and Rb alkali metal mixture, 1.9 atm of ^4He buffer gas, and 50 Torr N_2 quench gas. When the device operates as a SERF co-magnetometer sensitive to rotation, the cell contains a natural-abundance K and Rb alkali metal mixture, 2 atm of ^{21}Ne and 50 Torr N_2 quench gas. The circular-polarized laser light propagating along the z-axis is emitted by a distributed Bragg reflector (DBR) diode laser, interacting with the D1 resonance line of K atoms to realize the polarization of K atoms. The probe laser is a non-resonant linear-polarized light emitted along the x-axis to detect the atomic spin precession of the Rb atoms along the x-axis. A signal generator is used to control the triaxial magnetic field coil in order to apply excitation to the magnetic field. The ratio of alkali metal density inside the cell is K:Rb = 1:90 at 180 °C.

Figure 2 shows the inner magnetic shields of the magnetometer fabricated for this experiment. Amorphous and nanocrystalline shields are made of Fe-based amorphous and nanocrystalline films attached to a support frame made of non-magnetic PTFE. The thickness of the support frame made of PTFE is 5 mm, the outer radius of the barrel is 47 mm, and the barrel height is 120 mm. The test results of the inner shield layer of amorphous alloy and the inner shield layer of nanocrystalline alloy in SERF magnetometer was compared to analyze the influence of the inner shield magnetic noise on the measurement accuracy of SERF co-magnetometer.

- 2-layer amorphous inner magnetic shield
- 4-layer amorphous inner magnetic shield
- 2-layer nanocrystalline inner magnetic shield
- 4-layer nanocrystalline inner magnetic shield
- 2-layer amorphous+aluminum inner magnetic shield
- 4-layer amorphous+aluminum inner magnetic shield
- 2-layer nanocrystalline+aluminum inner magnetic shield
- 4-layer nanocrystalline+aluminum inner magnetic shield

Figure 2. Amorphous and nanocrystalline inner magnetic shield layers.

We designate these eight different inner magnetic shielding cylinder schemes as schemes 1–8.

As shown in Figure 3, the demagnetization coil structure is composed of six annular coils wound around the magnetic shield system. The coil can generate magnetic flux in a closed-loop mode in the shield layer, which can demagnetize the magnetic shield system. Further, Figure 3b is the size of the three-layer shielding cylinder. As the outer shield layer of the magnetic shield system, the main function of the two-layer permalloy is to shield the external magnetic field. The outermost layer of permalloy has a height of 180 mm and an outer diameter of 128 mm, and the second outer layer of permalloy has a height of 150 mm and an outer diameter of 108 mm. The innermost layer is an inner shielding layer composed of amorphous/nanocrystalline material and PTFE. PTFE cylinder with a thickness of 6 mm, amorphous/nanocrystalline material wound on the outside of the PTFE cylinder. The reason for using PTFE is that amorphous/nanocrystalline material cannot be fixed as ultrasoft magnetic materials, requiring a supporting skeleton. In addition, PTFE is a non-magnetic material and will not generate magnetic field interference.

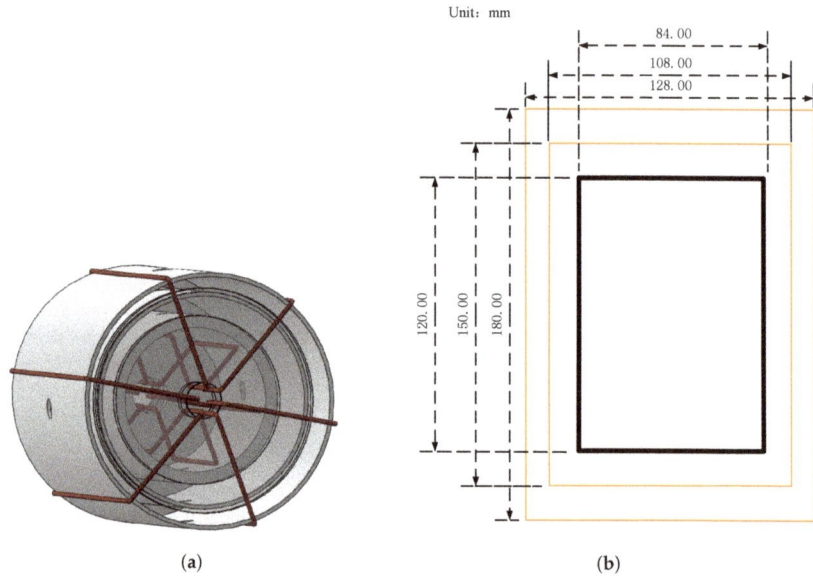

Figure 3. Internal structure of magnetic shield system composed of three shield layers of SERF co-magnetometer and method of wound demagnetizing coil. (**a**) 3D structure diagram (**b**) Plane dimension drawing

4. Results and Discussion

According to the analysis results in Section 2, the magnetic noise of the magnetic shield system mainly depends on the innermost magnetic shield layer. The low-frequency magnetic noise we focus on mainly includes the magnetic noise caused by hysteresis loss and eddy current loss. According to Equations (15) and (16), it is known that the noise due to hysteresis loss is similar to 1/f noise, and the noise due to eddy current loss is similar to white noise.

The magnetic noise characteristics of the inner shield layers of eight different schemes were measured in situ with a K-Rb-^{21}Ne SERF magnetometer. Table 1 shows the test results of the residual magnetic fields in the three directions of the center of the shielding system under the inner shield layer of Scheme 1 to Scheme 8 after the winding is demagnetized. The main purpose of the design of the inner shield is to suppress the magnetic noise felt by the central air chamber of the magnetic shielding system. On the one hand, it is necessary to suppress the magnetic noise caused by the outer magnetic shield. The second is to suppress the magnetic noise generated by the inner shielding material itself. This is also the reason why SERF co-magnetometers need to choose low noise and high permeability inner shields.

The degaussing current is generated by a laptop, a compact DAQ module, and a power amplifier. The degaussing current is about 60A. It can be seen that with the same degaussing method, the residual magnetic flux density of nanocrystals is significantly smaller than that of the amorphous material. The small residual flux density can make the angle measurement more sensitive and stable.

Table 1. Triaxial residual magnetic field after demagnetization of combined magnetic shielding system with different amorphous and nanocrystalline material as inner shield.

	B_x (nT)	B_y (nT)	B_z (nT)
without shielding	9385	29665	29850
Scheme 1	3	3.78	11.25
Scheme 2	1.3	2.4	12.45
Scheme 3	0.56	2.4	0.3
Scheme 4	0.86	1.08	0.375
Scheme 5	7.2	6.34	11.1
Scheme 6	13.14	5.54	0.9
Scheme 7	0.5	1.92	0.33
Scheme 8	0.46	0.46	0.855

Figure 4 shows the measurement results of magnetic noise of the SERF magnetometer under the inner magnetic shield layer of amorphous alloys and nanocrystalline alloys at frequencies below 100 Hz. It can be seen that the noise of the two amorphous inner shield layers is lower than the four amorphous inner shield layers. Similarly, the noise of the two nanocrystalline inner shield layers is lower than the four nanocrystalline inner shield layers. Therefore, the amorphous and nanocrystalline materials have similar properties when used as the inner magnetic shield. With the increasing number of layers, the magnetic noise of the amorphous and nanocrystalline inner magnetic shielding becomes greater. It can be seen from Figure 4 that the interference of the small peaks that appear is around 50 Hz, and is caused by power frequency fluctuations. Since the SERF co-magnetometer is used for rotation measurement, the further the magnetic noise is below 1 Hz, the higher the measurement accuracy of rotation. It can be seen that the magnetic noise level of nanocrystalline alloy is much better than that of amorphous alloy at frequencies lower than 1 Hz. Therefore, for rotation measurement, nanocrystalline alloys as the inner magnetic shield layer are better than amorphous.

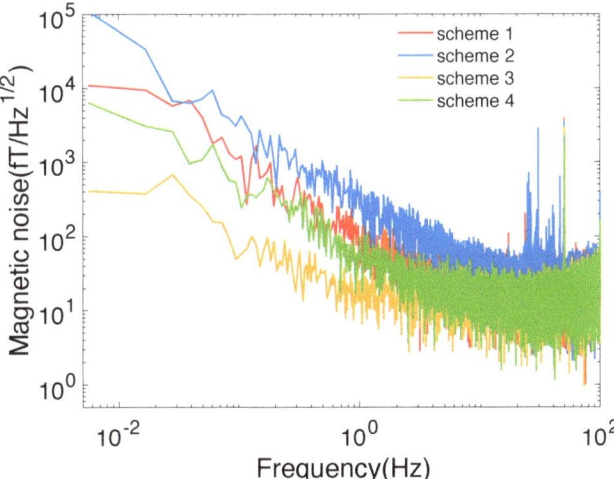

Figure 4. Magnetic noise measurement results of the SERF magnetometer under inner magnetic shield layers of different amorphous alloys and nanocrystalline alloys.

In addition, the magnetic noise characteristics of amorphous and nanocrystalline magnetic shield inner layers with aluminum addition were studied. Figure 5 shows magnetic noise measurement results of the SERF magnetometer under inner magnetic shield layers of different (Figure 5a) amorphous+aluminum alloys and (Figure 5b) nanocrystalline

+aluminum alloys. It can be seen from Figure 5 that with the increase of the number of layers, the magnetic noise of both amorphous and nanocrystalline material decreases, and the magnetic noise characteristics change. The magnetic noise of amorphous material after adding aluminum is lower than that without aluminum, but it is still higher than that of nanocrystalline material. However, after adding aluminum to nanocrystalline material, compared with no aluminum, the magnetic noise below 1Hz has increased; thus, adding aluminum to nanocrystalline material is not a good idea.

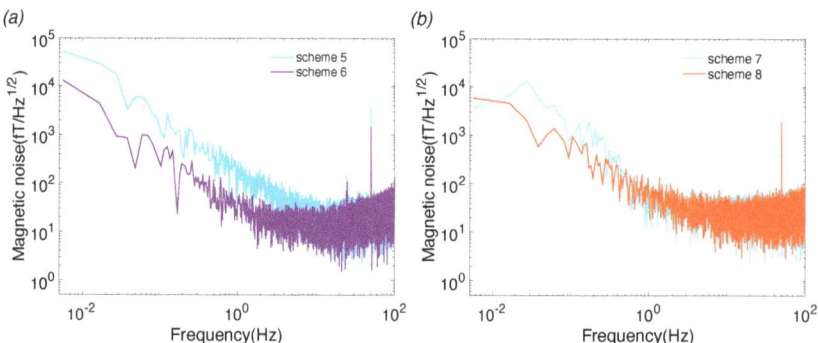

Figure 5. Magnetic noise measurement results of the SERF magnetometer under inner magnetic shield layers of different (**a**) amorphous + aluminum alloys, and (**b**) nanocrystalline + aluminum alloys.

Table 2 shows the magnetic noise values of the eight schemes at 1 Hz, from which it can be seen that the 2-layer nanocrystalline alloy is the scheme with the lowest noise, and the magnetic noise is 14.17 fT/Hz$^{1/2}$. Among the four kinds of amorphous inner magnetic shields, the best magnetic noise at 1 Hz is scheme 8, having a magnetic noise value of 54.13 fT/Hz$^{1/2}$.

Table 2. Magnetic noise value of schemes 1–8 at 1 Hz.

	1 Hz fT/Hz$^{1/2}$
Scheme 1	68.93
Scheme 2	239.65
Scheme 3	147.98
Scheme 4	28.62
Scheme 5	14.17
Scheme 6	46.59
Scheme 7	54.81
Scheme 8	54.13

According to the data measured thus far, nanocrystalline material with fewer layers is a better choice for inner magnetic shielding. However, the research on the magnetic noise characteristics of nanocrystalline is not adequate at the present time. Later, detailed magnetic noise modeling will be carried out to analyze its characteristics.

5. Conclusions

This paper analyzes the magnetic noise characteristics of amorphous and nanocrystalline inner shields used in SERF co-magnetometer combined magnetic shielding systems. Comparing the magnetic noise of amorphous and nanocrystalline materials with different layers, it is found that the magnetic noise increases with the increase of layers. Further, the magnetic noise of nanocrystalline material is markedly lower than that of amorphous material when the frequency is lower than 1 Hz. In addition, aluminum was added to amorphous and nanocrystalline materials. It was found that the magnetic noise decreases with the increased number of layers after adding aluminum, which requires further research into its characteristics. The research and findings in this paper are not only instructive for the exploration of miniaturized low-noise magnetic shield systems more suitable for high-precision SERF co-magnetometers, it also possesses important guiding significance for other high-precision atomic sensors that require magnetic shield systems, such as atomic clocks and nuclear magnetic resonance gyroscopes.

Author Contributions: Methodology, W.F.; Formal analysis, Y.L. and H.G.; Investigation, L.M.; Data curation, J.Q.; Writing—original draft, Y.L.; Writing—review & editing, X.X., Y.F., W.Q. and L.D.; Project administration, Y.L., Y.F. and W.Q. All authors have read and agreed to the published version of the manuscript

Funding: This research was funded by National Natural Science Foundation of China (grant number 62103026) and National Science Fund for Distinguished Young Scholars of China (grant number 61925301).

Institutional Review Board Statement: Not applicable.

Informed Consent Statement: Not applicable.

Data Availability Statement: Not applicable.

Conflicts of Interest: The authors declare no conflict of interest.

References

1. Flambaum, V.; Romalis, M.V. Limits on lorentz invariance violation from coulomb interactions in nuclei and atoms. *Phys. Rev. Lett.* **2017**, *118*, 142501. [CrossRef] [PubMed]
2. Kostelecký, V.A.; Russell, N. Data tables for lorentz and c p t violation. *Rev. Mod. Phys.* **2011**, *83*, 11. [CrossRef]
3. Sheng, D.; Li, S.; Dural, N.; Romalis, M.V. Subfemtotesla scalar atomic magnetometry using multipass cells. *Phys. Rev. Lett.* **2013**, *110*, 160802. [CrossRef]
4. Kominis, I.; Kornack, T.; Allred, J.; Romalis, M.V. A subfemtotesla multichannel atomic magnetometer. *Nature* **2003**, *422*, 596–599. [CrossRef]
5. Kornack, T.; Ghosh, R.; Romalis, M.V. Nuclear spin gyroscope based on an atomic comagnetometer. *Phys. Rev. Lett.* **2005**, *95*, 230801. [CrossRef]
6. Limes, M.; Sheng, D.; Romalis, M.V. He 3- xe 129 comagnetometery using rb 87 detection and decoupling. *Phys. Rev. Lett.* **2018**, *120*, 033401. [CrossRef]
7. Fan, W.; Quan, W.; Liu, F.; Xing, L.; Liu, G. Suppression of the bias error induced by magnetic noise in a spin-exchange relaxation-free gyroscope. *IEEE Sens. J.* **2019**, *19*, 9712–9721. [CrossRef]
8. Jiang, L.; Quan, W.; Liang, Y.; Liu, J.; Duan, L.; Fang, J. Effects of pump laser power density on the hybrid optically pumped comagnetometer for rotation sensing. *Opt. Express* **2019**, *27*, 420–430. [CrossRef]
9. Liu, J.; Jiang, L.; Liang, Y.; Li, G.; Cai, Z.; Wu, Z.; Quan, W. Dynamics of a spin-exchange relaxation-free comagnetometer for rotation sensing. *Phys. Rev. Appl.* **2022**, *17*, 014030. [CrossRef]
10. Fu, Y.; Wang, Z.; Xing, L.; Fan, W.; Ruan, J.; Pang, H. Suppression of non-uniform magnetic fields in magnetic shielding system for serf co-magnetometer. *IEEE Trans. Instrum. Meas.* **2022**, *71*, 9507008. [CrossRef]
11. Fan, W.; Quan, W.; Zhang, W.; Xing, L.; Liu, G. Analysis on the magnetic field response for nuclear spin co-magnetometer operated in spin-exchange relaxation-free regime. *IEEE Access* **2019**, *7*, 28 574–28 580, . [CrossRef]
12. Liu, Y.; Fan, W.; Fu, Y.; Pang, H.; Pei, H.; Quan, W. Suppression of low-frequency magnetic drift based on magnetic field sensitivity in k-rb-21 ne atomic spin comagnetometer. *IEEE Trans. Instrum. Meas.* **2022**, *71*, 1–8.
13. Lee, S.-Y.; Lim, Y.-S.; Choi, I.-H.; Lee, D.-I.; Kim, S.-B. Effective combination of soft magnetic materials for magnetic shielding. *IEEE Trans. Magn.* **2012**, *48*, 4550–4553. [CrossRef]
14. Lee, S.-K.; Romalis, M. Calculation of magnetic field noise from high-permeability magnetic shields and conducting objects with simple geometry. *J. Appl. Phys.* **2008**, *103*, 084904, . [CrossRef]
15. Sadovnikov, A.V.; Grachev, A.A.; Odintsov, S.; Martyshkin, A.; Gubanov, V.A.; Sheshukova, S.E.; Nikitov, S.A. Neuromorphic calculations using lateral arrays of magnetic microstructures with broken translational symmetry. *JETP Lett.* **2018**, *108*, 312–317. [CrossRef]
16. Sadovnikov, A.V.; Bublikov, K.; Beginin, E.N.; Sheshukova, S.E.; Sharaevskii, Y.P.; Nikitov, S.A. Nonreciprocal propagation of hybrid electromagnetic waves in a layered ferrite–ferroelectric structure with a finite width. *JETP Lett.* **2015**, *102*, 142–147. [CrossRef]
17. Li, J.; Quan, W.; Han, B.; Liu, F.; Xing, L.; Liu, G. Multilayer cylindrical magnetic shield for serf atomic co-magnetometer application. *IEEE Sens. J.* **2018**, *19*, 2916–2923. [CrossRef]

18. He, K.; Wan, S.; Sheng, J.; Liu, D.; Wang, C.; Li, D.; Qin, L.; Luo, S.; Qin, J.; Gao, J.-H. A high-performance compact magnetic shield for optically pumped magnetometer-based magnetoencephalography. *Rev. Sci.* **2019**, *90*, 064102. [CrossRef]
19. Freake, S.; Thorp, T. Shielding of low magnetic fields with multiple cylindrical shells. *Rev. Sci. Instrum.* **1971**, *42*, 1411–1413. [CrossRef]
20. Altarev, I.; Bales, M.; Beck, D.; Chupp, T.; Fierlinger, K.; Fierlinger, P.; Kuchler, F.; Lins, T.; Marino, M.; Niessen, B.; et al. A large-scale magnetic shield with 106 damping at millihertz frequencies. *J. Appl. Phys.* **2015**, *117*, 183903. [CrossRef]
21. Dang, H.; Maloof, A.C.; Romalis, M.V. Ultrahigh sensitivity magnetic field and magnetization measurements with an atomic magnetometer. *Appl. Phys. Lett.* **2010**, *97*, 151110. [CrossRef]
22. Smiciklas, M.; Brown, J.; Cheuk, L.; Smullin, S.; Romalis, M.V. New test of local lorentz invariance using a ne 21- rb- k comagnetometer. *Phys. Rev. Lett.* **2011**, *107*, 171604. [CrossRef] [PubMed]
23. Bevan, D.; Bulatowicz, M.; Clark, P.; Flicker, J.; Griffith, R.; Larsen, M.; Luengo-Kovac, M.; Pavell, J.; Rothballer, A.; Sakaida, D.; et al. Nuclear magnetic resonance gyroscope: Developing a primary rotation sensor. In Proceedings of the 2018 IEEE International Symposium on Inertial Sensors and Systems (INERTIAL), Lake Como, Italy, 26–29 March 2018; pp. 1–2.
24. Savukov, I.; Kim, Y.; Shah, V.; Boshier, M. High-sensitivity operation of single-beam optically pumped magnetometer in a khz frequency range. *Meas. Sci. Technol.* **2017**, *28*, 035104. [CrossRef]
25. Lu, J.; Sun, C.; Ma, D.; Yang, K.; Zhao, J.; Han, B.; Quan, W.; Zhang, N.; Ding, M. Effect of gaps on magnetic noise of cylindrical ferrite shield. *J. Phys. D Appl. Phys.* **2021**, *54*, 255002. [CrossRef]
26. Kornack, T.; Smullin, S.; Lee, S.-K.; Romalis, M. A low-noise ferrite magnetic shield. *Appl. Phys. Lett.* **2007**, *90*, 223501. [CrossRef]
27. Fan, W.; Quan, W.; Liu, F.; Pang, H.; Xing, L.; Liu, G. Performance of low-noise ferrite shield in a k-rb-21 ne co-magnetometer. *IEEE Sens. J.* **2019**, *20*, 2543–2549. [CrossRef]
28. Ma, D.; Lu, J.; Fang, X.; Yang, K.; Wang, K.; Zhang, N.; Han, B.; Ding, M. Parameter modeling analysis of a cylindrical ferrite magnetic shield to reduce magnetic noise. *IEEE Trans. Ind. Electron.* **2021**, *69*, 991–998. [CrossRef]
29. Herzer, G. Modern soft magnets: Amorphous and nanocrystalline materials. *Acta Mater.* **2013**, *61*, 718–734. [CrossRef]
30. Hasegawa, R. Advances in amorphous and nanocrystalline materials. *J. Magn. Magn. Mater.* **2012**, *324*, 3555–3557. [CrossRef]
31. Fish, G.E. Soft magnetic materials. *Proc. IEEE* **1990**, *78*, 947–972. [CrossRef]
32. Silveyra, J.M.; Ferrara, E.; Huber, D.L.; Monson, T.C. Soft magnetic materials for a sustainable and electrified world. *Science* **2018**, *362*, eaao0195. [CrossRef] [PubMed]
33. DeVore, P.; Escontrias, D.; Koblesky, T.; Lin, C.; Liu, D.; Luk, K.B.; Ngan, J.; Peng, J.; Polly, C.; Roloff, J.; et al. Light-weight flexible magnetic shields for large-aperture photomultiplier tubes. *Nucl. Instrum. Methods Phys. Res. Sect. A Accel. Spectrometers Detect. Assoc. Equip.* **2014**, *737*, 222–228. [CrossRef]
34. Fu, Y.; Sun, J.; Ruan, J.; Quan, W. A nanocrystalline shield for high precision co-magnetometer operated in spin-exchange relaxation-free regime. *Sens. Actuators A Phys.* **2022**, *339*, 113487. [CrossRef]
35. Kornack, T.; Romalis, M. Dynamics of two overlapping spin ensembles interacting by spin exchange. *Phys. Rev. Lett.* **2002**, *89*, 253002. [CrossRef] [PubMed]
36. Chen, Y.; Quan, W.; Duan, L.; Lu, Y.; Jiang, L.; Fang, J. Spin-exchange collision mixing of the k and rb ac stark shifts. *Phys. Rev. A* **2016**, *94*, 052705. [CrossRef]

MDPI
St. Alban-Anlage 66
4052 Basel
Switzerland
www.mdpi.com

Materials Editorial Office
E-mail: materials@mdpi.com
www.mdpi.com/journal/materials

Disclaimer/Publisher's Note: The statements, opinions and data contained in all publications are solely those of the individual author(s) and contributor(s) and not of MDPI and/or the editor(s). MDPI and/or the editor(s) disclaim responsibility for any injury to people or property resulting from any ideas, methods, instructions or products referred to in the content.